25th Salt Water Intrusion Meeting (SWIM 2018)

E3S Web of Conferences Volume 54 (2018)

Gdansk, Poland
17-22 June 2018

Editors:

A. Szymkiewicz
A. Sadurski
B. Jaworska-Szulc

ISBN: 978-1-5108-7122-9

Printed from e-media with permission by:

Curran Associates, Inc.
57 Morehouse Lane
Red Hook, NY 12571

Some format issues inherent in the e-media version may also appear in this print version.

Printed by Curran Associates, Inc. (2018)

For additional information, please contact EDP Sciences – Web of Conferences at the address below.

EDP Sciences – Web of Conferences
17, Avenue du Hoggar
Parc d'Activité de Courtabœuf
BP 112
F-91944 Les Ulis Cedex A
France

Phone: +33 (0) 1 69 18 75 75
Fax: +33 (0) 1 69 28 84 91

contact-edps@webofconferences.org

Additional copies of this publication are available from:

Curran Associates, Inc.
57 Morehouse Lane
Red Hook, NY 12571 USA
Phone: 845-758-0400
Fax: 845-758-2633
Email: curran@proceedings.com
Web: www.proceedings.com

TABLE OF CONTENTS

INFLUENCE OF TIDES, BATHYMETRY, LITHOLOGY AND REGIONAL FLOWS ON THE SALINIZATION PROCESS IN NATURE AREA THE RAMMEGORS 1
America Ilja, de Louw Perry, Bier George, van der Zee Sjoerd

HIGH FREQUENCY SALTWATER INTRUSION MONITORING USING BOREHOLE GEOPHYSICAL TOOLS (SMD) 7
Baïsset M., Neyens D.

AN APPROACH TO DELINEATE GROUNDWATER BODIES AT RISK: SEAWATER INTRUSION IN LIEPAJA (LATVIA) 13
Bikše J., Retike I.

MODELING SEAWATER INTRUSION TO COASTAL AQUIFERS IN SOUTH COAST OF LAIZHOU BAY, CHINA 19
Chang Yawen, Hu Bill X., Li Xue

SALINITY PROBLEMS IN MEDITERRANEAN AND ISLAND COASTAL AQUIFERS IN SPAIN 24
Custodio Emilio

THE ASCENSION AND INTRUSION PROCESSES OF SALT WATER INTO AQUIFERS ALONG THE TECTONIC DISCONTINUITIES IN THE ZARNOWIECKIE LAKE AREA 30
Damrat Mateusz, Warumzer Rafal

LAND SUBSIDENCE BY PEAT OXIDATION LEADS TO ENHANCED SALINIZATION THROUGH BOILS IN DUTCH POLDERS 36
de Louw Perry G. B., Bootsma Huite, Kooi Henk, Kramer Mark, Erkens Gilles

IT'S HYDROGEOLOGY BUT NOT AS WE KNOW IT: SUB-SEAFLOOR GROUNDWATER FLOW DRIVEN BY THERMAL GRADIENTS 42
Desens A., Post V. E. A., Houben G. J., Kuhn T., Walther M., Graf Thomas

LONG-TIME RESISTIVITY MONITORING OF A FRESHWATER/SALTWATER TRANSITION ZONE USING THE VERTICAL ELECTRODE SYSTEM SAMOS 47
Grinat Michael, Epping Dieter, Meyer Robert

LOCATION CHANGES OF "WYDRZANY" GROUNDWATER INTAKE IN POLISH PART OF THE UZNAM ISLAND AIMED AT GROUNDWATER STATE IMPROVEMENT 51
Hoc Ryszard, Sadurski Andrzej, Wisniowski Zenon

300 YEARS OF COASTAL SALINIZATION RESEARCH IN GERMANY – THE HOMANN (1718) MAP OF THE CHRISTMAS FLOOD OF 1717 56
Houben Georg J.

TECTONICALLY CONDITIONED BRINE LEAKAGE INTO USABLE FRESHWATER AQUIFERS – IMPLICATIONS FOR THE QUALITY OF GROUNDWATER EXPLOITED IN CENTRAL POLAND 59
Kaczor-Kurzawa Dorota

HYDROCHEMICAL CHARACTERIZATION OF VARIOUS GROUNDWATER AND SEEPAGE WATER RESOURCES LOCATED IN THE BAY OF PUCK, SOUTHERN BALTIC SEA 64
Klostowska Zaneta, Szymczycha Beata, Kulinski Karol, Lengier Monika, Leczynski Leszek

FOURIER SERIES SOLUTION FOR AN ANISOTROPIC AND LAYERED CONFIGURATION OF THE DISPERSIVE HENRY PROBLEM 70
Koohbor B., Fahs M., Belfort B., Ataie-Ashtiani B., Simmons C. T.

MAJOR GROUNDWATER RESERVOIR NR 112 IN THE COAST OF GDANSK BAY 76
Kordalski Zbigniew, Sadurski Andrzej

GROUNDWATER CHEMISTRY AND ORIGIN OF THE VISUTLA DELTA PLAIN 81
Krawiec Arkadiusz, Sadurski Andrzej

SALT WATER INTRUSION IN THE BREAKTHROUGH VALLEY OF THE RIVER AA BETWEEN THE FLEMISH COASTAL PLAIN AND THE SAINT OMER BASIN (FRANCE) 88
Lebbe Luc, Depret Devlin, Claus Jasper, Devriese Gert Jan

HYDROGEOLOGICAL RESEARCHES IN THE 4D CARTOGRAPHY PROGRAM IN THE COASTAL ZONE OF THE SOUTHERN BALTIC 94
Lidzbarski Miroslaw, Tarnawska Ewa

FERRUGINOUS GROUNDWATERS AS A SOURCE OF P, FE, AND DIC FOR COASTAL WATERS OF THE SOUTHERN BALTIC SEA: (ISOTOPE) HYDROBIOGEOCHEMISTRY AND THE ROLE OF AN IRON CURTAIN 99
Lipka Marko, Böttcher Michael E., Wu Zijun, Sültenfuß Jürgen, Jenner Anna-K., Westphal Julia, Dellwig Olaf, Escher Peter, Schmiedinger Iris, Winde Vera, Struck Ulrich

ON THE USE OF COMSOL MULTIPHYSICS FOR SEAWATER INTRUSION IN FRACTURED COASTAL AQUIFERS 104
Mozafari B., Fahs M., Ataie-Ashtiani B., Simmons C. T., Younes Rafic

MONITORING THE GROUNDWATER QUALITY/QUANTITY FROM YOUR DESKTOP – APPLICATION TO SALT WATER INTRUSION MONITORING EMI: ENVIRONMENTAL DATA MANAGEMENT INTERFACE 110
Neyens D., Baïsset M., Lovighi H.

CAN BOMB-PEAK TRITIUM PERSIST IN THE TRANSITION ZONE? A CASE STUDY FROM THE GERMAN ISLAND OF LANGEOOG 115
Post V. E. A., Houben G. J.

ASSESSING GROUNDWATER VULNERABILITY TO SEA WATER INTRUSION IN THE COASTLINE OF THE INNER PUCK BAY USING GALDIT METHOD 119
Potrykus Dawid, Gumula-Kawecka Anna, Jaworska-Szulc Beata, Pruszkowska-Caceres Malgorzata, Szymkiewicz Adam, Dzierzbicka-Glowacka L.

SEAWATER INTRUSION DUE TO PUMPING MITIGATED BY NATURAL FRESHWATER FLUX: A CASE STUDY IN WLADYSLAWOWO, NORTHERN POLAND 125
Pruszkowska-Caceres Malgorzata, Szymkiewicz Adam, Jaworska-Szulc Beata, Przewlócka Maria, Gomula-Kawecka Anna, Potrykus Dawid

FRESHENING OF SALINIZED GROUNDWATER IN GDANSK QUATERNARY AQUIFER 131
Przewlócka Maria, Jaworska-Szulc Beata, Kozerski Bohdan, Pruszkowska-Caceres Malgorzata, Szymkiewicz Adam, Kwaterkiewicz Andrzej

CHARACTERIZATION OF A REGIONAL COASTAL ZONE AQUIFER USING AN INTERDISCIPLINARY APPROACH – AN EXAMPLE FROM WESER-ELBE REGION, LOWER SAXONY, GERMANY 137
Rahman Mohammad Azizur, González Eva, Wiederhold Helga, Deus Nico, Elbracht Jörg, Siemon Bernhard

NEW DATA ON SEAWATER INTRUSION IN LIEPAJA (LATVIA) AND METHODOLOGY FOR ESTABLISHING BACKGROUND LEVELS AND THRESHOLD VALUES IN GROUNDWATER BODY AT RISK F5 144
Retike I., Bikše J.

DEEP GEOELECTRICAL INVESTIGATION TO BOUND A COASTAL THERMAL OUTFLOW AREA 150
Rizzo Enzo, Zuffianò L. E., Santaloia F., Polemio M.

SPREADING OF BRINE IN THE PUCK BAY IN VIEW OF IN-SITU MEASUREMENTS 155
Robakiewicz Malgorzata

DEVELOPMENT AND APPLICATION OF DIAGNOSTIC TOOLS FOR SEAWATER INTRUSION ANALYSIS IN HIGHLY HETEROGENEOUS COASTAL AQUIFERS 161
Adil Sbaï M., Amraoui N.

MODELING SALTWATER INTRUSION SCENARIOS FOR A COASTAL AQUIFER AT THE GERMAN NORTH SEA 167
Schneider A., Zhao H., Wolf J., Logashenko D., Reiter S., Howahr M., Eley M., Gelleszun M., Wiederhold H.

COMPARISON OF MANUALLY AND AUTOMATICALLY DERIVED FRESH-SALINE GROUNDWATER BOUNDARIES FROM HELICOPTER-BORNE EM DATA AT THE JADE BAY, NORTHERN GERMANY 173
Siemon Bernhard, Steuer Annika, Deus Nico, Elbracht Jörg

EVALUATION OF GROUNDWATER POTENTIAL AND SALINE WATER INTRUSION USING SECONDARY GEOPHYSICAL PARAMETERS: A CASE STUDY FROM WESTERN MAHARASHTRA, INDIA 179
Suneetha N., Gupta Gautam

HYDROGEOLOGICAL STRUCTURES AS BURRIED VALLEY ALONG THE EASTERN POMERANIA POLISH COAST OF THE BALTIC SEA 185
Szelewicka Anna

DEEP SUBMARINE GROUNDWATER DISCHARGE INDICATED BY PORE WATER CHLORIDE ANOMALIES IN THE GULF OF GDANSK, SOUTHERN BALTIC SEA 188
Szymczycha Beata, Klostowska Zaneta, Kulinski Karol, Winogradow Aleksandra, Jakacki Jaromir, Klusek Zygmunt, Grabowski Milosz, Brodecka-Goluch Aleksandra, Graca Bozena, Stokowski Marcin, Koziorowska Katarzyna, Rak Daniel

MODELING GROUNDWATER FLOW AND SALINITY EVOLUTION NEAR TSF ZELAZNY MOST. PART I – GROUNDWATER FLOW 192
Swidzinski Waldemar

MODELING GROUNDWATER FLOW AND SALINITY EVOLUTION NEAR TSF ZELAZNY MOST. PART II – CHLORIDE TRANSPORT 198
Swidzinski Waldemar

LABORATORY SCALE INVESTIGATION OF DISPERSION EFFECTS ON SALTWATER MOVEMENT DUE TO CUTOFF WALL INSTALLATION 204
Takahashi Masahiro, Momii Kazuro, Luyun Roger
MANTRA-O18: AN EXTENDED VERSION OF SUTRA MODIFIED TO SIMULATE SALT AND δ^{18}O TRANSPORT AMID WATER UPTAKE BY PLANTS 210
Teh Su Yean, De Angelis Donald L., Voss Clifford I., Sternberg Leonel, Koh Hock Lye
GROUNDWATER SALINITY MAPPING OF THE BELGIAN COASTAL ZONE TO IMPROVE LOCAL FRESHWATER STORAGE AVAILABILITY 217
Vandevelde D., Van Baaren E., Delsman J., Karaoulis M., Essink G. O., De Louw P., Vermaas T., Pauw P., De Kleine M., Thofte S., Teilmann R., Walraevens K., Van Camp M., Dominique H., Dabekaussen W., Gunnink J., Vandenbohede A.
SALTWATER INTRUSION UNDER CLIMATE CHANGE IN NORTH-WESTERN GERMANY - MAPPING, MODELLING AND MANAGEMENT APPROACHES IN THE PROJECTS TOPSOIL AND GO-CAM 223
Wiederhold Helga, Scheer Wolfgang, Kirsch Reinhard, Azizur Rahman M., Ronczka Mathias
Author Index

International Advisory Committee

Mark BAKKER, Delft University of Technology, Netherlands

Wesley DANSKIN, US Geological Survey, USA

Klaus HINSBY, Geological Survey of Denmark and Greenland, Denmark

Georg HOUBEN, Federal Institute for Geosciences and Natural Resources BGR, Germany

Gualbert OUDE ESSINK, Deltares, Netherlands

Vincent POST, Federal Institute for Geosciences and Natural Resources BGR, Germany

Christian SIEBERT, Helmholtz Centre for Environmental Research UFZ, Germany

Clifford VOSS, US Geological Survey, USA

Kristine WALRAEVENS, Ghent University, Belgium

Adrian WERNER, Flinders University, Australia

Helga WIEDERHOLD, Leibniz Institute for Applied Geophysics, Germany

Scientific Committee coordinator

Adam SZYMKIEWICZ, Gdańsk University of Technology, Poland

Scientific Committee

Behzad ATAIE-ASHTIANI, Sharif University of Technology, Iran

Giovanni BARROCU, University of Cagliari, Italy

Michael BÖTTCHER, Leibniz Institute for Baltic Sea Research, Germany

Kazimierz BURZYŃSKI, Gdańsk University of Technology, Poland

Jesus CARRERA, Institute of Environmental Assessment and Water Research IDAEA-CSIC, Spain

Jean-Christophe COMTE, University of Aberdeen, UK

Emilio CUSTODIO, Universitat Politecnica de Catalunya, Spain

Lidia DZIERZBICKA-GŁOWACKA, Institute of Oceanology PAS, Poland

Maria-Dolores FIDELIBUS, Polytechnic University of Bari, Italy

Thomas GRAF, Leibniz University Hannover, Germany

Rainer HELMIG, University of Stuttgart, Germany

Bill HU, Jinan University, Guangzhou, China

Joseph HUGHES, US Geological Survey, USA

Beata JAWORSKA-SZULC, Gdańsk University of Technology, Poland

Jimmy JIAO, Hong Kong University, China (Hong Kong)

Bohdan KOZERSKI, Gdańsk University of Technology, Poland

Christian LANGEVIN, US Geological Survey, USA

Luc LEBBE, Ghent University, Belgium

Perry de LOUW, Deltares, Netherlands

Chunhui LU, Hohai university, Nanjing, China

Leszek ŁĘCZYŃSKI, University of Gdańsk, Poland

Teresa de MELO, Instituto Superior Tecnico Lisboa, Portugal

Holly MICHAEL, University of Delaware, USA

Suzana Maria Gico de Lima MONTENEGRO, Federal University of Pernambuco, Brazil

Leanne MORGAN, University of Canterbury, New Zealand

Mike MÜLLER-PETKE, Leibniz Institute for Applied Geophysics, Germany

Namsik PARK, Dong-A University, South Korea

Maurizio POLEMIO, Research Institute for Geo-Hydrological Protection, Italy

Maria POOL, Institute of Environmental Assessment and Water Research IDAEA-CSIC, Spain

Małogrzata PRUSZKOWSKA-CACERES, Gdańsk University of Technology, Poland

David PULIDO VELAZQUEZ, Geological and Mining Institute, Spain

Clare ROBINSON, University of Western Ontario, Canada

Andrzej SADURSKI, Polish Geological Institute, Poland

Frans SCHAARS, Artesia, Netherlands

Jiri ŠIMŮNEK, University of California - Riverside, USA

Peter SINCLAIR, Pacific Community

Shaul SOREK, Ben Gurion University, Israel

Torben SONNENBORG, Geological Survey of Denmark and Greenland, Denmark

Pieter STUYFZAND, KWR Watercycle Research Institute, Netherlands

Waldemar ŚWIDZIŃSKI, Institute of Hydro-Engineering PAS, Poland

Alexander VANDENBOHEDE, De Watergroep, Belgium

Yoseph YECHIELI, Geological Survey of Israel, Israel

Statement of Peer review

In submitting conference proceedings to *Web of Conferences*, the editors of the proceedings certify to the Publisher that

1. They adhere to its **Policy on Publishing Integrity** in order to safeguard good scientific practice in publishing.
2. All articles have been subjected to peer review administered by the proceedings editors.
3. Reviews have been conducted by expert referees, who have been requested to provide unbiased and constructive comments aimed, whenever possible, at improving the work.
4. Proceedings editors have taken all reasonable steps to ensure the quality of the materials they publish and their decision to accept or reject a paper for publication has been based only on the merits of the work and the relevance to the journal.

Title, date and place of the conference

25th Salt Water Intrusion Meeting (SWIM)
17 - 22 June, 2018 Gdańsk, Poland

Proceedings editor(s):

Adam Szymkiewicz
Andrzej Sadurski
Beata Jaworska-Szulc

Date and editor's signature 21.08.2018

17, avenue du Hoggar - PA de Courtabœuf – BP 112 - 91944 Les Ulis Cedex A (France)
Tél. : 33 (0)1 69 18 75 75 - Fax : 33(0)1 69 07 45 17 – www.edpsciences.org

Influence of tides, bathymetry, lithology and regional flows on the salinization process in nature area the Rammegors

Ilja America[1], Perry de Louw[2], George Bier[1], Sjoerd van der Zee[1]
1 Environmental Sciences, Wageningen University, Wageningen, Netherlands
2 Deltares

ABSTRACT

Nature area Rammegors, which has recently been transformed from a fresh inner-dyke nature area to a salt tidal area. Due to this transformation, salt water is infiltrating in a fresh waterlens. This salinisation process is investigated in more detail by two- and three dimensional models together with mearsurements in the area. Zeeland project FRESHEM has provided detailed isohaline maps of the area and Deltares is making transient isohaline maps based on measurements made by an ERT-cable which is situated in Rammegors. These data has been and will be used to investigate which factors; bathymetry, lithology, tides or regional groundwater flow, will have the largest impact on the salinization process in Rammegors. This investigation shows that discretization size has an influence on the speed and spatial distribution of salt plumes. Lithology has the largest influence on the salinization process, followed by bathymetry. Spring and neap tides do differ from the normal tides situation only when bathymetry is not taken into account.

INTRODUCTION

Rammegors (figure 1) was still an inner-dike nature reserve near the island of Tholen, wedged by the Scheldt-Rhine channel on the one hand and the Easternscheldt on the other, 40 years ago. It took part of a dynamic tidal system consisting of deep channels, tidal flats and salt marshes (De Louw et al., 2016). This changed with the construction of the Scheldt-Rhine canal, 40 years ago. Rammegors was isolated for that part of the Easternscheldt and changed from an open tidal area to a fresh inner-dike nature area. Typical salty tidal nature that is unique in Europe, was lost by the creation of Rammegors. This unique salty tidal nature is still present in the Easternscheldt. However, since the construction of the storm surge barrier in 1986, less water has been flowing into and out of the Easternscheldt which result in a decrease in tidal amplitudes, volumes and velocities of the water but most of all, a decrease in sediment exchange. A situation is created where erosion and sedimentation are not in equilibrium with each other. This is called the sand deficit problem or either 'sand hunger'. The ecological value of the area decreases because the surface of intertidal areas is reducing 50 ha per year, on average (De Louw et al., 2016). In order to give this valuable nature more space, the connection between Rammegors and Easternscheldt is restored. This creates a salt tidal area, where salt marshes can thrive again and birds can find food on the mud flats. The Rammegors is therefore completely given back to the Easternscheldt. The recovery of the original Zeeland landscape, but now controlled behind a safe dike. This MSc-study focuses on nature area Rammegors which experiences an unstable density stratification, as heavier seawater lies on top of lighter freshwater, since the connection with the Eastern Scheldt in December 2016. This freshwater forms a lens which was created by precipitation-excess when the Rammegors was embanked by dikes (Pauw, 2015). The unstable density stratification can give rise to free convection (fingering) were salinization occurs much more rapid because advective flow is driven by density differences (Post and Kooi, 2003).

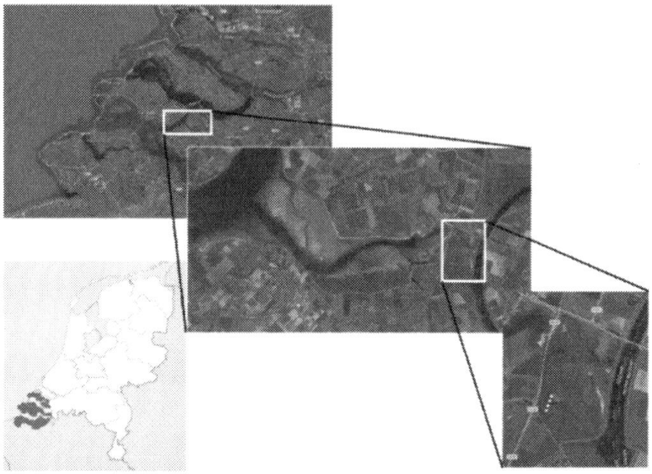

Figure 1. Study area: nature area Rammegors, the Netherlands (Google Maps). Dotted white line is transect where 2 dimensional model will be based upon.

The general scarcity of high-resolution field data is the main cause that the role of density driven fingering triggered by seawater inundation, has not yet fully assessed under field conditions. Many studies have investigated the process of free convection in unstable conditions in the past (Wooding et al. 1997a,b; Simmons et al., 1999; Kinzelbach, 2000; Post et al., 2004). Most of them have focused on small-scale problems and almost no aquifer scale investigations have been done, due to the strict discretization requirements (Kooi et al., 2000).

Many areas in the world (will) experience salinization because of sea level rise. This is mainly caused by lateral seawater intrusion or vertical infiltration during ocean surge inundation (Yu et al., 2016). Especially low-lying estuaries will experience higher grades of salinization in the future because those areas will be inundated earlier (FitzGerald et al., 2008; Ketabchi et al., 2016). Unfortunately, these future inundations will not be controlled and could therefore damage large parts of the world. It is therefore of great importance to understand the salinization process in Rammegors, where inundations are controlled, in order to minimalize damage in possible future inundations all over the world.

The previous mentioned studies have provided much understanding in the free convection process on small scale. The scarcity of high-resolution field data and the low computation ability prevented investigation on larger scales, which could be validated with trustworthy data. Nature area Rammegors is currently changing to a controlled dynamic tidal system consisting of tidal flats and salt marshes. It is therefore the perfect situations to investigate how an area transforms from a freshwater nature area to a saltwater nature area. In the period 2014 and 2015, several electromagnetic measurements has been made commissioned by the province of Zeeland, Deltares, BGR and TNO for the Zeeland-project FRESHEM. These electromagnetic measurements gives information about the fresh, brackish and saline groundwater. An ERT-cable has been installed by Deltares to measure the salinization process in more detail during time. These isohaline images are still in development and are hopefully finished in June, so they can be showed during the SWIM.
The main goal of this MSc-studyis to investigate which factors; bathymetry, lithology, tides or regional groundwater flow, will have the largest impact on the salinization process in Rammegors. The influence of discretization size, is also be investigated.

 https://doi.org/10.1051/e3sconf/20185400001

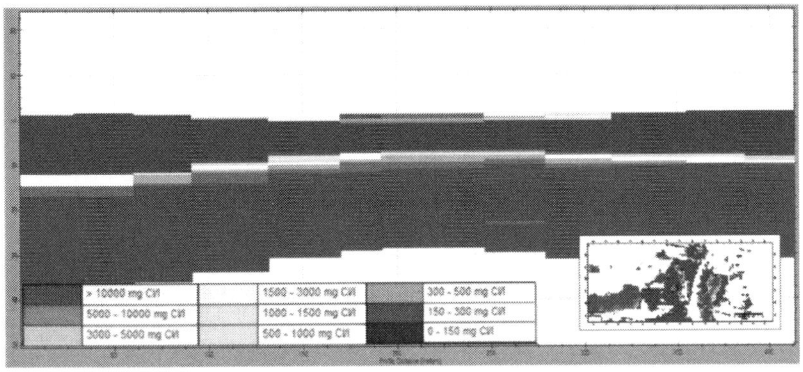

Figure 2. Isohaline FRESHEM map of 2D transect.

METHODS

In order to investigate how tides, bathymetry, lithology and regional groundwater flow influence free convection and salinization processes in nature area the Rammegors, a 2 dimensional model has been created in a generic MODFLOW/MT3DMS-based computer program, named SEAWAT in a GMS environment. A first impression of the spatial distribution of the chloride concentrations in March 2015 is made with the FRESHEM data of Rammegors. Figure 2 shows the isohaline map of the transect (figure 1) made with the FRESHEM data. The thin salt layer which is located at the top of the salt layer is exactly located in the gully where salt water is flowing. It also shows that the fresh water lens which is created in 40 years is 15 meter thick, on average.

A 2 dimensional model has been made to understand the process of free convection in more detail. First, the influence of discretization size on the spatial and temporal scales on the free convection process are investigated. Six different discretization sizes (0.5, 1, 2, 5, 10 and 25 meter) has been used. The model is based on a transect in the south-west of Rammegors (figure 1). This transect is situated perpendicular to the hydraulic heads, therefore no sideways fluxes are expected. Some parts of this transect will be inundated always/periodically/never due to its bathymetry and tides. Table 1 shows the implementation of the different factors. The lower elevated areas, which are present due to bathymetry, will always be inundated whereas higher elevated areas will periodically be inundated due to the influence of tides. The discretization size with the most detail and relative lowest computational burden will eventually be used for nine different versions where the impact of regional groundwater flow, tides, bathymetry and lithology on the salinization process will be investigated. In the first version (a) the variable density flow will be switched off and in the second version (b) it will be turned on. This second version will be the reference case to which all other versions are compared to . In the third version (c), a simplified bathymetry transect is implemented. Lithology is implemented in the fourth version (d). In fifth till eighth version (e-h), tides, springtide, tides with bathymetry and springtide with bathymetry are implemented, respectively. In the last version (i), all factors are implemented in a combined version.

In the last stage, a 3 dimensional model will be created where all factors will be implemented. In this model, the future perspective of Rammegors will be created. The 3D-model is under construction and therefore not shown here.

Table 1. Implementation of different factors in 2 dimensional model.

Factor	Implementation
Bathymetry	5 different levels ranging from +1.42 m NAP (left) to +0.36 m NAP (right).
Lithology	7 lithological layers, with poorly permeable layers around -13 m NAP and -25 m NAP.
Tides	Sinusoidal shape of $0.98 + 0.4\sin(\frac{2\pi}{\frac{12.4}{24}} * t)$.
Spring and neap tides	Sinusoidal shape of $0.98 + 0.2\sin(\frac{2\pi}{14.75*24} * t) + 0.4\sin(\frac{2\pi}{12.4} * t)$.

RESULTS

Discretization size

In figure 3, the concentration contours after 10 years are shown with different discretization sizes, ranging from 0.5 till 25 meter. A clear distinction in plume shapes between the smaller discretization sizes (0.5 till 2 meter) and the larger discretization sizes (5 till 25 meter) can be observed, whereby the smaller discretization sizes show more detailed plumes. With increasing discretization size the amount of plumes with concentrations higher than 20 kg m^{-3} increases within the scales till 1 meter. The free convection plumes which are developed with a discretization size of 2 meter, are more emerged and broader than the plumes that were formed at smaller scales. The areas where fresh water flows in the direction of surface are clearly visible when the discretization size is small. This indicates that the heavier salt water can flow downwards and lighter freshwater will flow upwards. The start locations of the different plumes do differ which indicate that there are no preferential start locations.
The concentration contours of the larger discretization sizes show large salt containing areas, which do not look like plumes. The areas become larger when discretization size increases. The locations where no salt is observed are more or less the same but decreases with increasing discretization size. The influence of the boundaries are clearly visible in the models when smaller discretization sizes have been used. The plumes which have formed in the models with larger discretization sizes migrate faster downwards. The salt mass, which is present in earlier stages, is higher and with that the maximum salt mass is reached earlier.

Factors influencing salinization process

In figure 4, the concentrations contours for the different factors after 10 years can be observed. By comparing figures 4a and 4b with each other, the influence of density and thus convection becomes clear. In figure 4a, no plumes are present. The salt infiltrates very gradually downwards. The infiltration is influenced by both boundaries as the infiltration goes faster at the left and right boundaries. When only bathymetry is implemented in the model, the plumes have changed to larger salt containing areas and the shape of the original plumes have diminished completely (figure 4c). The model where only lithology is implemented, shows an increase in salt concentration in the upper most layer which forms a dense salty layer (figure 4d). Figures 4e till 4h show how spring and neap tides differ from a normal tide situation where bathymetry is taken into account and not. In the situations where bathymetry is not implemented the salinization process shows the same pattern as the initial situation. The plumes are only smaller, thinner and less dense. The situation where springtides are implemented (figure 4f) shows more plumes than the situation where normal tides are implemented (figure 4e). The positions of these plumes are not the same. The

 https://doi.org/10.1051/e3sconf/20185400001

plumes in the situations where bathymetry (figures 4g and 4h) is implemented have more or less the same location and their shapes are almost the same. In the situation where all factors are combined, the influence of bathymetry is clearly visible by the 4 different section of plumes. The plumes are larger when the surface is covered by water for a longer period which is influenced by tides (most right plume). The most left plume, which has no water levels on the left hand side, is also large. The influence of lithology is clearly visible by an abrupt change in concentration distribution.

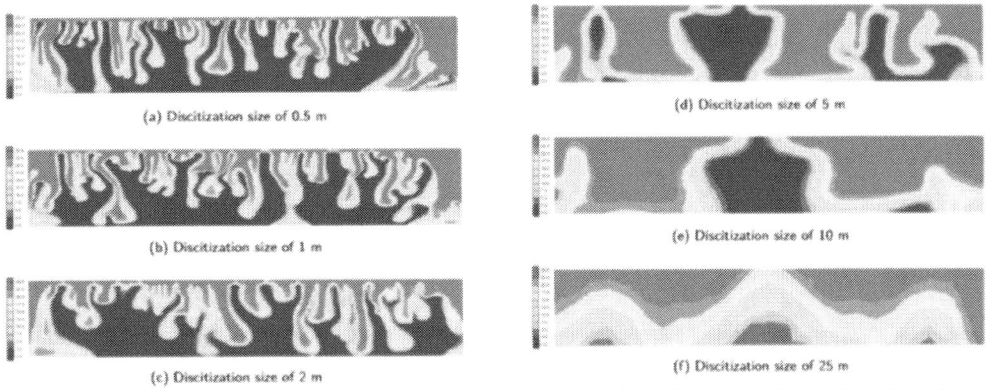

Figure 3. Concentration contours after 10 years with different discretization sizes.

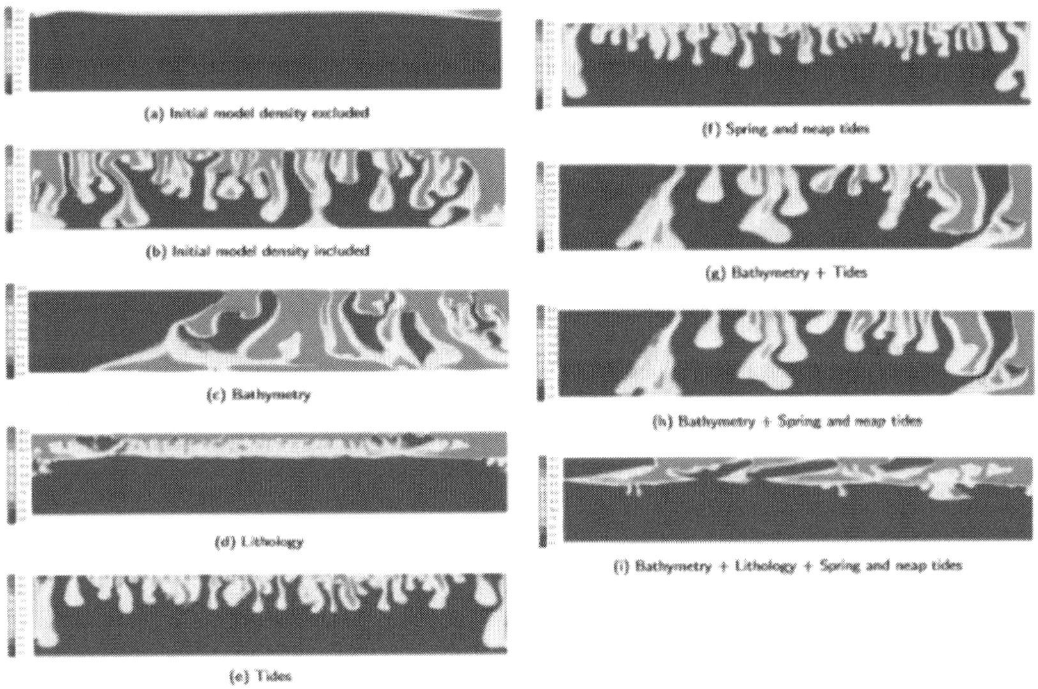

Figure 4. Concentration contour after 10 years for different factors on the salinization process, with a discretization size of 1 meter.

DISCUSSION AND CONCLUSIONS

In conclusion, discretization size has a major influence if free convection wants to be measured and recorded. There is a large difference between small and large discretization sizes, whereas small discretization sizes show free convection plumes and large discretization sizes not. Lithology has the largest influence on the salinization process, followed by bathymetry. The influence of bathymetry is mainly caused by preferential start location, which are located at the transition zones where higher elevated areas begin. Spring and neap tides do differ from the normal tides situation only when bathymetry is not involved.

REFERENCES

De Louw, P., Ysebaert, T., Bouma, T., B.-M., Van Dalen, J., Van Belzen, J.-S., 2016. Rammegors tidal restoration.

FitzGerald, D. M., Fenster, M. S., Argow, B. A., Buynevich, I. V., 2008. Coastal impacts due to sea-level rise. Annu. Rev. Earth Planet. Sci. 36, 601-647.

Ketavbchi, H., Mahmoodzadeh, D., Ataie-Ashtiani, B., Simmons, C. T., 2016. Sea-level rise impacts on seawater intrusion in coastal aquifers: Review and integration. Journal of Hydrology 535, 235-255.

Kinzelbach, W., 2000. A three-dimensional physical model for verification of variable-density flow codes. In: Calibration and Reliability in Groundwater Modelling: Coping with Uncertainty; Proceedings of the ModelCare'99 Conference Held in Zurich, Switzerland, form 20 to 23 September 1999. No. 265. IAHS, p. 399.

Kooi, H., Groen, J., Leijnse, A., 2000. Modes of seawater intrusion during transgressions. Water resources research 36 (12), 2581-3589.

Pauw, P. S., 2015. Field and model investigations of freshwater lenses in coastal aquifers. Wageningen University.

Post, V. E., Kooi, H., 2003. Rates of salinization by free convection in high-permeability sediments: insights form numerical modelling and application to the dutch coastal area. Hydrogeology Journal 11 (5), 549-559.

Post, V. E. A., et al., 2004. Groundwater salinization processes in the coastal area of the netherlands due to transgressions during the holocene. Ph.D. thesis.

Simmons, C. T., Narayan, K. A., Wooding, R. A., 1999. On a test case for density-dependent groundwater flow and solute transport models: The salt lake problem. Water Resources Research 35 (12), 3607-3620.

Wooding, R., Tyler, S. W., White, I., 1997a. Convection in groundwater below an evaporating salt lake: 1. onset of instability. Water Resources Research 33 (6), 1199-1217.

Wooding, R., Tyler, S. W., White, I., Anderson, P. 1997b. Convection in groundwater below an evaporating salt lake: 2. evolution of fingers or plumes Water Resources Research 33 (6), 1219-1228

Yu, X., Yang, J., Graf, T., Koneshloo, M., O'Neal, M. A., Michael, H. A., 2016. Impact of topography on groundwater salinization due to ocean surge inundation. Water Resources Research 52 (8), 5794-5812.

Contact Information: Ilja America, Wageningen University, Environmental Sciences Department, Email: ilja.america@wur.nl;

High frequency saltwater intrusion monitoring using borehole geophysical tools (SMD)

M. Baïsset[1], D. Neyens[1]
[1]imaGeau, Montpellier, Hérault, FRANCE

ABSTRACT

In February 2016, two remote controlled geophysical monitoring tools (SMD) have been installed for the first time in the Reunion Island. Settled into two piezometers drilled into a basaltic coastal aquifer, between the ocean and a production well, they allow the record of groundwater electrical conductivity (ECw) logs on a 30 min basis. Thanks to those two tools, water operator continuously knows the shape and the position of the SWI as data are available online on a secured web application designed especially for SWI data management.

During the observation period a 5,15 m rise of SWI interface has been recorded. Knowing the average porosity, water table elevation and SWI interface position it is possible to estimate available fresh groundwater volume. Along a 1 km band between extraction well and the ocean, available fresh groundwater volume was found to be 1 259 000 m3 in June 2016. In June 2017, due to SWI progression this volume was found to be 777 000 m3, that to say a 480 000 m3 volume of freshwater replaced by brackish water.

SMD network will now be spread in the Reunion Island to improve coastal extraction well management knowing SWI shape and position on a high frequency basis.

INTRODUCTION

Due to a high demographic pressure, Reunion Island coastal aquifers are more and more subject to saltwater intrusion. Some coastal wells settled too close from the shore, in basaltic aquifers has to be abandoned, and new ones have to be drilled farthest from the coastline inducing financial damaged. As saltwater intrusion is increasing years after years on this island, it demands strong investments to precisely monitor fresh/saline interface position and its evolution according freshwater abstractions.

To achieve this objective, two remote controlled geophysical monitoring tools (SMD) have been installed into two piezometers settled in between the coastline and a production wells. They allow the record of hourly EC vertical logs to know on an hourly basis SWI interface position. Those geophysical tools are connected to a Web application (EMI) used by water operator to adapt fresh water extraction flow rate according to SWI.

METHODS

An SMD is an autonomous geophysical tool installed in a screened PVC piezometer used to constantly record water's electrical conductivity along the aquifer's vertical axis. It provides a real-time picture of the position and evolution of the saltwater intrusion available online on a web application. Two SMD have been installed in piezometers settled rescpectively at 125 m and 355 m from the Indian Ocean. They continuisly record EC logs every 30 min from 2 m absl to 27 m absl with a 100 cm vectival resolution.

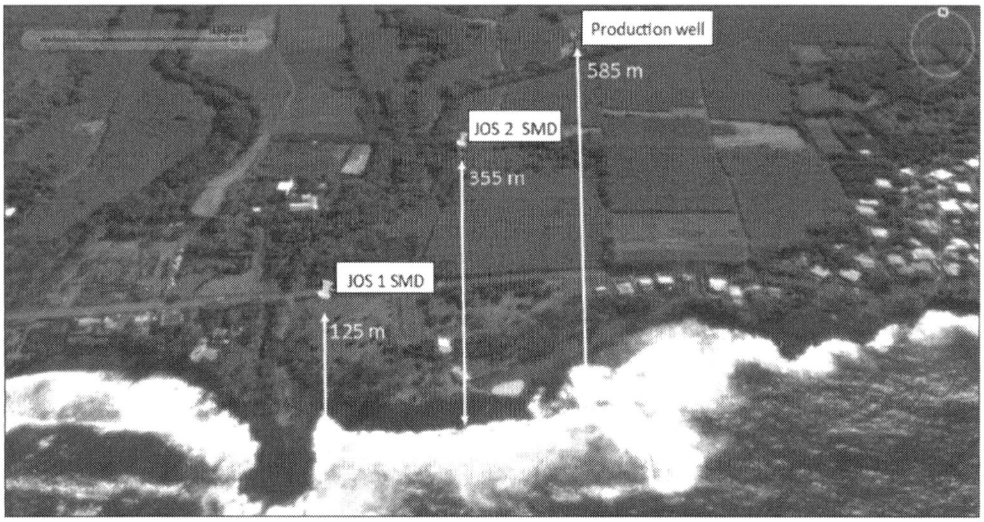

Figure 1. SMD localization named JOS1 and JOS2.

RESULTS

During the observation period, from February 2016 to July 2017, more than 20 000 ECw logs have been recorded. This huge data amount allows analyzing SWI movement according different tide events, different rain events and different well extraction rates. Furthermore, it allows accurate calculation of monthly SWI interface mean position without artefact. Indeed, comparison of punctual logs do not allows apprehending low frequency SWI interface movements as they are not recorded in comparable tide event, rain event or swell event.

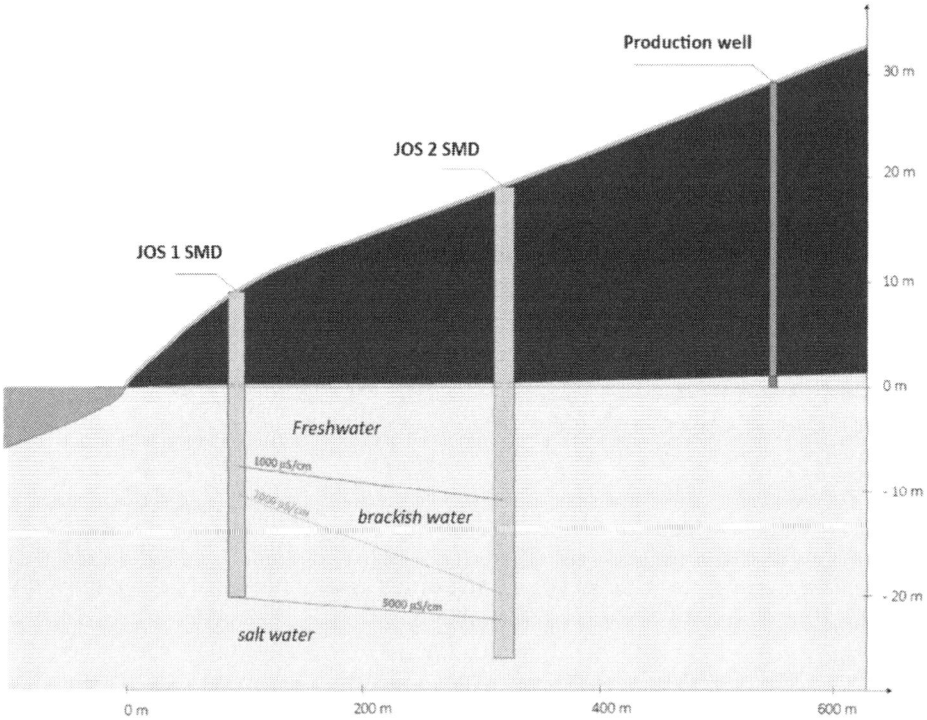

Figure 2. SWI shape measured by SMD.

 https://doi.org/10.1051/e3sconf/20185400002

During the observation period at JOS 1 SMD, a 17% water conductivity increase has been recorded between the two dry seasons and a 22% water conductivity increase has been record between the two rainy seasons (Figure 3). Mean SWI interface position (1mS/cm limit) is going 30 cm up from June 2016 to June 2017.

At JOS 2 SMD located 200 m further inland, a 35% water conductivity increase has been recorded between the two dry seasons and a 32% water conductivity increase has been recorded between the two rainy seasons (Figure 4). Mean SWI interface position (1mS/cm limit) is going 5,15 m up from June 2016 to June 2017.

Figure 5 presents comparison of SWI interface position between June 2016 and June 2017 (1mS/cm limit).

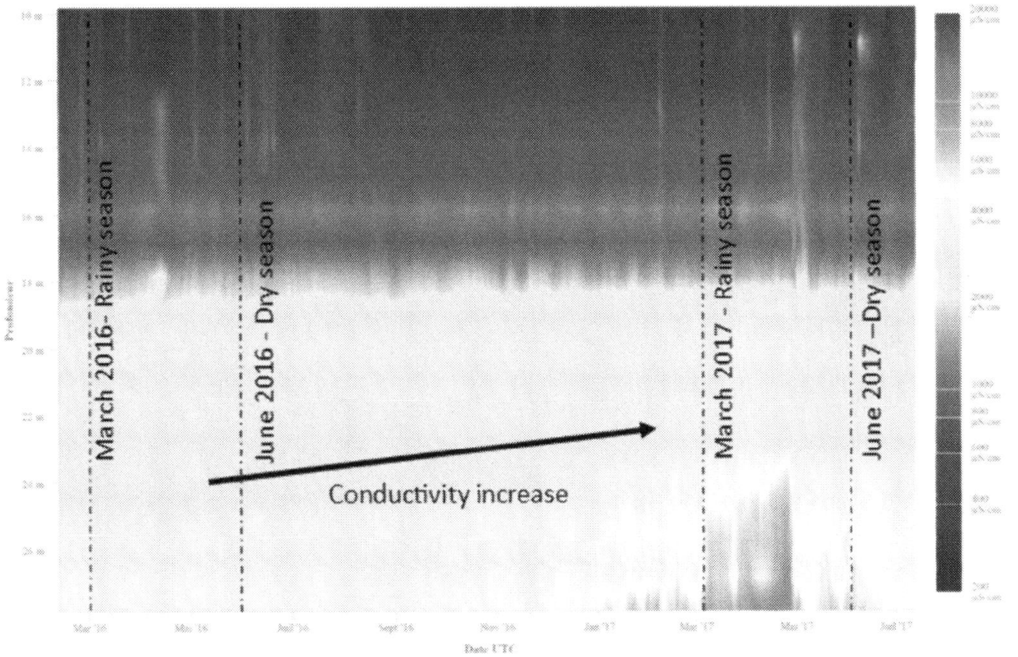

Figure 3. SMD logs measured by JOS 1 from July 2016 to July 2017.

Available fresh groundwater volume along 1 km coastline can thus be estimate knowing position of SWI interface (1mS/cm limit), water table elevation and mean porosity according the following equation:

Available fresh groundwater volume = porosity x (Area between Water table and SWI interface) x 1000

As water table elevation and SWI position are measured on a 30 min base, it is possible to know available fresh groundwater volume at this range of time. Monthly average can also be calculated to know the impact of SWI on available fresh groundwater amount. Figure 6 presents fresh groundwater volume balance from June 2016 to June 2017. This total balance gives a 480 000 m3 volume of freshwater lost, replaced by brackish water during the observation period.

https://doi.org/10.1051/e3sconf/20185400002

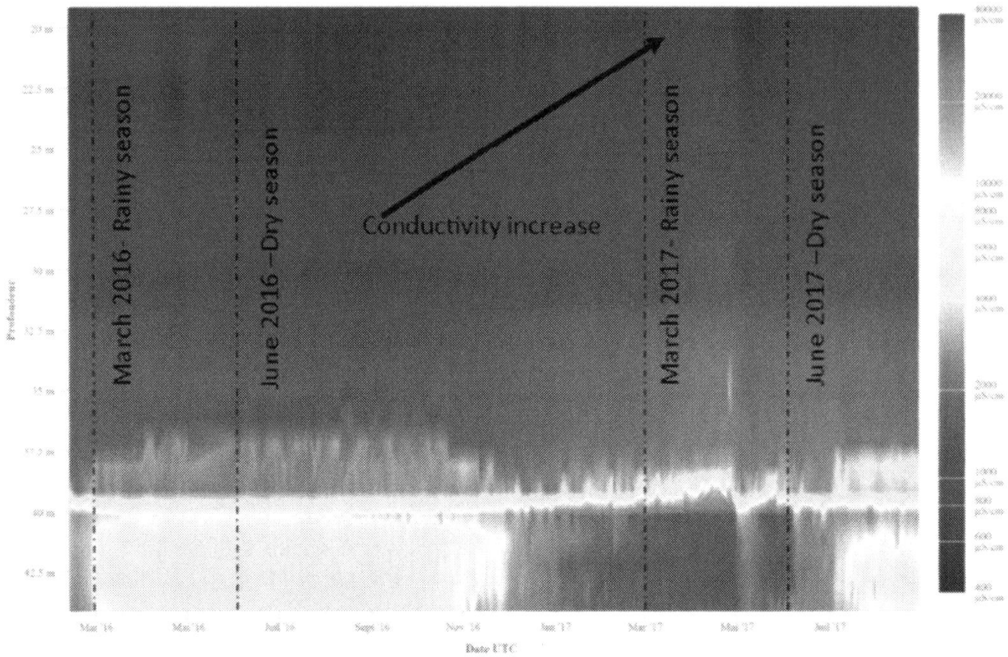

Figure 4. SMD logs measured by JOS 2 from July 2016 to July 2017.

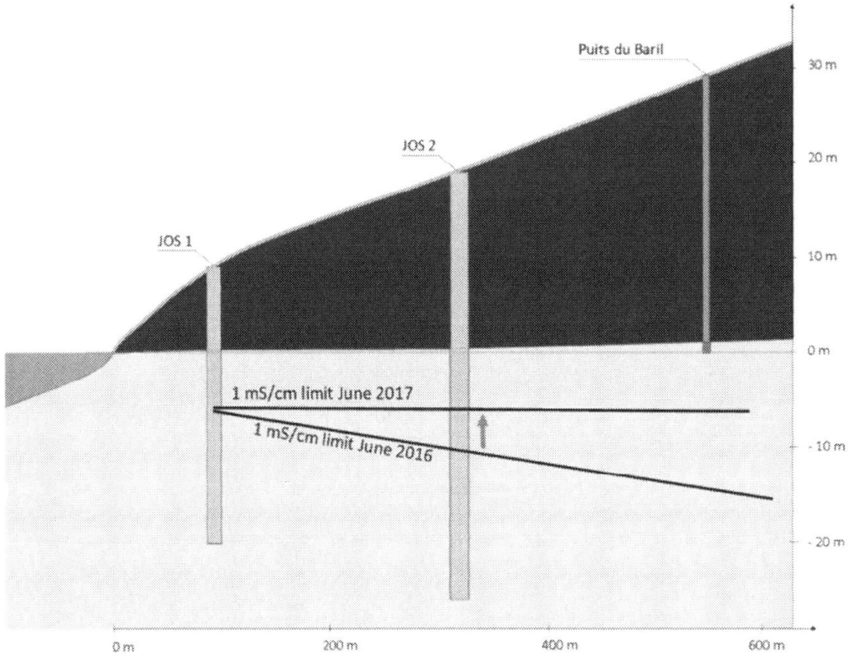

Figure 5. SWI interface position in June 2016 and June 2017 (1mS/cm limit)

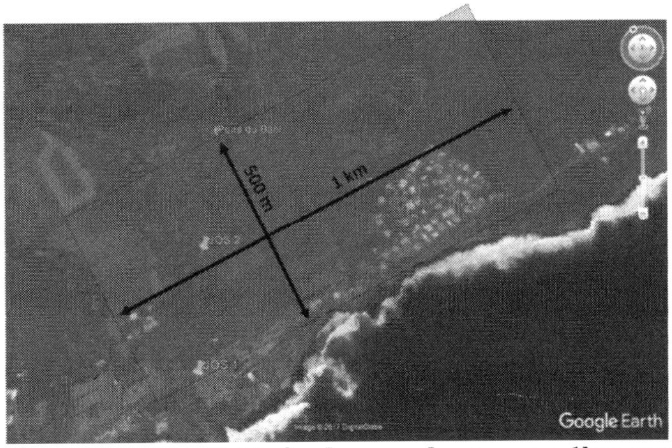

Figure 6. Calculation of available fresh groundwater on a 1km coastline band.

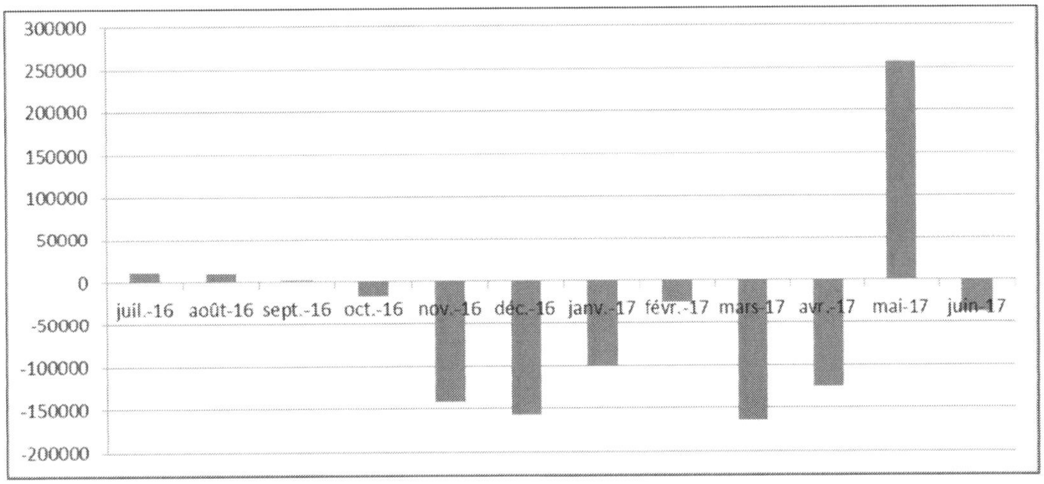

Figure 7. Monthly freshwater volume loss or gain between June 2016 and June 2017 in cubic meter on a 1km coastline band.

DISCUSSION AND CONCLUSIONS

This is the first SMD network that has been installed in the Reunion Island. Since the beginning of the measurement period, more than 20 000 EC logs have been recorded allowing to know SWI interface position on a 30 min basis.

These measurements have showed a 5,15 m rise of SWI position (1mS/cm limit) between June 2016 and June 2017. Knowing the average porosity, water table elevation and SWI interface position it is possible to estimate available fresh groundwater volume. Along a 1 km band between extraction well and the Ocean, available fresh groundwater volume was found to be 1 259 000 m3 in June 2016. In June 2017, due to SWI progression this volume was found to be 777 000 m3, that to say a 480 000 m3 volume of freshwater replaced by brackish water.

In the Reunion Island, SMD network will now be spread to improve coastal extraction well management knowing SWI shape and position on a high frequency basis.

REFERENCES

Aunay B, Broch d'Autelans R (2011) Montée du niveau marin induite par le changement climatique. Phase 1 - Diagnostic préalable aux conséquences sur l'intrusion saline dans les aquifères côtiers de La Réunion - Rapport final - BRGM/RP-59049-FR. BRGM, La Réunion

Bourhane A (2014) Méthodes d'investigation de l'intrusion marine dans les aquifères volcaniques : La Réunion et La Grande Comore. Thèse de Doctorat, Université de La Réunion

La Réunion. Rapp. BRGM/RP-63818-FR, 90p. BRGM, La Réunion

An Approach to Delineate Groundwater Bodies at Risk: Seawater Intrusion in Liepāja (Latvia)

J. Bikše[1], I. Retike[1,2]
[1]Faculty of Geography and Earth Sciences, University of Latvia, Riga, Latvia
[2]Hydrogeology division, Latvian Environment, Geology and Meteorology Centre, Riga, Latvia

ABSTRACT

Groundwater quality in coastal areas is frequently affected by seawater intrusion as a consequence of intensive water consumption. To achieve "good chemical status" of a groundwater body according to Water Framework Directive the effects of saline or other intrusions should not be observed. Groundwater pumping in former decades has caused a significant seawater intrusion into confined aquifer in Liepāja and has led to deterioration of relatively wide coastal area of the third largest city in Latvia. However, the area affected by seawater intrusion is a small part of groundwater body F1 which overall chemical status is good. Thus, no specific management measures have been applied to explore and control seawater intrusion. A political decision was made to delineate the area affected by seawater intrusion as new groundwater body at risk- F5. This study demonstrates simple approach for delineation of groundwater bodies at risk in coastal areas. Delineation process was based on chloride concentration gradient along the well profile and gradient based buffers. Finally, the worst-case scenario was selected for delineation of boundaries.

INTRODUCTION

The key purpose of the Water Framework Directive (WFD) is to protect and improve the quality of all European waters, including groundwater. Another purpose of maintaining good groundwater status is to eliminate any significant damage to ecosystems which directly depend on groundwater body. Member States already had to reach good groundwater status (both chemical and quantitative) by 2015 but no later than 2027 (WFD 2000).

Groundwater body is a reporting unit set to estimate its quantitative and chemical status, and to which environmental objectives under Article 4 of the directive should apply. Groundwater body should be delineated in such a way to ensure proper status assessment and monitoring, effective management and future treatment (Kovács et al. 2012). According to WFD (2000) groundwater body is in bad chemical status if the chemical composition of the body is such that the concentration of pollutants exhibits the effects of saline or other intrusions. Groundwater Directive (GWD 2006) goes on to state that the chemical status assessment shall be carried out for all groundwater bodies at risk of not meeting WFD Article 4 objectives in relation to each of the pollutants which contribute to the groundwater body being so characterized.

Currently the area affected by seawater intrusion in Liepāja is relatively small part of groundwater body F1 (total area of F1 is 2974 km^2) which is in good chemical status. Latvia used proposed procedure for general assessment of the chemical status of the whole groundwater body and applied 20% criterion. It follows that the proportion of the total area or volume of the groundwater body affected by seawater intrusion (in this case) was compared to the total area or volume of the groundwater body. An exceedance of less than

20% of the area does not lead to a poor status of the groundwater body (European Commission 2009).

The boundaries of 22 groundwater bodies in Latvia were mainly delineated by watersheds using modelled groundwater levels as input data. On the one hand, it is impossible to exceed 20% criterion considering the size of F1 groundwater body. One the other hand, it would be inappropriate to set whole groundwater body in bad status considering the size of the area affected by seawater intrusion. Thus, a political decision was made to delineate the area affected by seawater intrusion in Liepāja as a separate groundwater body at risk- F5. This study demonstrates developed simple approach for delineation of groundwater bodies at risk in coastal areas using long term chloride data.

Hydrogeological setting

Seawater intrusion takes place in freshwater aquifer at city Liepāja - Upper Devonian Mūru-Žagares (D_3*mr-žg*) partly confined aquifer which is formed of weakly cemented sandstones, siltstones and dolomites in total thickness of 44 - 47 m and at depth of 38 - 43 m. D_3*mr-žg* is covered by Upper Devonian clayey formations and Quaternary till and sand around Liepaja city. Deposits of D_3*mr-žg* aquifer are exposed at the bottom of the Baltic Sea - approximately 5 km from the coast. Cause is the dipping of Devonian deposits towards S-SE (and outcropping at N-NW) and the lack of Quaternary sediments at some areas at the sea.

Underlying formations consist of dolomitic marls, clays, dolomite and sandstones forming several aquitards and minor aquifers. At the depth of 230 - 241 m lies Upper Devonian Gaujas and Middle Devonian Burtnieku formation (D_2br+D_3gj) - significant hydraulically connected aquifer with total thickness of more than 100 m consisting of sandstones and clays. The D_2br+D_3gj aquifer has no direct connection to uppermost aquifers and the Baltic Sea, however, D_2br+D_3gj aquifer is mainly used for industrial water supply due to elevated mineralization and high sulphate content from gypsum dissolution.

Thus, the most important aquifer for water supply needs in Liepāja surroundings is shallow D_3*mr-žg* freshwater aquifer. It contains good quality drinking water and has been extensively used for decades.

Historical evolution of seawater intrusion

In the beginning of the 20th century and intensive groundwater abstraction from Upper Devonian Mūru-Žagares (D_3*mr-žg*) aquifer took place in Liepāja surroundings. First evidence of water level decrease and quality change (high chloride concentrations) in water supply wells were reported in early 1930's. However, regular groundwater monitoring started only in 1961 and already formed depression cone was identified. A decision was made to switch to centralized water supply and the new well field "Otaņķi" was created in 1961. Still, it abstracted groundwater from the same aquifer - D_3*mr-žg*. As a result, depression cone expanded southeast and in 1976 reached "Otaņķi". In 1986 the center of depression cone was located in "Otaņķi" area and groundwater levels were reported as 14 m below sea level in D_3*mr-žg* aquifer. In ten years the seawater intrusion has moved 1 km southeast and reached the northern part of the lake "Liepāja". A specific measure was established to prevent further movement of saltwater - an abstraction of the affected water for technical needs (Janikins et al. 1993).

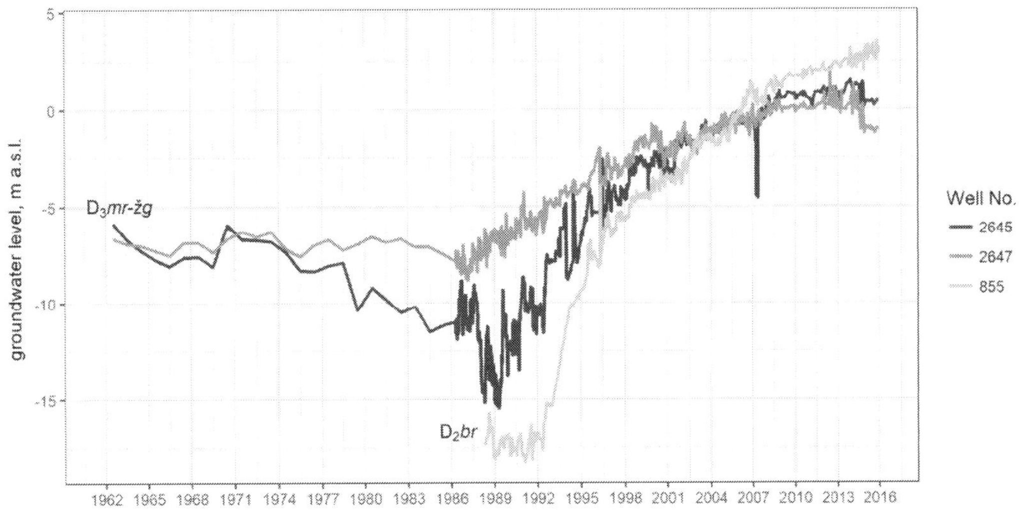

Figure 1. Historical groundwater levels in D_3*mr-žg* and D_2*br* aquifers.

It was decided to exploit Middle Devonian Burtnieku and Upper Devonian Gaujas (D_2*br*+D_3*gj*) aquifer to reduce the pressure on Upper Devonian Mūru-Žagares (D_3*mr-žg*) aquifer. New wells in "Otaņķi" in D_2*br*+D_3*gj* aquifer were installed in 1967. Other aquifers in the area contain high sulphate content and mineralization, and therefore are unsuitable for water supply.

Depression cone started to decline in the beginning of 1990's when consumption rate dramatically decreased because of the collapse of USSR. In about 15 years groundwater level in D_3*mr-žg* aquifer restored and exceeded the Baltic Sea level (Figure 1). Chloride concentrations decreased in marginal zone of the area affected by seawater intrusion, still in central part of the zone chloride values remain high.

METHODS

Information about monitoring and abstraction wells was gathered from the largest Latvian hydrogeological database "Wells" (limited access) (Urbumi 2017). Records about chloride concentrations are dated from 2017 back to 1960's.

Delineation of groundwater body at risk was made based on chloride concentrations from all wells with available data. The profile consisting of four groundwater wells located at the edge of the intrusion was used to determine chloride concentration gradient in D_3*mr-žg* aquifer (Figure 2). Six groundwater sampling campaigns from 2003 to 2017 have been made in these wells thus, permitting calculation of chloride concentration gradient along the profile for several time frames. Linear gradients along the well No.2647 and rest of the wells on the profile was calculated. Natural background level for chloride (13.2 mg/l) was subtracted from saltwater impacted samples to obtain gradient that represents only seawater impact (Bikse et al. 2016).

Chloride ion gradient was used to calculate buffer around each groundwater well based on chloride concentration. If time series from a single well were available, then sample with highest chloride concentration was used as it accounts for worst case scenario. As a result, gradient based buffer shows the maximum distance around each groundwater well where seawater intrusion could take an effect. Thus, the boundary of the groundwater body at risk was delineated as a combination of outermost buffers (Figure 3).

RESULTS

Calculation of chloride concentration gradient along the profile (Figure 2) yielded results ranging from 0.65 mg/l/m to 1.01 mg/l/m with median value of 0.83 mg/l/m. To account for worst case scenario, the value of 0.65 mg/l/m was selected for calculation of gradients buffers.

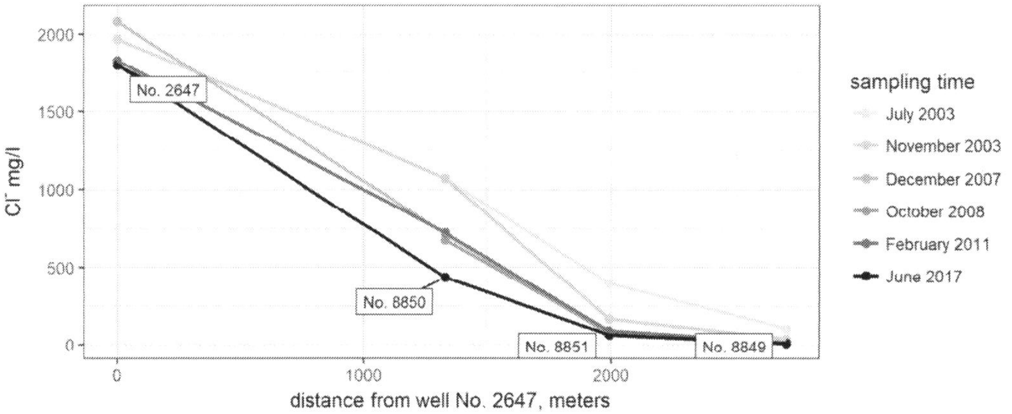

Figure 2. Chloride concentration variations in groundwater well samples along the profile.

Constructed buffers cover large part of the city Liepāja, especially at the central and northern part of the city (Figure 3). The largest buffer has a radius of 3385 m that corresponds to chloride concentration of 2200 mg/l.

The boundary of the groundwater body was delineated along the outermost buffers and along the Baltic Sea shoreline, however, minor corrections were made to smoothen the boundary and to include closest wells from groundwater well field "Otaņķi" which is the main water supply in study area and must be protected from seawater intrusion. Total area of the new groundwater body at risk (F5) is 46.42 km^2 and it includes most of the city Liepāja, western part of lake "Liepāja" and southern part of lake "Tosmare".

DISCUSSION AND CONCLUSIONS

The methodology described in this paper permit simple approach to delineate groundwater bodies at risk in coastal areas, however, two preconditions must be met: (1) large number of groundwater samples from seawater affected area with acceptable spatial coverage must be available and (2) profile of groundwater wells covering central and marginal part of seawater intrusion must be sampled simultaneously for several times to account for temporal variability of chloride concentration gradient. A simplification was made - that seawater intrusion has equal impact on all directions from each groundwater well, therefore an approach that would include seawater intrusion dynamics would yield boundary with different extension and therefore may improve this approach.

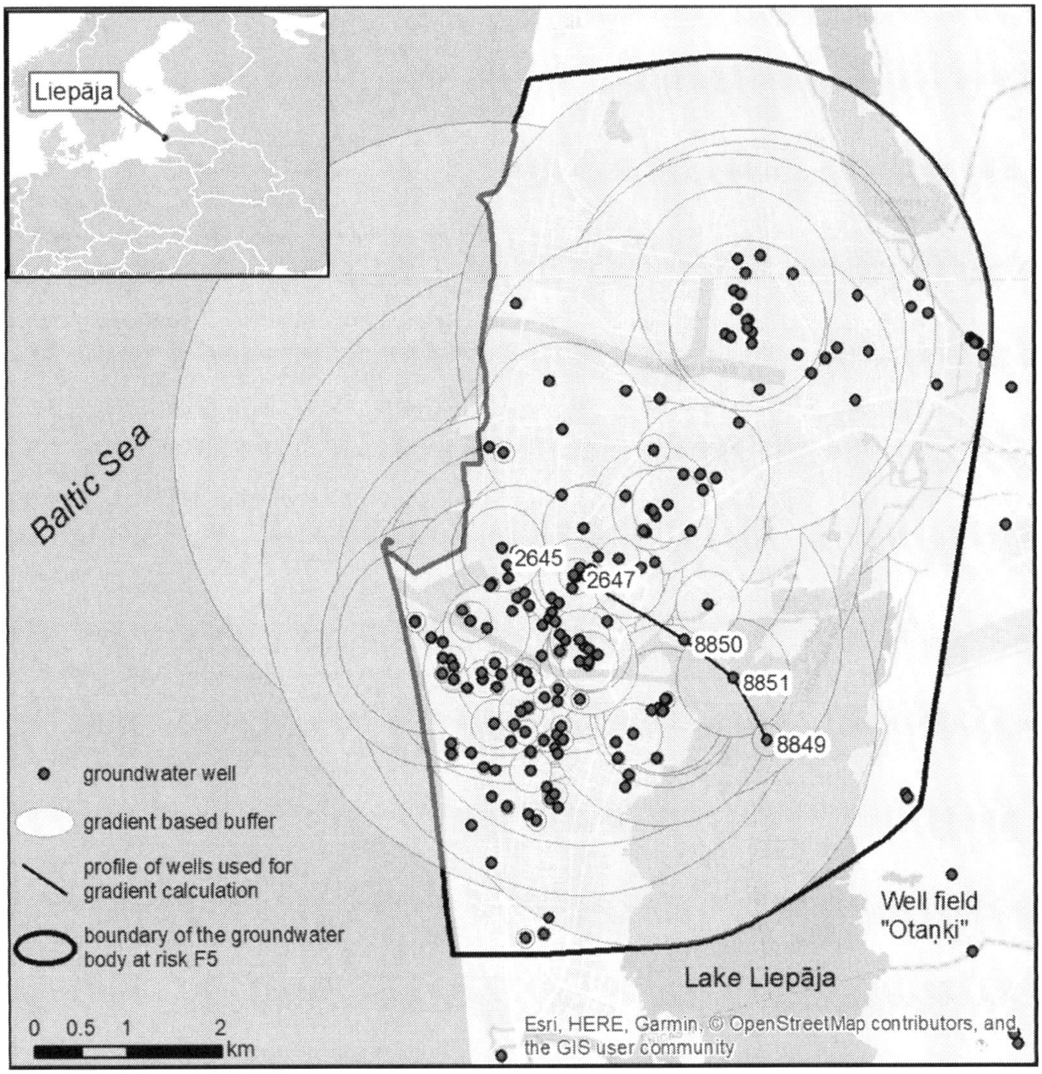

Figure 3. The boundaries of delineated groundwater body at risk - F5.

ACKNOWLEDGEMENT

The study was supported by performance-based funding of University of Latvia Nr.AAP2016/B041//ZD2016/AZ03 within the "Climate change and sustainable use of natural resources" programme.

REFERENCES

Bikse, J., Retike, I., Kalvans, A. 2016. Historical evolution of seawater intrusion into groundwater at city Liepaja, Latvia. In Proceedings of XXIX Nordic Hydrological Conference v.1: 29. Aleksandras Stulginskis University.

European Commission 2009. Guidance Document No 18: Guidance on Groundwater Status and Trend Assessment. European Communities, Luxembourg.

GWD 2006. Directive 2006/118/EC of the European parliament and of the Council of 12 December 2006 on the protection of groundwater against pollution and deterioration.

Janikins, J., Levina, N., Levins, I., Prols, J., Straume, J. 1993. Pazemes ūdeņu monitorings Latvijā (Groundwater monitoring in Latvia). Valsts ģeoloģijas fonda inventāra Nr.14674 (In Latvian).

Kovács, J., Nagy, M., Czauner, B., Kovács, I.S., Borsodi, A.K., Hatvani, I.G. 2012. Delimiting sub-areas in water bodies using multivariate data analysis on the example of Lake Balaton (W Hungary). Journal of Environmental Management. 110: 151–158.

Urbumi 2017. Derīgo izrakteņu atradņu reģistrs (Register of Mineral Resources). Latvian Environment, Geology and Meteorology Centre. Available from: www.meteo.lv (In Latvian).

WFD 2000. Directive 2000/60/EC of the European parliament and of the Council Establishing a Framework for Community Action in the Field of Water Policy.

Contact Information: J.Bikše, University of Latvia, Faculty of Geography and Earth Sciences, Jelgavas Street 1, LV-1004, Riga, Latvia, Email: janis.bikse@lu.lv

Modeling Seawater Intrusion to Coastal Aquifers in South Coast of Laizhou Bay, China

Yawen Chang[1,2], **Bill X. Hu[1,2,3,*]**, Xue Li[1,2]
[1]School of Water Resources and Environment, China University of Geosciences (Beijing), No. 29 Xueyuan Road, Haidian District, Beijing 100083, P. R. China
[2]MOE Key Laboratory of Groundwater Circulation and Environmental Evolution, China University of Geosciences (Beijing), Beijing 100083, P. R. China
[3]Institute of Groundwater and Earth Sciences, Jinan University, Guangzhou 510632, P. R. China

ABSTRACT

In this study, a two-dimensional SEAWAT 2000 model is developed to simulate the seawater intrusion to coastal aquifers and brine water/fresh water interaction in the south of Laizhou Bay, Shandong Province, China and forecast the seawater intrusion and brine water/freshwater interface development in the coming years. The model profile is perpendicular to the coastal line, about 40 km long and 110 m in depth, and consists of two interfaces, freshwater-saline water interface and brine water-saline water-seawater interface. The parameters of aquifers in the SEAWAT-2000 model are calibrated by trial-error method repeatedly to fit the head and salinity measurements. Based on the historical groundwater and brine water exploration and natural precipitation condition, the prediction results indicate that equivalent freshwater head in shallow freshwater-saline water area will decrease year by year and decline 2.0 m in the forecasting period, caused by groundwater over-pumping for irrigating farmlands. The groundwater head in the brine-saline water area will also decrease about 1.8 m in forecasting period. A larger depression cone appears in the brine area, with smaller funnels in other areas. The salinity in the brine area finally drops below 105g/l. In the meanwhile, the salinity increases in other areas, damage fresh groundwater resources.

INTRODUCTION

Excessive exploitation of groundwater, climate change, sea-level rise, variation of land use haveleaded to serious seawater intrusion to coastal aquifers (Werner et al. 2013).To meet water needs in the agriculture, lives and industry, groundwater resources have been seriously over exploited in the last several decades (Datta et al. 2009).

Physically, the seawater intrusion is a density-dependent issue (Guo et al.2002). Mathematical modeling a seawater intrusion process needs to couple groundwater flow equation with solute (salt) transport equation (Priyanka et al.2015). Some studies were focused the freshwater and seawater interaction areas. For example, Li et al. (2009) applied a density-variable numerical code, SEAWAT 2000, to simulate a submarine groundwater discharge (SGD) caused by tidally induced sea water recirculation and a terrestrial hydraulic gradient. Some other studies were on freshwater-saline or brine water areas. Xue et al. (1997) developed a three-dimensional numerical model of seawater intrusion and brine water and freshwater interaction in Longkou-Laizhou area to study the interface between freshwater and saltwater, evolution and development of the transitional zone, and evaluate seawater intrusion caused by pumping.

The above studies focus either on fresh water-saline water interaction area or brine-saline water area, without considering both interaction areas. In this paper, a large scale of a two-dimensional cross-section, perpendicular to the coastal line in the Laizhou bay, China is chosen to study the variations of both fresh water-saline water interface and brine-saline water interface with time. The cross-section is about 40 km in length and 110 m in depth. Along the 40 km cross section, there exist a freshwater-saline water interaction area and a brine water and saline water interaction area. The two interaction areas will influence each other.

OVERVIEW OF THE STUDY AREA

The study area is located in the south of Laizhou bay, specifically in Changyi, Weifang City, Shandong Province(Figure 1).

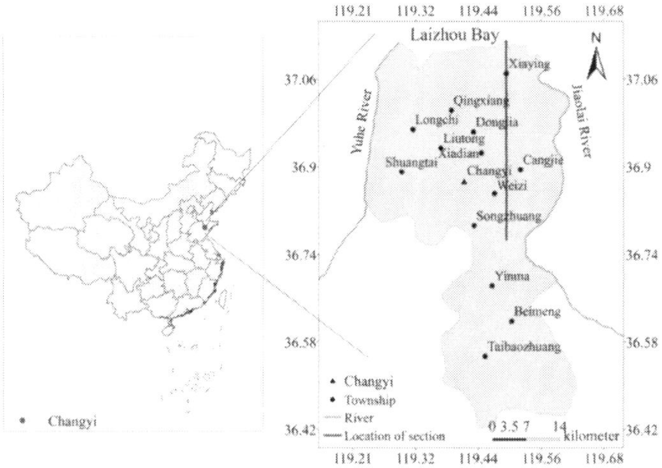

Figure 1. Geographical location of the study area.

The average annual precipitation from 1956 to 2012 is 612.5-660.1 mm, and 72.4-76.2 % of the annual precipitation is from June to September. The evaporation in April to July is relatively large. The average monthly evaporation is l91.4-371.6 mm.

The groundwater in the Quaternary loose sediments is widely distributed in the coastal plain area. The lithology of the water-bearing sand strata is from medium coarse sands and gravels (thickness 5-30 m) to silt, fine sands and medium sands (sand thickness 10-30 m).

Since the late 1970s, due to the over exploitation of fresh groundwater, the pressure difference between salt water head and freshwater head decreased, and the brine water flowed through the ancient alluvial sand layer to the south quickly. In the Laizhou Bay coastal areas, groundwater depth in the sea-land interlaced layers is shallow and it is easy to be evaporated or to receive the supply of precipitation. So the regional water levelis closely related with weather condition and affected by seasonal change, showing a certain periodical characteristics.

GROUNDWATER FLOW MODEL IN THE STUDY AREA

According to hydrogeological conditionsin the section, the model area is ranged from Songzhuang Township-Weizi Town in the south to Xiaying Town-Wei River Estuary in the north. The total length is about 39200 m with 1000 m extended beyond the shoreline. The model boundaries are defined as: The south piedmont boundary of the model is a constant flow boundary with 40 m^3/d since several small rivers recharge the aquifer; the north boundary extended into sea can be defined as a constant head and constant salt concentration boundary with 0 m and 3500 mg/l, respectively; the bottom boundary is treated as no-flow boundary; Water exchange between the aquifer and atmosphere takes place at the upper boundary, including atmospheric precipitation and evaporation. Atmospheric precipitation and evaporation during the period from January to December, 2013 are measured by Changui weather station. The values of atmospheric precipitation recharge are products of precipitation and precipitation infiltration coefficient of 0.15. Evaporation limit depth is 4 m. In the freshwater-saline water area, the pumping rate of with 207,200 m^3/year are estimated according agricultural irrigation amounts for various crop requirements during the period from January to December, 2013, because the pumped groundwater is mainly used for irrigation. The pumping rate divided by the area is0.3 m^3/m^2·year, which is used in the simulation. Brine exploitation is mainly concentrated in April, May and June with a total exploitation of 486,000 m^3/year during the period from January to December, 2013, which, divided by the area, is 0.43 m^3/m^2·year. In the freshwater-saline water area pumping rates are distributed to the grids of 0 m - 27200 m, and in the brine-saline water area pumping rates are distributed to the grids of 30600-35400 m. The lateral recharge in the piedmont is 15,330 m^3/year. The hydraulic structure of the sediment cross-section is generated according to hydraulic properties of the sediment layers. The section is divided into 15 layers. The first, third, fifth, seventh, ninth, eleventh, thirteenth, fifteenth layer are aquifers. The lithology of the aquifers is from silt to coarse gravel. The remaining layers consist of sandy clay and clay with poor permeability, and are classified as aquitards.

In the freshwater-saline water area, the groundwater circulation in the south coast of Laizhou Bay is mainly controlled by precipitation, runoff and pumping processes. In recent years, due to the reduction of precipitation and the over-pumping, the groundwater level declines rapidly. In the brine-saline water area, the middle and deep layers in the brine funnel has a weak hydraulic connection with precipitation and shallow aquifers. The brine groundwater is mainly affected by the hydrostatic pressure and mining intensity in the recharge area.

In this paper, the mathematical model of groundwater flow with variable density is applied to simulate the seawater and brine water intrusion processes. The numerical simulation model of solute transport in groundwater flow is established using SEAWAT 2000, which couples MT3D with MODFLOW-2000, and considers the effect of density on groundwater flow.

This model consists of rectangular grid with 500 × 100 m. There are 392 columns and 15 layers with 5880 cells in total. The simulation period is from January 1, 2013 to December 31, 2013, and each month is taken as a stress period, and the number of days every stress period is usedas a time step. According to the lithology, the parameters areca liberated.

In general, the simulated groundwater levels and salt concentrations match the observations well. The groundwater flow field (Figure2(a)) and concentration field (Figure2(b)) could well describe the groundwater level and salinity distributions in the study area, in which there are freshwater-saline area and brine-saline water area. In freshwater-saline area, the groundwater head is relatively high, basically greater than 0 m, and salinity is less than 20,000 mg/l. Pumping groundwater occurs in shallow aquifers and the pumped water is used for irrigation. In brine-saline water area, there is depression cone with about a depth of 7.0 m below sea level, and its salinity is about 10,0000 mg/l. The brine exists in the deep aquifer and the brine water is pumped for salt industry, which has been practiced for several decades.

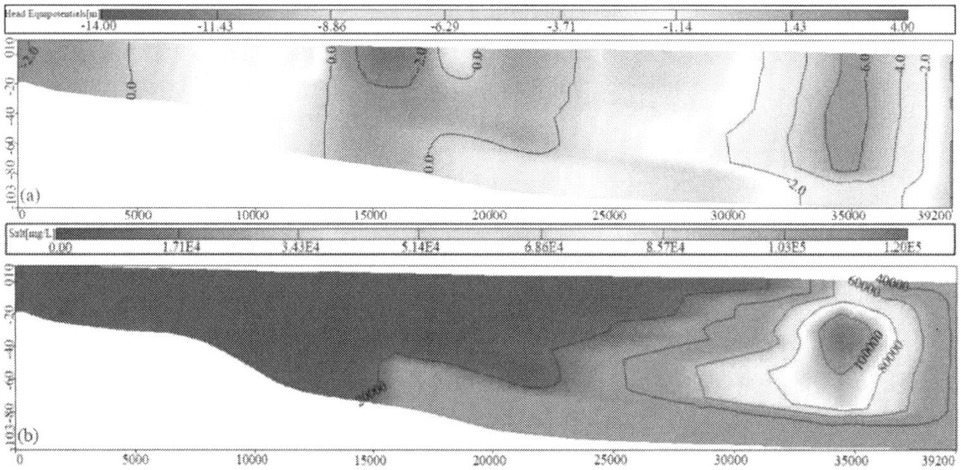

Figure 2. Groundwater flow field(a) and salinity field (b) at the end of simulation.

PREDICTION RESULTS AND ANALYSIS

The prediction simulation period for groundwater flow and solute transport in the cross-section area is 7 years (from January, 2014 to December, 2020).The rainfall has the periodicity, and the rainfall period during the period from January, 1983 to December, 1989 is a rainfall cycle, and assumed to be the cycle for the period from January, 2014 to December, 2020 in the prediction model. In the seven rainfall years, there are three consecutive dry years (> 75% cumulative frequency, rainfall < 483.1 mm), which can allow our model to meet the requirements for predicting groundwater level and quality trends under special climate conditions. Also, the evaporations from 1983 to 1989 are taken as the evaporations inputs in the prediction simulation. Pumping rates from 2014 to 2020 are taken as the same as those in 2013 in the freshwater-saline water area and brine-saline water area.

In the shallow freshwater area, the groundwater level is declining year by year. In a hydrological year, the amount of groundwater pumping increases with less precipitation and more evaporation when spring irrigation season starts. Therefore, the water level declines 0.3 m to the lowest in June. Then, the groundwater table will rise slowly with the precipitation increase and groundwater exploitation decrease, and reach the highest level in September or October. Due to large water demand for winter wheat, the water level declines from November to February in the following year. For many years, the groundwater level will decrease year by year due to the continuous drought condition, the reduction of the precipitation recharge and the excessive exploitation of groundwater. A large hydraulic cone has formed in the coastal plain, induced seawater intrusion to the freshwater aquifer. The salinity of the shallow freshwater area increases, the main reason is that the exploitation amount of fresh water is higher than recharge, which decreases hydraulic head and increases salt water inflow into the shallow layers.

In the deep layers of funnel area, the depth of the deep water recharge zone is large, and the hydraulic head is slow to restore, the water level has been decreasing for many years. In the forecast periods, the annual variation is about 6 m, and the mean hydraulic head decreases about 1.8 m in the next 7 years. The hydraulic gradient in the northern brine zone will increase, which would increase seawater intrusion to the groundwater. The salinity of the deep brine funnel is also reduced, mainly due to the pumping of the brine with high salinity, while the recharge of seawater with relatively low salinity or fresh water and no brine recharge will make the brine water salinity decrease.

SUMMARY AND CONCLUSIONS

The groundwater head at the freshwater-saline water area is decreasing year by year under drought condition, and the annual variation is about 0.3 m, and the hydraulic head decrease in the forecasting period is about 2 m. The groundwater head in the brine mining area is also decreasing year by year, and the annual variation is about 6 m, and the hydraulic head decrease in forecasting period is about 1.8 m.

The salinity of the brine decreases gradually, and finally drops below 105 g/l and the salinity of the other areas increases.The overpumping in the brine area will significantly decrease the brine salinity and deepen the brine water depth, which means the production will be less effective and cost will increase. The overpumping in the fresh water area will increase water and soil salinity, significantly affect the biological environment.

ACKNOWLEDGEMENTS

This work was partly supported National Natural Science Foundation of China (41530316) and National Key Research Project(2016YFC0402805).

REFERENCES

Datta, B., Vennalakanti, H., Dhar, A.2009. Modeling and control of saltwater intrusion in a coastal aquifer of Andhra Pradesh, India. Journal of Hydro-environment Research, no. 3:148-159.

Guo, W., Langevin, C.D.2002. User's guide to SEAWAT: a computer program for simulation of three-dimensional variable-density groundwater flow. US Geological Survey, 1-434.

Li, X.Y., Bill, X.H., Burnett, W.C., Santos, I.R., Chanton, J.P. 2009.Subarine groundwater discharge driven by tidal pumping in a heterogeneous aquifer. Ground Water, 47(4): 558-568.

Priyanka, B.N., Maheshab, A. 2015. Parametric studies on saltwater intrusion into coastal aquifers for anticipate sea level rise. Aquatic Procedia, no. 4: 103-108.

Werner, A.D., Bakker, M., Post, V.E.A., Vandenbohede,A., Lu, C.H., Ataie-Ashtianiab, B., Simmonsab, C.T., Barry, D.A. 2013. Seawater intrusion processes, investigation and management: recent advances and future challenges. Adv. WaterResour. 51 (1):3-26.

Xue, Y.Q., Wu, J.C., Xie, C.H., Zhang, Y.X. 1997. Research of seawater and salt water intrusion of Laizhou Bay. Chinese Science Bulletin,11(22):2360-2368.

Contact Information: bill.x.hu@gmail.com

Salinity problems in Mediterranean and island coastal aquifers in Spain

Emilio Custodio[1]
[1]Royal Academy of Science (Spain). Emeritus Professor, Technical University of Catalonia, Barcelona, Spain

ABSTRACT

The Spanish coastal aquifers are often complex and bounded by or inside mountainous areas. Most of them are in Quaternary and Miocene littoral sediments or highly karstified carbonate formations, or in the case of the Canary Islands in volcanic formations. Along the Spanish Mediterranean coast and the Balearic and Canarian archipelagos coasts, 95 groundwater bodies have been identified, often including several aquifers. Some kind of marine salinization problems have been identified in 70 groundwater bodies (20 with generalized problems) but only a few aquifers have detailed, specific studies on sea water intrusion. Information on salinization is often based on scarce and occasional data.

Table1. Synthesis of the conditions on the significant Mediterranean and island aquifers in Spain (elaborated from SASMIE, 2017).

Coastal aquifer	Salinization problems	Current situation	Action done
Explanations	*WS, water supply* *IF, impact on factories* *IA, impact on agriculture* *GW, groundwater* *WR, water resources* *IWRM, integrated WR management* *CUAS, GW user's association*		*GWAR, GW abstract. reduction* *WI, water importation* *RW, relocation of wells* *SWD, seawater desalination* *WA, well abandonment* *EI, efficiency improvement* *WWR, waste water reclamation*
Tordera Delta (Catalonia)	Serious in the 1980s. WP, IF, IA	Partial recovery; New use	GWAR, SWDP
Besós Delta (Barcelona)	Serious in the 1960s. Ws, IF	Reasonable recovery; Current partial use	WA
Llobregat Delta (Barcelona)	Serious in the 1970s. WS,IF, (IA)	Under control and managed. Full use. Key water reserve	GWAR, IWRM, CUAS, WI, SWD Inj.. barrier. Artif. recharge, WWR
Tarragona–Reus (Catalonia)	Serious in the 1960s WS, IF	Recovery	WA, WI
Vall d'Uixò (Plana de Castelló)	Serious in the 1990s. IA	Recovering	RW, EI
Marina Alta (Alicante)	Serious in the 1990s. WS	Partial recovery	SWD
Campo de Cartagena (Murcia Region)	Since the 1990s Brackish return irrigation flows. Poor disposal of "debrackishing" plant rejection	Uncertain. Water mixing, "debrackishing". Mar Menor eutrophication	Control of "debrackishing" WWR, WI Uncertain WI → enhances using brackish GW use
Campo de Níjar (Almeria)	Since the 1980s. IA Brackish natural GW (climatic) Saline irrigation return flow.	Uncertain	SWD (low use). CUAS (local goals) Mixing water
Campo de Dalías (Almeria)	Since the 1990s. Currently serious. IA Large greenhouse agricultural area in risk	Uncertain No clear decisions	CUAS (several), SWD, mixing Incipient IWRM
Mallorca Island (Balearic Is.)	Serious since the 1970s WS, touristic areas supply	Improvement around	WR, SWD Local management

		Palma de Mallorca	
Menorca Island (Balearic Is.)	Since the 1980s, in the two extremes	Uncertain, risk limited	SWD (not operative)
Eivissa–Ibiza (Balearic Is.)	Serious since the 1980s	Uncertain	SWD (available not operative)
Gran Canaria Island (Canary Is.)	Locally serious, since the 1970s	Uncertain No clear decisions	SWD, WR (partial), WA, WWR Local incipient IWRM
Tenerife Island (Canary Is.)	Local, since the 1980s	Excessive local GW exploitation restrictions	SWD, WR, WWR Restricted coastal aquifer use

The high human pressure near the coast for urban, industrial and tourism water supply and especially for irrigation has produced serious salinity problems in the coastal aquifers of Spain. These problems were acute in the 1950s in the Besós Delta and in the 1960s in the Llobregat Delta, near Barcelona. Afterwards, salinity problems showed up progressively in many other continental and island areas between 1970 and 1990. Up to 60 affected aquifers were identified in a survey carried out by the Geological and Mining Institute (MIMAM, 2000; López-Geta and Gómez-Gómez, 2008; Lópex-Geta and Fernández-Ruiz, 2012). Some of these salinization problems persist today while others have dwindled due to well abandonment after surface, reclaimed, and sea desalinated water was made available or to displacement of groundwater exploitation to areas further inland.

The public water administration, and the study and research institutes and the Academia made a great knowledge and monitoring effort on coastal aquifers in the period 1960–1990, and this is partly reflected in the SWIM and many other meeting proceedings, as summarized in SASMIE (2017), in which the existing scientific, technical and socio-economic information has been summarized. But the effort decayed since two decades ago, so current monitoring and study activity is much less that what it was. Several older than 30 years coastal aquifer public monitoring networks exist, although there are periods without data. Data become more scarce and irregular since the early 2000s, with some exceptions, due to growing economic difficulties in the public water authorities. Groundwater users' are rarely involved in monitoring and paying for the expenses of the public organizations in charge.

MANAGEMENT ACTIONS IN COASTAL AQUFIERS

Public management action is dominantly directed to new infrastructures to bring new water to coastal areas with seawater intrusion instead of non-structural action. Infrastructural action is easier, allows continuing existing development and is well accepted by people, but is expensive and often involves direct, indirect and hidden subsidies. Non-structural action is more difficult, may include closing abstractions, redefining water rights, taxation and limitations, is less prone to be accepted by people, and needs trained personnel for management, which is scarce in old organizations with a high inertia, and not prone to adapt to new requirements, as is the case of the water authorities.

Import of water from other areas is done of southeastern Spain, mainly in the Segura Basin, but the availability of this water is not secured in dry years, even transactions at a price, as this involves inter-regional problems and some. The most expensive seawater desalination was adopted two decades ago in the Mediterranean and Balearic Islands and more than four decades ago in the Canary Islands. Reclamation of treated sewage water is developing and well established in the Segura Basin, in southeastern Spain, where about 95% of urban sewage water is used for irrigation and part of it is applied in the coastal area. Private action is limited and often it reduces to voluntarily closing salinized wells and drilling new, more expensive ones in areas farther from the coast, or to "debrackish" local coastal aquifer groundwater.

Under favorable circumstances, coastal aquifers are often used to abstract seawater to feed a part of the desalinations plants when hydrogeological conditions favor the possibility of obtaining the needed saline water flows Getting seawater from the aquifer avoid many of the hydraulic and water quality problems associated to direct uptake from the sea. However, ill-designed uptakes from the coastal aquifer may entrain a consumption of aquifer freshwater that mixes with seawater. This explains why the water authorities are not favorable to grant permits to abstract saline and brackish water to avoid increasing the consumption of continental water as well as saline affections to other groundwater users and to nearby wetland areas. Mixed water may also create problems to the plants if alkalinity and hardness increases, as well as to high SO_2 concentration in some cases in the Canary Islands.

ENVIRONMENTAL CONSEQUENCES OF COASTAL AQUIFER DEVELOPMENT

Many small- to medium-size groundwater related wetlands are found in coastal Spain, presenting a wide spectrum of natural and man-made typologies. They were and are significant for local economies, especially in semiarid areas, and they have to be preserved by law. But hydrological studies are scarce. The role of coastal groundwater discharge to the sea to maintain important ecosystems and their services has been rarely considered up to now. According to current water planning norms, a given coastal discharge has to be allowed and its amount is subtracted before assigning water resources to the different demands. The erroneous and biased thinking that groundwater discharge to the sea is a "loss of freshwater" is still deeply rooted in population and groundwater users living in water scarce areas, and this influences coastal groundwater management policies.

Negative externalities associated to coastal aquifer abstraction are not generally considered. Such are the deterioration of ecosystems and the salinization of other wells. In some cases, the outflow of high nitrate concentration groundwater creates serious eutrophication problems to coastal lagoons and littoral waters. This is currently a serious concern in the Mar Menor, near Cartagena, southeastern Spain, which is ecologically important, yield valuable ecological services and have around large tourism developments.

The numerous large coastal seawater desalination plants dispose safely their reject brines into the sea. This is not always the case of the up to 2000 small and medium size groundwater "debrakishing" plants applying membrane technology. This creates serious environmental problems in numerous places, mostly in the Campo de Cartagena–Mar Menor (Murcia) and in eastern Gran Canaria Island.

ADMINISTRATIVE AND LEGAL ASPECTS OF COASTAL AQUIFER MANAGEMENT

In the current Spanish water law and water planning regulations, seawater intrusion is mainly considered a water quality issue. Except in some water plans (Xùquer-Jucar and Catalonia), it is not taken into account that the freshwater flow that can be sustainably abstracted form a coastal aquifer is less than recharge. The relative scarce consideration of coastal aquifers in the Spanish water legislation is due to be local situations or affecting a small area of the

 https://doi.org/10.1051/e3sconf/20185400005

territory the water authorities have to control and manage, even if they are densely inhabited and economically relevant areas. This situation is enhanced by the fact that despite being all water a public domain since 1985, in practice most of groundwater rights remain private as they correspond to previous legal regulations. The numerous actors in each case hinder the effective action capacity of water authorities.

There is not a clear policy on the application of the European Water Framework Directive further than trying that the groundwater bodies in bad status change into good status. In some cases, getting the good status implies a disproportionate effort, which may be not socially justified. So, less strict conditions are proposed, but not always supported by studies and a wide space and time vision of consequences and externalities.

ECONOMIC ASPECTS OF COASTAL AQUIFER DEVELOPMENT

In many Spanish coastal areas there is no other water resource than groundwater. When water is scarce in the coastal area, water importation from other areas and seawater desalination for urban and tourism supply and wastewater reclamation for irrigation are now common solutions, but at an increasing cost. The water cost at the place of use can be afforded by urban population, non-water intensive industry and tourism, but not by common farmers. Consequently, these farmers use as much as possible local resources, which often is groundwater abstracted by means of deep wells that are depleting aquifer water reserves or enhancing seawater intrusion. This situation is extremely difficult to be controlled and action is often politically unsupported. For urban and tourism supply, water of the general supply network is preferred to decrease quality hazards associated, although water from local coastal aquifers is often used a reserve to deal with water demand peaks and breakdowns. However this is mostly a private action as water planning do not generally take this into account

In much of the Spanish Mediterranean coastal area, common groundwater costs/prices at the production site range from 0.3 to 0.5 €/m^3, depending on circumstances. In the Canary Islands the most common prices are about 0.5 €/m^3, although in high water demand moments prices may exceed 1 €/m^3. These costs/prices are close to the farmers' willingness to pay for permanent supply, although they are higher for peak supplies to save a crop. In other cases costs are too high and crop fields become temporally or permanently idle.

Highly efficient irrigated crops, mostly to supply the European markets, can afford temporarily or permanently expensive water, such as desalinated seawater, every time more. But seasonal variability of water demand and the preferred use of local groundwater resources is the cause of the low use factor of most of the Mediterranean desalination plants. This increases the cost of produced water.

In the Spanish coastal areas, local or imported surface water may be cheaper than local groundwater, especially when only maintenance and operation costs are charged. But if all costs are taken into account, including transportation to the place of use and subsidies are cancelled, it is possible that the cost/price difference is reduced and even disappears. This may be also true even for desalinated seawater, especially when the water has to be pumped to high elevation and to faraway areas of use. In the case of "debrakished" groundwater, the cost of safe disposal of reject brines may cancel the economic advantages. This has been ignored until present.

Efficient water use seems a sound economic goal but it is not as simple as there are many side and perverse effects. In many Spanish coastal aquifers recharge depends on infiltration of water coming from inner areas, as local return irrigation flows. Using allochtonous irrigation water more efficiently decreases recharge, which may seriously affect local coastal aquifer resources and the groundwater discharge into the sea that keeps the saline wedge to penetrating excessively and also maintain some special littoral environments.

A common practice in water-short intensive Spanish agricultural areas is producing every time the cheapest mixture of water with the appropriate salinity for the stage of the plants and to avoid soil deterioration. Water sources include available local or imported surface water, reclaimed wastewater, desalinated seawater and local groundwater. When groundwater or reclaimed water is brackish, a part of it may be "debrackished" with membrane technology.

What is paid for water in irrigated intensive agriculture is a small fraction of the total cost of the agricultural activity. However, as much of the needed inputs are out of the farmer's control, the water cost affects significantly the net margin that can be controlled by the farmer. Thus, the farmer looks for the cheapest water. Thus, water from the coastal aquifer is preferred if there cheap imported water available is not available, even if this implies mining groundwater reserves and an increase of salinization or the associated risk

SOCIAL ASPECTS OF COASTAL AQUIFIER DEVELOPMENT

Intensive exploitation of coastal aquifers has been the driver of economy in numerous areas. This allowed a more stable and evolved society, but with progressive cost increase and some environmental damage. A negative result is that in many places local economy depends excessively on crop irrigation. In successful areas, the high economic relevance achieved propitiates the false social and media feeling that these areas are essential to the country's economy and thus they deserve public non-refundable investment and support paid by others, and that the employment have to be secured at any price. This is reflected in subsidized action by governmental agencies, fostered by political pressure. Providing more water easies management, but sustainability is unclear and economically unsound if permanently done. This socio-economic model seems exhausted and a change of paradigm is needed. The change is demanded by the civil society, but still it is not in the public administration and politician minds. In fact, there is a strong reluctance to increase water supply tariffs and prices, although there is a consensus on non-increasing and even reducing agricultural irrigated areas. However, groundwater resources will continue to be abstracted in spite of serious problems in some areas (MASE, 2015).

The development of civil society institutions is still poor. The Academia has a moderate role and decreasing activity as economic resources are progressively dwindling due to the general economic crisis and to poor vision of medium- to long-term issues. This is especially true for the Spanish coastal aquifers.

Groundwater Users' Associations to protect their water resources are scarce in the Iberian Peninsula, even more in coastal aquifers. However, a successful groundwater users' association was launched in 1975 in the Llobregat delta, Barcelona, to join efforts with the water authority to reduce and control the conspicuous sea water intrusion in the aquifers (Custodio, 2012; Niñerola et al., 2009). This has been a successful enterprise that has converted the coastal aquifer in one of the key pieces of the integrated water resources management in the Barcelona Metropolitan Area. In this case, urban water suppliers and industrial enterprises dominate. Other 9 associations exist in coastal areas with variable goals and success. Forming them in agriculture dominated areas is difficult due to the large number of actors and the fact that owning a well right is a comparative advantage relative to other farmers, when surface water supply fails. These groundwater associations have to be different from the common irrigators associations, which are very numerous in Spain and deal mostly with the operation and management of public water concessions but not the water resource. In the Mediterranean and Balearic areas, groundwater water transactions are done in the private domain, but are local and of limited importance. However, In the Canary Islands, especially in Gran Canaria and Tenerife, water and private water rights trade was well established, approaching free markets. Currently they are dwindling due to the public offer of water at controlled prices. This public sector interference has positive and negative aspects, but it holds back private investments just when public economic resources are scarce. In Spain, the attempts to set specific taxes to compensate for environmental and other negative externalities have failed. Farmers do not pay the cost of management of groundwater. Current legal rules make difficult that these taxes have a final destination, so their effect and the public acceptability are doubtful. Moreover there is no experience on using the generated funds for spatial and time compensations

The cost of investments to avoid salinization and for management is normally not paid by the users, especially the farmers who use 60 to 80% of coastal water resources, depending on the area.

REFERENCES

Custodio, E. (2012). Low Llobregat aquifers: intensive development, salinization contamination and management. In: S. Sabater, A. Ginebreda, D. Barceló. The story of a Polluted Mediterranean River. The Handbook of Environmental Chemistry, 21: 27–50. DOI: 10.1007/698_2011_138.

López–Geta, J.A., Gómez–Gómez, J.D. (2008). La intrusión marina y su incidencia en los acuíferos españoles [Marine intrusión and its implications in Spanish aquifers]. Enseñanza de las Ciencias de la Tierra, Madrid, 153: 266–273.

López–Geta, J.A., Fernández Ruiz, L. (2012). Importancia de los acuíferos costeros en España y el papel del TIAC [Relevance of coastal aquifers in Spain and the role of TIAC]. In: IV TIAC (Technology of Intrusion in Coastal Aquifers), Instituto Geológico y Minero de España, II: 7–14.

MASE (2015). Aspectos hidrológicos, ambientales, económicos, sociales y éticos del consumo de reservas de agua subterránea en España [Hydrological, environmental, economic and ethical aspacts of groundwater reserves consumption in Spain]. Prepared by E. Custodio, for UPC and AQUALOGY–CETAQUA. Barcelona. UPC e-book: 1–487. http://hdl.handle.net/2117/111272.

MIMAM (2000). Libro blanco del agua en España [White book of water in Spain]. Ministerio de Medio Ambiente. Madrid.

Niñerola, J.M.; Queralt, E.; Custodio, E. (2009). Llobregat delta aquifer. In: Ph. Quevauviller, A–M. Fouillac, J. Grath, R. Ward (eds.), Groundwater Monitoring. John Wiley & Sons: 289–301.

SASMIE (2017). Salinización de las aguas subterráneas en los acuíferos costeros mediterráneos e insulares españoles [Groundwater salinization in Mediterranean and island coastal aquifers in Spain]. Prepared by E. Custodio, for UPC and Suez Solutions–CETAQUA, Barcelona. UPC e-book. Barcelona: 1–852. http://hdl.handle.net/2117/111515

The ascension and intrusion processes of salt water into aquifers along the tectonic discontinuities in the Żarnowieckie Lake area

Mateusz Damrat[1], **Rafał Warumzer**[2]
[1]Mineral Raw Materials and Fossil Fuels Department, Polish Geological Institute - National Research Institute, Warsaw, Poland
[2]Marine Geology Branch in Gdańsk, Polish Geological Institute - National Research Institute, Gdańsk, Poland

ABSTRACT

In this article, the authors want to pay attention to the intensity of the phenomenon of ascension of salt water from the Permo-Mesozoic formation in the geological structures of the Dębki - Żarnowiec trough. This is a unique area in Poland. The sediments (up to 320 m in thickness) of the deep Pleistocene trough, cross the Tertiary, Cretaceous and Jurassic deposits, and reach the Triassic layers in the deepest place. An additional aspect is the unquestionable presence of older structures and tectonic loosening, which conditioned the formation of such a large erosion structure. Younger tectonic (neotectonic) activity, both large-scale and resulting from isostatic movements, certainly uses the existing fault network, which in combination with the deeper layer of relatively well-permeable Pleistocene sediments is a system extremely conducive to ascension brines.

INTRODUCTION

The issue of the occurrence of mineralized waters in the aquiferous layers of the Polish Baltic coast was topic of several dozen works in the field of hydrogeology. At present, opinions about the origin of salty waters in the Quaternary aquifer and older formations along the South Baltic coast are clashing. The main processes affecting this phenomenon are ascension and ingression. The area of Żarnowckie Lake trough (Fig. 1), due to its complex geological structure and location in the coastal zone, is an excellent research ground for a better understanding of the regularities at the interface between the two processes.

The geological setting of the Żarnowieckie Lake area is well-known. Apart from the standard recognition works of the geological structure of the country (Ostaficzuk et al. 1976; Ostaficzuk 1978),, it was an object of interest due to the occurrence of a hydrocarbon deposit in Cambrian sandstones and due to geotechnical investigation for the planned nuclear power plant operated by PGE EJ 1 in the "Żarnowiec" area (Gniewino and Krokowa municipalities).

General geological and geophysical information about the marine part of the described region is provided by the geological Atlas of the Southern Baltic in the scale 1:500,000 (Mojski eds. 1995). The recognition of the sub-Quaternary bedrock was ensured by the development of a Geological Map of the Baltic bottom without Quaternary deposits (Kramarska (eds.) 1999). The coastline zone was studied by numerous authors from the Polish Geological Institute, which issued in 2003 65 sheets in the scale 1:10 000 of the geodynamic map of the Polish coastal zone of the Baltic Sea. The location of subglacial forms in the sub-Quaternary relief of the Polish coast was analyzed by the Mojski and Tomczak (1994; Mojski (eds.) 1995).

Hydrogeological diagnosis within the sheet of the Hydrogeological Map of Poland (in 1: 50,000 scale) was performed by Sierżęga & Chmielowska (2000). Chlorides anomalies occurring in the groundwaters of aquifers of the Polish coast have been a topic of many works over the last several decades. Works related to the process of ascension and ingression were carried out, among others Dowgiałło (1971), Płochniewski & Sierżęga (1980), Kawaterkiewicz & Sadurski (1986), Kozerski & Pruszkowska (1996), and Krawiec (2013, 2015).

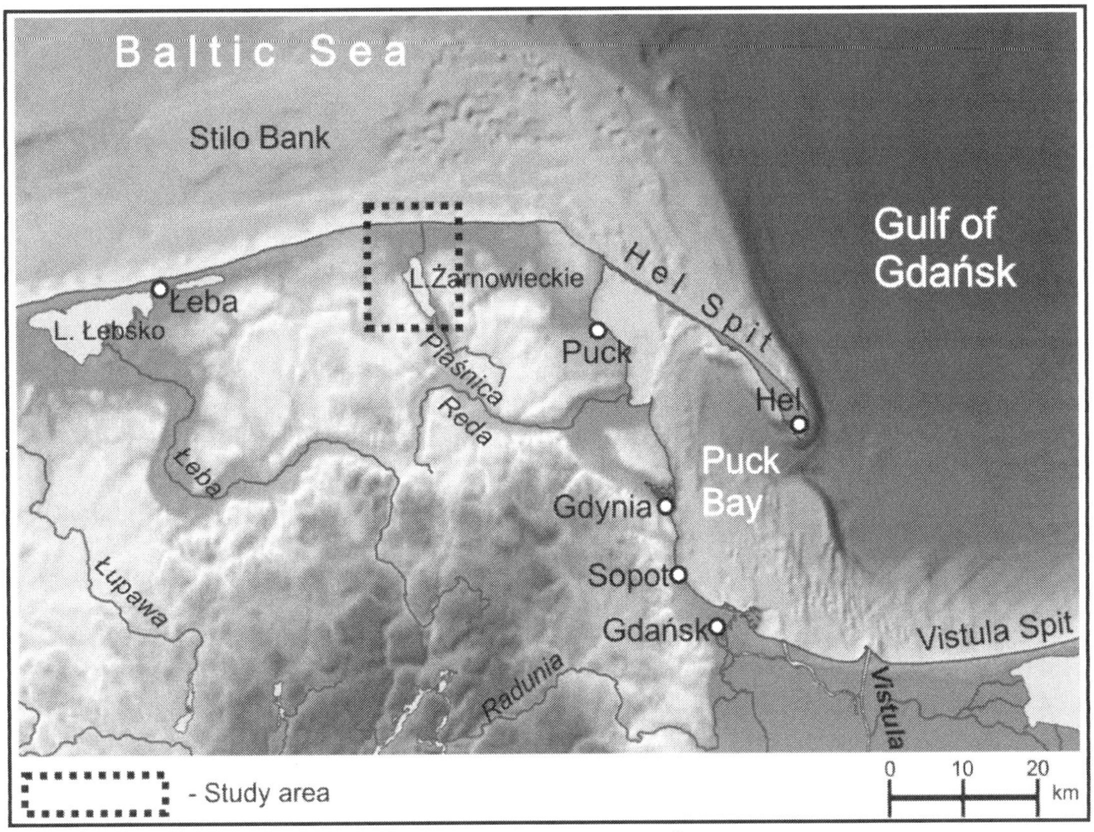

Figure. 1. Localization of the Dębki and Żarnowiec area.

METHODS

Despite the geological surveys of the Dębki and Żarnowiec region presented above, there is a lack of the synthetic study aimed at taking into account the occurrence of hydrogeological processes, against the background of tectonic and geophysical processes. An additional difficulty is the location and structure of the Dębki - Żarnowiec trough both in the land and sea areas, which results in the lack of continuous geophysical reconnaissance and differences in results. In the case study below, information and data relevant to the migration of salt water in the discussed region are extracted from available sources. A synthetic model was made, including known hydrogeological conditions. An analysis was also made of what research should be done in order to identify the processes occurring on the unique research area, namely the area of the gutter of Żarnowieckie Lake as best as possible.

RESULTS

The area of the Dębki - Żarnowiec trough is located in the area of the Peribaltic Syneclise in the NE part of the Łeba elevation. The Quaternary basement is built by the cover of the Paleozoic, Mesozoic and Tertiary sediments. The structural stages: Old Paleozoic, Permo-Mesozoic and Cenozoic were distinguished. Paleozoic (2.6 km in thickness) build Cambrian sandstones and silt-clay sediments, Silurian claystones and the Zechstein evaporates in the top (dolomites, anhydrites and rock salts). Above there are Mesozoic sediments: clays, mudstones and sandstones of the Triassic and the middle Jurassic. The youngest in this series are quartz-glauconite sands and sands with phosphates (upper Cretaceous), (Fig. 2). The area of Żarnowieckie Lake is characterized by a diverse geomorphology. The main form is the buried Pleistocene trough, the development of which was associated with the processes of erosion of glacial waters as well as strong exaration during the Weichselian glaciation. The remaining geomorphological forms occurring in the discussed area are: kame terraces and moraine plateaus. The north part of the Żarnowieckie Lake adjacent to forms associated with lacustrine and marine sedimentation such as lagoons and lake plains, spits and sea terraces.

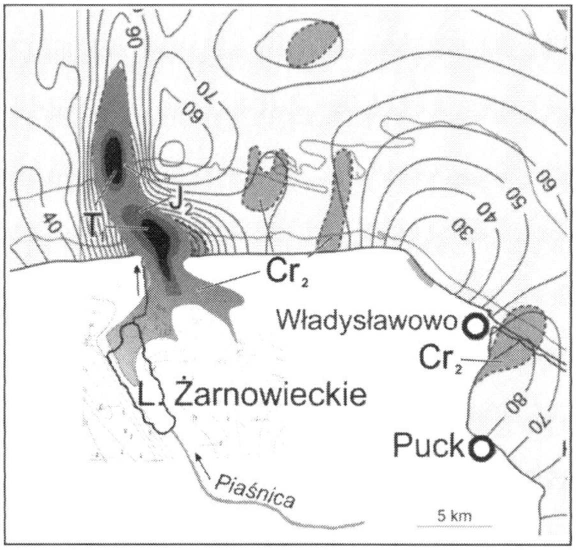

Figure. 2. Reconstruction of pre-Quaternary surface along the Dębki - Żarnowiec trough area. Isolines of Quaternary thickness in meters; after Ostaficzuk (1978) and Kramarska (edit. 1999) modified.

The trough course is close to the meridian (NNW-EES). Quaternary sediments within the research hole in Dębki (Dębki IG-1) were found up to 320 m - the highest thickness of Q in Poland. At the same time being an erosive surface reached the Triassic (sandstone) formation. Such a large geomorphological form certainly required tectonic conditions and long-lasting and repetitive erosion. The tectonic movements of the Laramiian phase in the upper Cretaceous and Paleogene were certainly an additional factor intensifying erosion, and this tectonic predisposition had a direct impact on the modern location of the Piaśnica Valley and the Żarnowieckie Lake (Tyski 1973). The prolongation of structures observed on land was also documented during geophysical work at sea in the elaboration of the Geological Map of the Baltic bottom without Quaternary formations (Kramarska (edit.) 1999), (Fig. 3).
Within the Dębki - Żarnowiec trough two aquifers belonging to the Quaternary can be distinguished. The first (upper) aquifer is associated with sand-gravel sediments of upper Pleistocene, under the bottom of the lake is spreading to the area of coastal lowlands. In the

marginal parts of the lake aquifer is not isolated and recharged through the lateral inflow from the plateau. The water table is not strained. The second (bottom) aquifer includes a base part of the trough. It occurs at a depth of 70.0 to 93.5 m and is underlined with tills of older glaciations and in the top there are Eemian clay and silts (Fig. 4). The contact between both quaternary layers takes place by interaquifer drainage through poorly permeable structures. Outside the Dębki - Żarnowiec trough - in the upland, there is also an aquifer that is associated with Oligocene and Miocene deposits. These aquifers are in direct hydraulic contact with the deeper Quaternary aquifer and form a common quaternary-tertiary regional aquifer. The processes of ascension of mineralized waters within the Dębki - Żarnowiec trough can be facilitated by the direct contact of the quaternary layer (lower) with the aquifer associated with the Triassic sandstones belonging to the Permo-Mesozoic stage (Fig. 4). This hydrogeological window was found based on the drilling hole in Dębki. The layers are contacted at a depth of 306 m. There is an increase in the salinity of groundwater in the range from 11598 mg/dm^3 at depth 40 m, to 13617 mg/dm^3 (200 m), (Fig. 4). The course of the contact area includes a narrow zone with a longitudinal course between the northern part of Żarnowieckie Lake and Dębki.

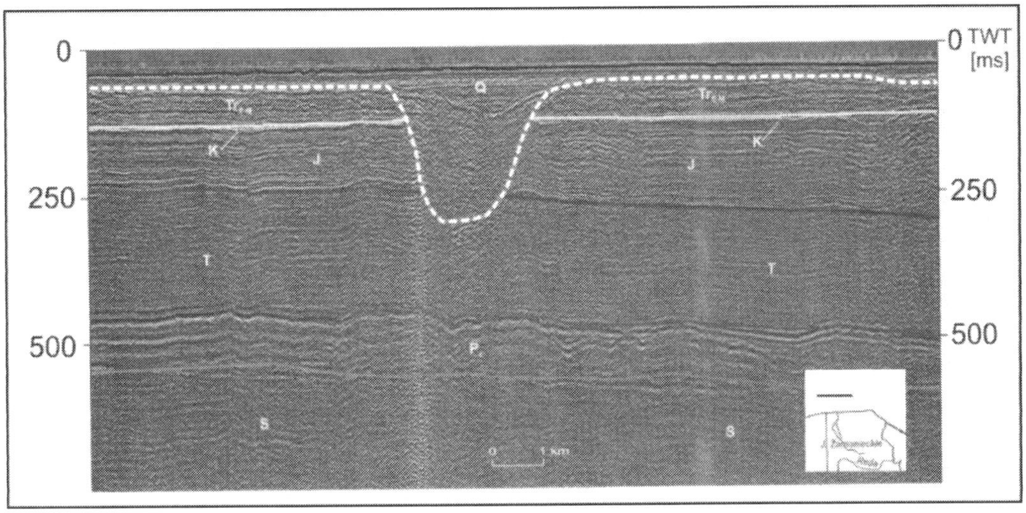

Figure. 3. The offshore geophysical record of Dębki - Żarnowiec trough reaching the Triassic layers; after Kramarska (edit. 1999) modified.

DISCUSSION AND CONCLUSIONS

It should be noted that according to Płochniewski and Sierżęga, Krawiec (Krawiec 2015, Płochniewski & Sierżęga 1980), the presence of salt and brackish waters in the analyzed area is related to the ascension of brines from the Mesozoic formations. This ascension takes place along the zones of tectonic (neotectonic) discontinuities and deep erosion in the Quaternary basement. The aforementioned zones of faults are present in the Mesozoic complex both in the land zone and under the sea. The identification of these zones is a serious issue. A series of studies from various branches of geology should be used for this task. Especially hydrogeochemical saline water ingression tests, isotope studies, noble gas determinations and geophysical surveys, both on- and offshore. New drilling hydrogeological wells located in the area of deep structures, buried troughs, with the possibility of sample deep aquifers that may be in contact with the Permo-Mesozoic layers. The occurrence of chlorides anomalies in the groundwaters of the Dębki - Żarnowiec trough and the majority of the Polish Baltic coast has a polygenetic nature connected with the processes of ascension, mixing of waters and ingression. Ascension processes as previously mentioned occur in contact zones of Permo-Mesozoic aquifers with younger layers, while the processes of ingression last from approximately 5.5 thousand years when the Littorina Sea reached a sea level similar to today's (Uścinowicz 2003) and its salinity reached above 12 ‰ (Bianchi et al. 2000) and even 15-20 ‰ (Hyvärinen et al. 1988, Witkowski 1994). An important process affecting the presence of saline and brackish waters was the washing out of saline waters during lower base level of groundwater drainage, which was influenced by ice sheet loads and relaxation during interglacials.

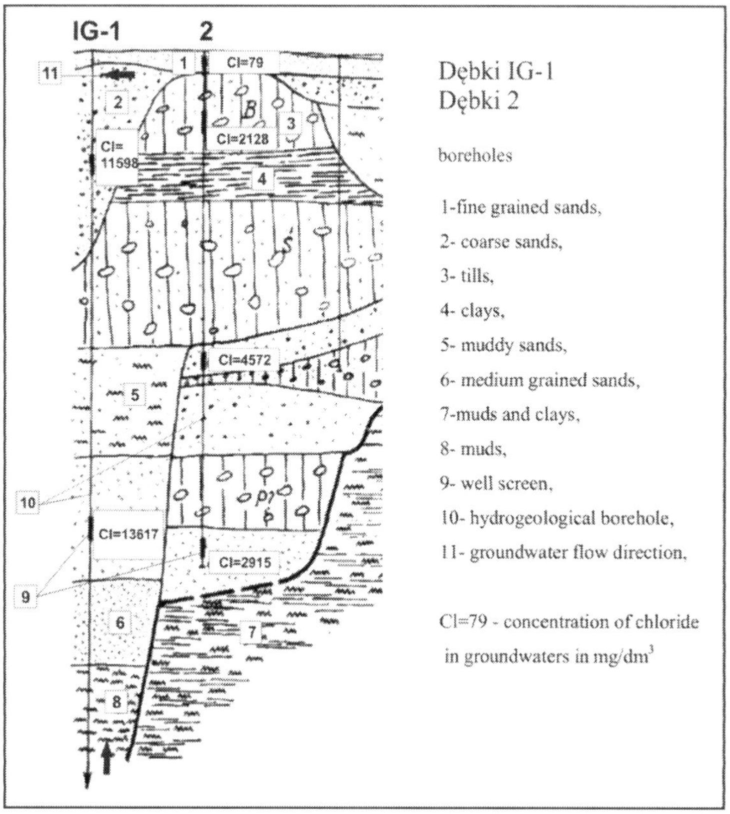

Figure. 4. Hydrogeological cross section of the Dębki - Żarnowiec trough after Kwaterkiewicz & Sadurski (1986) modified.

REFERENCES

Bianchi, T., P. Westman, C. Rolff, E. Engelhaupt, T. Andrén, and R. Elmgren. 2000. Cyanobacterial blooms in the Baltic Sea: Natural or Human-induced? Limnol. Oceanogr., 45(3), 716–726,

Dowgiałło, J., 1971, Studium genezy wód zmineralizowanych w utworach mezozoicznych Polski północnej, Biul. Geol. Wydz. Geol. UW, Warszawa

Graniczny, M., 1994, Strefy nieciągłości tektonicznych w świetle korelacji wielotematycznych danych geologicznych, na przykładzie Żarnowca i Ziemi Kłodzkiej, IMBG, 54, PIG

Graniczny, M., Mizerski W., 2003, Lineamenty na zdjęciach satelitarnych Polski – próba podsumowania, Przegląd Geologiczny, nr 6, 51: 474–482.

Hyvärinen, H., J. Donner, H. Kessel, and A. Raukas. 1988. The Litorina and Limnaea Sea in the northern and central Baltic, in Problems of the Baltic Sea History, edited by J. Donner, and A. Raukas, Ann. Acad. Sci. Fenn., Ser. A3, 148, 25–35.

Kozerski, B. and Pruszkowska, M. 1996. O pochodzeniu zasolenia wód podziemnych polskiego wybrzeża Bałtyku, Inżynieria Morska i Geotechnika

Kramarska, R. (eds.). 1999. Mapa geologiczna dna Bałtyku bez utworów czwartorzędowych w skali 1:500 000, Państw. Inst. Geol., Gdańsk – Warszawa.

Krawiec, A., 2013. Pochodzenie anomalii chlorkowych w wodach podziemnych wybrzeża Bałtyku, UMK, Toruń

Krawiec, A., 2015. Ingresje i ascenzje wód słonych na Pobrzeżu Słowińskim, Przegląd Geologiczny, vol. 63, nr 10/1, 2015, Warszawa

Kwaterkiewicz, A.and Sadurski, A. 1986. Problemy genezy wód zmineralizowanych w sąsiedztwie jeziora Żarnowieckiego, Annales Societatis Geologorum Poloniae

Mapa geodynamiczna polskiej strefy brzegowej w skali 1 : 10 000. 2003. Praca zbiorowa PIG-PIB

Mojski, J.E. and Tomczak, A. 1994. Większe formy subglacjalne w rzeźbie podczwartorzędowej polskiego wybrzeża. Acta Univ. Nicolai Copernici. Geografia XXVII. Nauki Mat–Przyrod. z. 92. Toruń.

Mojski, J.E., (eds). 1995. Atlas geologiczny Południowego Bałtyku (1:500 000). Państwowy Instytut Geologiczny.

Mojski, J.E. 2000. The evolution of the Southern Baltic Coastal Zone. Oceanologia 42 (3), Gdańsk.

Ostaficzuk, S., Jakubicz, B. and Skompski, S. 1976. Szczegółowa mapa geologiczna Polski w skali 1:50 000 Arkusz Sławoszyno. Wyd. Geol., Warszawa.

Ostaficzuk, S. 1978. Objaśnienia do Szczegółowej mapy geologicznej Polski w skali 1:50 000, Arkusz Sławoszyno. Wyd. Geol., Warszawa.

Płochniewski, Z. and Sierżęga, P. 1980. Warunki hydrogeologiczne w rejonie Elektrowni Jądrowej Żarnowiec, Mater. Konf. Bezpieczeństwo Elektrowni Jądrowych i Ochrona Środowiska, Bydgoszcz

Sierżęga, P. and Chmielowska, U. 2000. Mapa hydrogeologiczna Polski w skali 1: 50 000, arkusz Sławoszyno, PIG, Warszawa

Uścinowicz, S. 2003. Relative sea level changes, glacio-isostatic rebound and shoreline displacement in the Southern Baltic. Polish Geol. Inst. Sp. Papers, 10: 1–79.

Witkowski, A. 1994. Recent and fossil diatom flora of the Gulf of Gdansk, southern Baltic Sea, Origin, composition and changes of diatom assemblages during the Holocene, Bib. Diatomologica, 28, 315 pp.

Contact Information: Rafał Warumzer, Polish Geological Institute – National Research Institute, Marine Geology Branch, Kościerska 5 str., 80-328 Gdansk, Poland, Tel. +48 58 554 29 09 ext. 213, Email: rwar@pgi.gov.pl

Land subsidence by peat oxidation leads to enhanced salinization through boils in Dutch polders

Perry G.B. de Louw[1,2], Huite Bootsma[1], Henk Kooi[1], Mark Kramer[3], Gilles Erkens[1,4]
[1]Deltares, Utrecht, The Netherlands
[2] Wageningen University
[3]Rijnland Waterboard, Leiden, The Netherlands
[4]Utrecht University, The Netherlands

ABSTRACT

Peat oxidation in deep Dutch polders leads – in addition to subsidence - to the development of new saline boils, enhancing the salinization of these polders. This on-going process is studied in detail in the Middelburg-Tempelpolder. The objective of the study was to get more in-depth knowledge about this process and to assess it for the present situation and for future landscapes (after 10, 50, 100 and 500 years).

INTRODUCTION

Salinization of deep polders of the Netherlands (reclaimed lakes) happens mainly through boils, contributing more than 50% to the total salt load (De Louw, 2013). Boils occur as conduits in the upper aquitard, connecting the underlying aquifer to the surface and allowing groundwater to discharge at high velocities (Figure 1). Concentrated types of groundwater discharge at higher rates like boils, discharge groundwater from deeper strata with more salty groundwater than diffuse types of seepage at low rates. Due to this natural saltwater upconing process (De Louw et al., 2013) the salinity of groundwater discharged by boils is much higher than diffuse types. The saline groundwater in the aquifers originates from marine transgressions during the Holocene.

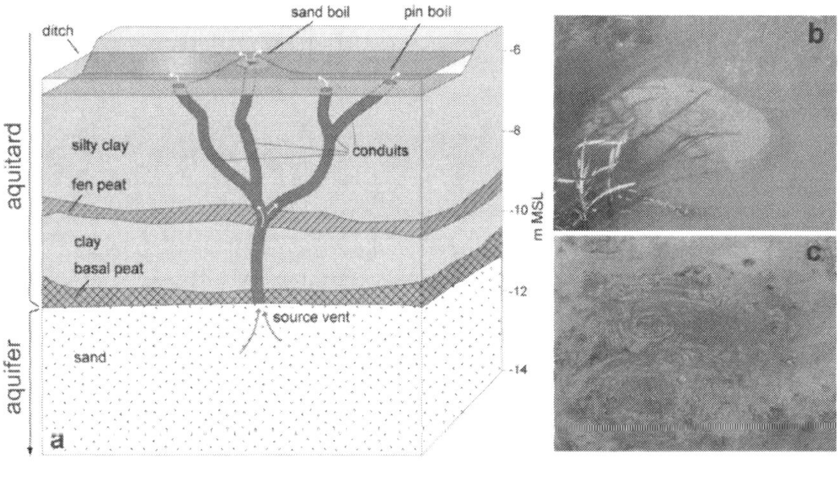

Figure 1. Boils in deep polders: (a) diagram of boils with several conduits in aquitard, (b) sand boil transporting sand from the aquifer (c) a boil emitting water and methane (adapted from De Louw et al., 2013).

https://doi.org/10.1051/e3sconf/20185400007

Boils develop when the pressure of water in the aquifer is greater than the pressure exerted by the weight of the overlying stratum. Due to the higher water pressure, heaving and cracking of the soil occur, creating flow pathways that lead to the development of boils. This is the reason why boils mainly occur in deep polders where large hydraulic gradients exist and where hydraulic heads of the upper aquifer exceed ground levels (see Figure 2).

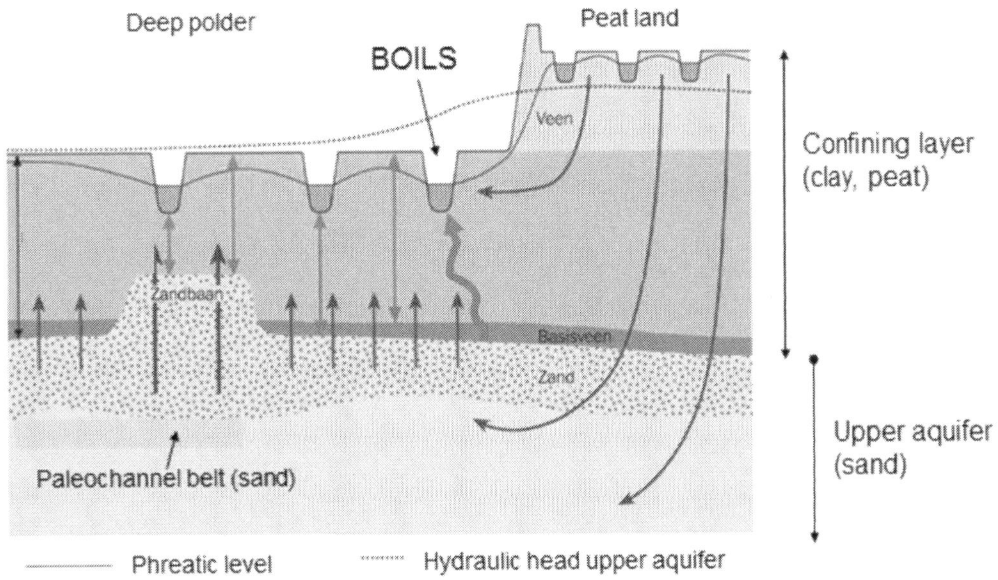

Figure 2. Hydrogeological cross section through a typical deep polder and adjacent elevated peatland. Heaving and cracking of the soil occurs when the water pressure (hydraulic head) in the upper aquifer (blue arrows) exceeds the weight of the overlying confining layer (green arrows).

The topsoil of most Dutch deep polders consists of clay, as most peat has been mined or degraded over the last centuries. However, some polders still have shallow peat remaining. Peat oxidizes when it comes in contact with air (oxygen), resulting in a reduction of peat volume,(oxidation) and consequently land subsidence. The average land subsidence in the Netherlands is about 0,8 mm per year (Van den Born et al., 2015). In oxidation prone areas, typically water levels are maintained by the Water Boards as high as possible to slow down subsidence. This means an average surface water level of about 20 to 60 cm below ground level, which is just deep enough to facilitate dairy farming. To sustain the aforementioned drainage depths under conditions of progressive subsidence, surface water levels must be periodically lowered, evoking new subsidence. This forms a self-perpetuating circle, which lead to meters of subsidence in the Dutch peatlands over the last 10 centuries (Erkens et al., 2016). With ongoing subsidence and decreasing confining layer weights, the risk of new boils development increases and consequently also the salinization of the polder will increase.

The process of land subsidence, the development of boils and related salinization is studied in detail for a deep polder called Middelburg-Tempelpolder (MT-polder). The objective of the study was to get more in-depth knowledge about this process and to assess it for the present situation and for future landscapes (after 10, 50, 100 and 500 years).

METHODS

We have combined modeling with an extensive field survey to collect detailed information about geology, hydraulic heads, land subsidence, heaving and cracking, occurrence and development of boils and the history of boil development.

The field survey contained the following elements:

- Interview of farmers: to collect data about (1) agricultural activities in subsiding areas with boils, (2) location of boils and history of boil development, (3) their view regarding future developments and solutions.
- Detailed geological borehole descriptions (in total 31 geological drillings until 5 to 11 meter depth).
- Weighting of wet bulk density samples of different lithology.
- Installation of 15 piezometers in the aquifer to measure the hydraulic head in the upper aquifer as well as the salinity of the groundwater. At most locations also phreatic piezometers were installed.
- EC (salinity) and temperature routing to map boils and quantify the salinization of the polder.

The modeling exercises involved:

- Groundwater modeling of hydraulic head in upper aquifer using Seawat (present situation)
- Construction of 3D-geological model (present situation)
- Producing maps of the risk index (for heaving/cracking and, hence development of saline boils) (present situation), the index being the ratio of overburden weight and hydraulic head at the base of the confining layer.
- Modeling land subsidence using land subsidence model Phoenix (Geisler, 2015) due to peat oxidation and resulting landscapes after 10, 50, 100 and 500 years.
- Modeling hydraulic heads for future landscapes resulting from land subsidence after 10, 50, 100 and 500 years.
- Calculating the risk index of the development of saline boils for the future landscapes 10, 50, 100 and 500 years after present.
- Quantifying the salinization of the polder through boils for the different future landscapes.

Finally, potential policy actions for the Water Boards concerning water management of the polder and for the farmers will be formulated based on this combined field and modeling study. The results of this regional study will be extrapolated to all other deep polders with shallow peat occurrence, which are undergoing significant land subsidence by peat oxidation.

RESULTS

This study started in October 2017 and will finish in May 2018 and the most significant results will be presented at the SWIM. In this extended abstract, merely some highlights of the results will be presented since both the field survey and modeling exercises are halfway.

The modeled hydraulic head exceeds ground level for about 50% of the MT-polder with the largest hydraulic heads up to 0.75 cm above ground level at the edges of the polder (Figure 3). The confining layer is 3 to 6 m thick, consists of peat and clay, and serves as an aquitard on top of the upper aquifer. Saturated peat has a much lower weight than clay, 0.9-1.1 g cm^{-3} and 1.3-1.6 g cm^{-3} respectively. The combination of large hydraulic heads and the subsoil consisting largely of peat, results in high risks for saline boil developments (Figure 3). These preliminary calculations are based on regional and national scale data.

The collected data of lithological composition of the confining layer and hydraulic heads during the field surveys will be used to improve the reliability of the model calculations. With the improved models, the different hydraulic heads and risk for saline boils developments due to land subsidence will be calculated for the future landscapes (10, 50, 100 and 500 years). Two water management scenarios will be assessed: (1) surface water levels follow the land subsidence to maintain favorable conditions for agriculture, (2) surface water levels will be fixed to slow down land subsidence. For the first water management scenario, areas with peat at the surface will subside with a rate of approximately 5 mm/year leading to ~0.25 meter of land subsidence in 50 years. Land subsidence will continue until all peat has been oxidized. With the fixed level scenario soil subsidence will slow down due to increasingly wet conditions (less oxygen intrusion). The subsidence will eventually stop when the land surface subsides to the fixed water levels causing anoxic conditions preventing the oxidizing of peat. However, this will lead to the disappearance of traditional dairy farming agriculture due to wet conditions.

Figure 3. Left: The modeled hydraulic head in the upper aquifer, referenced to ground level. At the edges of the polder, the hydraulic head rises up to 80 cm above ground level causing high risk for boil development. Right: The risk for the development of saline boils for the MT-polder. A value lower than 1.1 indicates a high risk for boil development, a value higher than 1.1 indicates a low risk.

First calculations show that land subsidence will increase the risk for boil development and consequently enhances salinization for most areas. However, for some areas the opposite is true. Due to regional land subsidence, the hydraulic head will be lowered too and when the reduction of the hydraulic head outranges the weight reduction by peat oxidation, the risk of boil development decreases. The development of new boils will therefore be stagnated for these areas when only land subsidence is taken into account. However, all activities that reduce the weight of the confining layer, such as excavations and lowering of surface water levels, may lead to the development of new boils and consequently enhanced salinization. The field surveys result in new insights and confirms existing theory about boil developing and salinization. According to the farmers, most boils are old and only a few have developed recently. However, the old boils still discharge saline groundwater and when they occur on land, the boils limit the agricultural production due to wet and saline conditions and in-accessibility of the land for machines due to the low bearing capacity of the soil. Most boils developed directly after the formation of the polder (reclamation of lake) at the end of the 19th century. But also during periods with increased hydraulic head conditions (e.g. due to reduced groundwater extraction), lowered surface water levels, constructing ditches and canals or other activities when soil is removed, have caused the development of boils. Figure 4 shows pictures of boils in the MT-polder. During a short frost period, the boils in the ditches were clearly visible as holes in the ice (Figure 4c-d), due to the constant temperature of the groundwater of 10.5 °C being discharged via boils. For two ditches with a total length of 950 m at the north-eastern edge of the MT-polder in total 60 large boils and 66 small boils were visually mapped during the frost period, resulting in a boil density of 1 boil every 7.5 meter. About 75% of the boils found in the field occur in the zones where the calculated risk for boils development is high (Figure 3). With the data locally collected in the field, the risk calculations will be improved significantly.

Figure 4. (A) Height difference of 3 meter between peat land (left) and deep polder with shallow peat occurance (right) causing large hydraulic gradients. (B) Boil on land surface causing wet saline conditions and soft soil. (C) and (D) Boils in surface water which don't freeze due to constant temperature of 10.5 °C of discharging groundwater.

From the 15 installed piezometers, groundwater from the upper aquifer just below the confining layer was sampled, and the salinity was measured. According to earlier findings (De Louw, 2013), almost all groundwater directly below the confining layer was fresh (EC < 1.200 μS/cm) except for three piezometers in the centre of the polder, and most boils were saline (EC 4.000 – 10.000 μS/cm). Boils have a much higher salinity than the diffuse types of groundwater seepage due to saltwater upconing whereas diffuse seepage only discharges groundwater from the top of the aquifer. Hydraulic head measurements in the installed piezometer confirm the modelled larger hydraulic heads at the edge of the polder (0.20-0.6 m above ground level) than in the centre of the polder (just above or below ground level). The measured hydraulic heads in the south-eastern part of the MT-polder are lower than the modelled ones which is probably due to the increased discharge of groundwater as a result of a large number of boils in this part of the polder. These boils were not incorporated in the model and this effect on the hydraulic head was therefore not accounted for.

SUMMARY

From the results so far the following preliminary conclusions can be made:

- Collecting local data increased the knowledge about the boil development process and improved the models (groundwater, geology, land subsidence).
- The combination of large hydraulic heads and shallow peat in the subsurface results in high risks for saline boil development.
- Land subsidence by peat oxidation result in both a reduction of the hydraulic head and weight of the confining layer. The ratio determines whether this would lead to an increase or decrease of the risk of boils development.

REFERENCES

Erkens, G., van der Meulen, M.J., Middelkoop, H., 2016. Double trouble: subsidence and CO2 respiration due to 1,000 years of Dutch coastal peatlands cultivation. Hydrogeol J 24, 551–568, DOI 10.1007/s10040-016-1380-4

Van den Born, G.J., Kragt, F., Henkens, D., Rijken, B. van Bemmel, B., van der Sluis, S., 2016, Dalende bodems, stijgende kosten. PBL, Den Haag, 96 pp.

Geisler, L., 2015 Improving the land subsidence model Phoenix, MSc thesis Utrecht University, 74 pp.

De Louw, P.G.B., 2013. Saline seepage in deltaic areas. Preferential groundwater discharge through boils and interactions between thin rainwater lenses and upward saline seepage. PhD thesis, Vrije Universiteit Amsterdam. ISBN/EAN 9789461085429.

De Louw, P.G.B., Oude Essink, G.H.P., Stuyfzand, P.J., Van der Zee, S.E.A.T.M., 2010. Upward groundwater flow in boils as the dominant mechanism of salinization in deep polders, The Netherlands. Journal of Hydrology 394, 494-506.

De Louw, P.G.B., Vandenbohede, A., Werner, A.D., Oude Essink, G.H.P., 2013. Natural saltwater upconing by preferential groundwater discharge through boils, Journal of Hydrology 490, 74-87.

Contact Information: Perry G.B. de Louw, Department of Soil and Groundwater, Deltares, Utrecht, The Netherlands Phone: +31-6-3054800, Email: perry.delouw@deltares.nl

It's hydrogeology but not as we know it: Sub-seafloor groundwater flow driven by thermal gradients

A. Desens[1], **V.E.A. Post**[1], G.J. Houben[1], T. Kuhn[1], M. Walther[2,3], Thomas Graf[4]
[1]BGR, Hannover, Germany
[2]Helmholtz-Centre for Environmental Research GmbH – UFZ Leipzig, Department of Environmental Informatics, Leipzig, Germany
[3]Technische Universität Dresden, Professorship of Contaminant Hydrology, Dresden, Germany
[4]Leibniz Universität Hannover, Institute of Fluid Mechanics and Environmental Physics in Civil Engineering, Hannover, Germany

ABSTRACT

Groundwater flow beneath the oceans plays an important role for cooling the earth's crust and geochemical cycles, yet it remains an understudied subject in hydrogeology. This contribution focuses on the circulation of seawater through basalt covered by deep-sea sediments in the equatorial northeast Pacific Ocean. Numerical model simulations are used to infer the factors controlling the flow patterns that develop between basalt outcrops. The energy to drive the flow is derived from the crustal heat flux. It is found that the sediment thickness plays a key role in determining the development of hydrothermal siphons, i.e. the flow between two adjacent seamounts where one acts as a recharge point and the other as a discharge point for seawater. Amongst the various factors tested, the outcrop width was an important factor as well.

INTRODUCTION

It is now firmly established that submarine groundwater discharge plays a major role in the delivery of nutrients and other chemical substances to the ocean. This process has been studied primarily in the near-shore environment (Moore, 2010). Flow processes around mid-oceanic ridges, evidenced e.g. by "black smokers", have also received a lot of attention. Much less remains known, however, about the role of groundwater flow processes in the deeper oceanic crust between the continents and the mid-oceanic ridges. Large parts of the ocean floor are of this type. In the absence of topographic driving forces, flow here is driven by processes like tectonic compression, sediment loading, mineral transformation, and heat flow. The latter is the focus of this contribution, which deals with the flow in basalt layers at the floor of the Pacific Ocean.

The study area is located 1700 km southwest of Manzanillo, Mexico in the equatorial northeast Pacific Ocean. The depth of the ocean ranges between 1460 and 4680 m (Kuhn *et al.*, 2017) and the seabed is formed by basaltic crust, which is draped by a sediment cover of up to 100 m thickness. Locally, the crust protrudes through the sediment layer in the form of seamounts and N-S oriented ridges, which are about 100 - 300 m high, a few kilometers wide, and 10s of kilometers long. The seamounts are extinct submarine volcanoes, which rise a few hundred to almost 3000 m over the surrounding abyssal plains. The circulation of seawater though the basaltic crust has been confirmed previously by heat flow measurements, and pore-water geochemical profiles (Kuhn *et al.*, 2017). The outcrops of the protruding seamounts and N-S ridges are entry and exit points for seawater. Similar flow systems have been described elsewhere and the phenomenon of seawater entering one seamount and exiting from another has been termed a hydrothermal siphon (Fisher, 2005).

Figure 1 shows a conceptual model of such a hydrothermal siphon. A net flow between two outcrops develops only when a potential difference exists between them. The hydrostatic pressure exerted by the overlying ocean water is assumed to be the same everywhere, so the flow can only be initiated by density differences that are caused by the crustal heat flux, which is around 0.1 W/m^2 for oceanic crust of the working area (Kuhn *et al.*, 2017). The high permeability of the basalts aids in the formation of upwelling plumes of relatively warm and downwelling plumes of relatively cold groundwater. Because of their fine-grained nature, the sediments are much less permeable than the basalts and thus mass transport in them is diffusion-dominated.

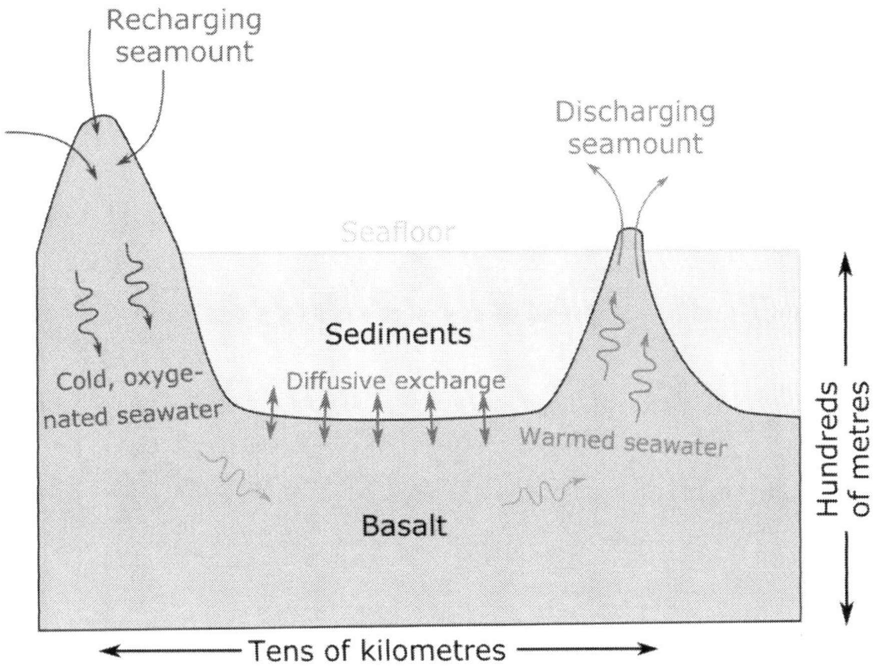

Figure 1. Conceptual model of the flow between two seamounts (modified fromFisher & Wheat, 2010).

Little is known about hydrothermal siphons but they can be expected to play an important role in the cooling of the oceanic lithosphere, the solute budgets of the oceans (Fisher, 2005) and, potentially the formation of mineral resources such as manganese nodules. In the study area, for example, it has been found that dissolved oxygen is entrained by the seawater circulating through the basalt layer and enters the sediment layer from below by upward diffusion. To better understand the flow dynamics, as well as the heat and solute transport process, a numerical model was constructed that simulates the flow of seawater through a sediment-covered basaltic aquifer between a pair of basalt outcrops, which could either be seamounts or N-S ridges.

MODEL SETUP

Figure 2 shows a schematic cross-section with the key features of the numerical model. The model was constructed using the software OpenGeoSys5 (Kolditz *et al.*, 2012), which solves the system of equations for coupled fluid flow and heat transport in an iterative coupling scheme. The model domain as a vertical cross-section had a width of 10 km and the basalt layer beneath the sediment was 500 m thick. The thickness of the sediment layer was varied between 10 and 100 m and the height of the seamounts, which equaled the thickness of the sediment layer, was varied accordingly. The width of the outcrops was 200 m.

The pressure at the seafloor was set to $P_0 = 4 \times 10^7$ Pa (400 bar), and the temperature to $T_0 =$ 1.48 °C, based on measurements. The temperature at the right outcrop was not specified though and was left to be calculated by the model instead. This allowed for the development of a pressure difference between the two basalt outcrops. A subsurface heat flux of q_h 0.1 W/m^2 was assumed along the bottom boundary of the model. The pressure and viscosity of water thus varied as a function of temperature. Other model parameters are shown in Figure 2. Initial runs were conducted to test the effect of grid discretization and time stepping on the simulation results. The mesh of the model with a sediment thickness of 100 m had 60,061 nodes and 60,000 elements, and tie time step used was 10^9 s.

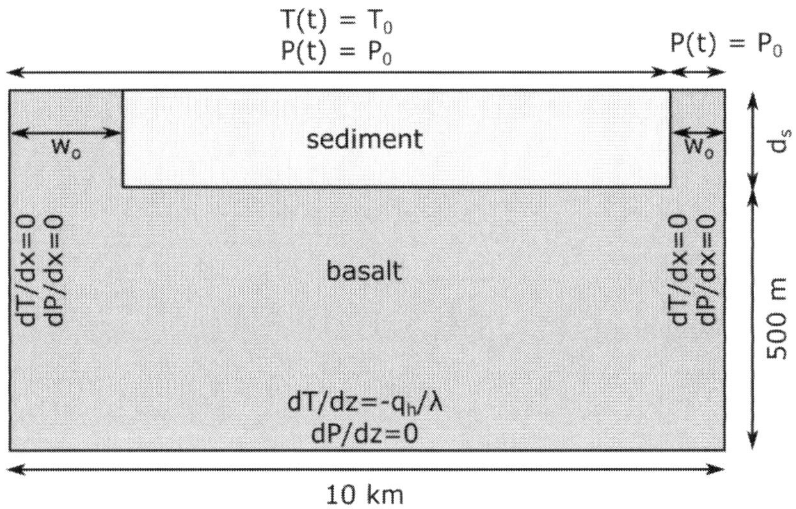

Figure 2. Cross-section showing the geometry of the model domain, lithological units and boundary conditions for temperature T and pressure P. w_o is the width of the outcrop, which was 200 m for the case presented here but varied and could be different between left and right for asymmetrical models. d_s was the sediment thickness, i.e. 100 m for the case presented here. Initial conditions were set to a constant temperature T = 1.48 °C and a hydrostatic pressure distribution. q_h is the crustal heat flux (0.1 W/m^2) and λ is the thermal conductivity.

RESULTS

Figure 3 shows the temperature distribution after about 30,000 years for the model with a sediment thickness of 100 m. The corresponding flow field is shown in Figure 4. These results clearly show that the system is unstable and that a convective flow system developed which is driven by the bottom heat flux.

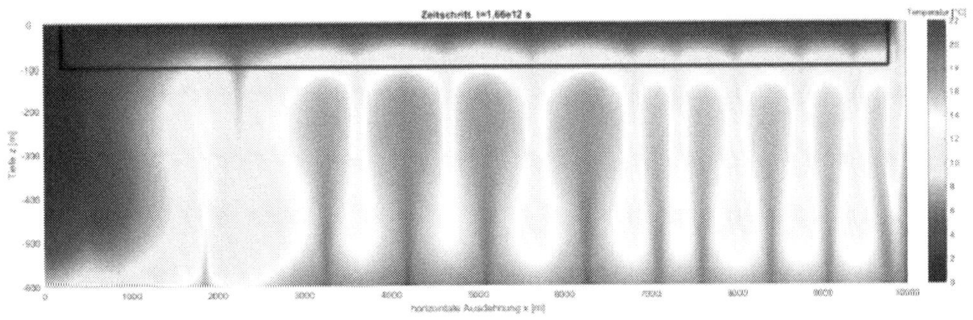

Figure 3. Temperature distribution for the model with a sediment layer thickness of 100 m after 9.15×10^{11} s (~30,000 y). The sediment layer is indicated by a black line. The corresponding flow field is shown in Figure 4.

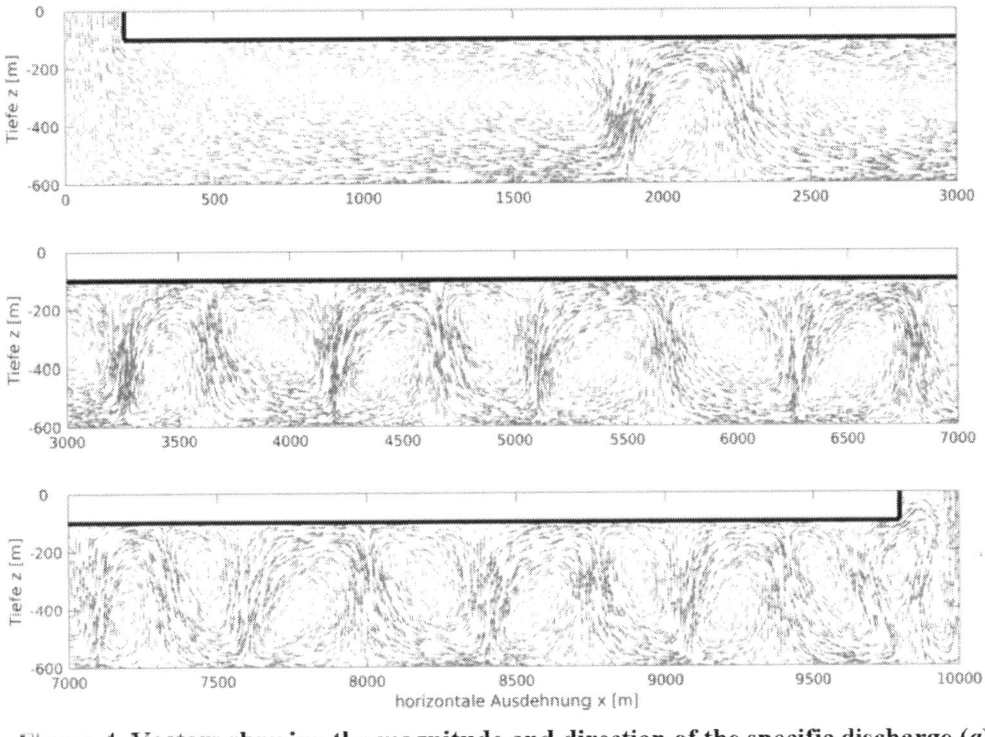

Figure 4. Vectors showing the magnitude and direction of the specific discharge (q) field after 9.15×10^{11} s (≈ 30,000 y). Note that the model domain has been subdivided into three parts to improve the visibility of the vectors. The vectors were normalized with respect to the maximum specific discharge and the largest arrow represents a value of $q = 5.7\times10^{-8}$ m/s.

The basalt outcrop on the left acts as the entry for seawater, whereas in the basalt outcrop on the right, there is both down- and upwelling of seawater (Figure 4). Mass balance calculations show that there is a net flow of seawater through the basalt from left to right. Running simulations with different sediment thicknesses showed that this net flow did not occur in the models with a sediment thickness ≤ 40 m. Additional simulations (not shown here) indicate that the width of the basalt outcrops is another important parameter.

Asymmetry of the outcrop width promotes the development of a siphon. Simulations yet to be conducted will also consider the distance between outcrops.

DISCUSSION AND CONCLUSIONS

The development of an unstable, free convection flow system was expected as the Rayleigh number (*Ra*) for the initial temperature profile is $Ra = 516$, which is much larger than the critical value of $Ra_{cr} = 17.65$. The model results indicate that, for the range of conditions considered in this study, a hydrothermal siphon does not develop when the sediment thickness ≤ 40 m. Our preliminary interpretation of this observation is that this is because the transport of heat via conduction through the thinner sediment layers becomes more effective than via convection through the basalts. In models with a sediment thickness ≥ 60 m, the thickness of the sediment layer prevented this, and thus a lateral flow component develops that conveys the energy out of the system. These modelling results provide support for the inferred circulation of seawater through the crustal basalt in the study area, as was suggested by Kuhn *et al.* (2017) based on heat flow and pore water concentration data. It is clear that pore waters beneath the ocean flow need not be stagnant but that there can be conditions where vigorous flow develops, provided that the permeability and heat flux are high enough. Hydrothermal siphons develop under favorable conditions, which include a heat flux high enough to cause potential differences, a thick sediment layer, and asymmetric basalt outcrops widths. The circulation of seawater through the basaltic crust does not only lead to its eventual cooling but also to a decrease of its permeability due to hydration reactions of the basaltic minerals, which lead to a decrease of porosity. Moreover, such hydration reactions may also lead to the mobilization of elements into solution and their transport back into the ocean. Considering the wide-spread existence of this type of off-axis ventilation, the net-effect of the return flow of mobilized elements on element fluxes and oceanic budgets may be very large.

REFERENCES

Fisher A.T. 2005. Marine hydrogeology: recent accomplishments and future opportunities. Hydrogeology Journal 13: 69-97.

Fisher A.T. and C.G. Wheat 2010. Seamounts as Conduits for Massive Fluid, Heat, and Solute Fluxes on Ridge Flanks. Oceanography 23: 74 - 87.

Kolditz, O., S. Bauer, L. Bilke, N. Böttcher, J.O. Delfs, T. Fischer, U.J. Görke, T. Kalbacher, G. Kosakowski, C.I. McDermott, C.H. Park, F. Radu, K. Rink, H. Shao, H.B. Shao, F. Sun, Y.Y. Sun, A.K. Singh, J. Taron, M. Walther, W. Wang, N. Watanabe, Y. Wu, M. Xie, W. Xu, B. Zehner 2012. OpenGeoSys: an open-source initiative for numerical simulation of thermo-hydro-mechanical/chemical (THM/C) processes in porous media. Environmental Earth Sciences 67: 589-599.

Kuhn T., G.J.M. Versteegh, H. Villinger, I. Dohrmann, C. Heller, A. Koschinsky, N. Kaul, S. Ritter, A.V. Wegorzewski and S. Kasten 2017. Widespread seawater circulation in 18–22 Ma oceanic crust: Impact on heat flow and sediment geochemistry. Geology 45: 799-802.

Moore W.S. 2010. The Effect of Submarine Groundwater Discharge on the Ocean. Annual Review of Marine Science 2: 59-88.

Contact Information: Vincent Post, Federal Institute for Geosciences and Natural Resources (BGR), Stilleweg 2, 30655 Hannover, Germany, Phone: +49-511-6432393, Email: vincent.post@bgr.de

Long-time resistivity monitoring of a freshwater/saltwater transition zone using the vertical electrode system SAMOS

Michael Grinat[1], Dieter Epping[1] and Robert Meyer[1]
[1]Leibniz Institute for Applied Geophysics, Hannover, Germany

ABSTRACT

In September 2009 two newly developed vertical electrode systems were installed in boreholes in the water catchment areas Waterdelle and Ostland at the North Sea island Borkum to monitor possible changes of the transition zone between the freshwater lens and the underlying saltwater. The vertical electrode systems, which were both installed between 44 m and 65 m below ground level, are used for geoelectrical multi-electrode measurements carried out automatically several times per day; the measurements are still ongoing. The whole system consisting of a vertical electrode system in a borehole and the measuring unit at ground level is called SAMOS (Saltwater Monitoring System).

At both locations the data show a clear resistivity decrease that indicates the transition zone between freshwater and saltwater. The depth of the transition zone as well as the kind of resistivity decrease is very stable since 2010.

Temporal changes are visible if single depths are considered. In 2015 Miriam Ibenthal used a vertical 2D density-dependent groundwater flow model to explain the long-term resistivity measurements and showed that the temporal changes at CLIWAT 2 (Ostland) could be explained by variations of the groundwater level, changing groundwater recharge rates and changing pumping rates of the nearby located drinking water supply wells.

INTRODUCTION

At the North Sea island Borkum the necessary freshwater is completely taken from the freshwater lens. The water is extracted in the two water catchment areas Waterdelle and Ostland. Within these water catchment areas up-coning of saltwater caused by high pumping rates in the wells is a permanent threat. This up-coning should lead to changes of the depth and shape of the transition zone between the freshwater lens and the underlying saltwater; therefore a system that monitors the transition zone acts as an early-warning system.

In September 2009 newly developed vertical electrode systems were installed in boreholes in each of the two water catchment areas Waterdelle and Ostland to monitor possible changes of the transition zone between the freshwater lens and the underlying saltwater. The transition zone is located in depths between 44 m and 65 m below ground level. The installation as well as the first measurements were carried out in the framework of the EU Interreg project Climate & Water (see http://cliwat.eu/, Sulzbacher et al. 2012, Wiederhold et al. 2013). Therefore the boreholes were called CLIWAT 1 (Waterdelle) and CLIWAT 2 (Ostland).

Each of the two vertical electrode systems is about 20 m long and includes 78 stainless steel ring electrodes. The spacing between adjacent electrodes is 0.25 m. The electrodes are mounted on an isolating PVC rigid pipe. After the installation of the vertical electrode systems the boreholes were refilled with material similar to the sediments that were found in the correspondent depths. Moreover, two water gauges with filters in different depths were installed above each of the vertical electrode systems to monitor the groundwater level.

The measuring system is installed at ground level. The geoelectrical measurements are carried out automatically several times per day using a modification of the commercial resistivity meter 4point light 10W (see www.l-gm.de). The power is supplied by batteries recharged by solar panels. Since December 2009 the data are regularly transmitted to Hannover by telemetry. The automated measurements are still ongoing. The whole system consisting of the vertical electrodes in a borehole and the measuring unit at ground level is called SAMOS (Saltwater Monitoring System).

Recently, similar systems were also used by Ogilvy et al. (2009) and Poulsen et al. (2010) for monitoring coastal aquifers.

RESULTS

The geoelectrical multi-electrode measurements are carried out automatically using the Wenner-alpha array (electrode spacings between 0.25 m and 6.25 m). Each multi-electrode measurement includes 975 different four-point arrays.

In the beginning the measurements were strongly influenced by moisture effects and showed a lot of outliers, but the use of newly developed active electrode switchboxes increased the quality of the data very much. In the last years a maintenance of the system at ground level was necessary once to twice a year only.

At both locations CLIWAT 1 and CLIWAT 2 the data show a clear decrease of the apparent resistivity from about 80-90 Ωm in depths around 45 m to about 1-2 Ωm around 65 m depth (spacing a = 0.25 m). This decrease indicates the transition zone between freshwater and saltwater. In CLIWAT 1 only sand is found in the transition zone, but in CLIWAT 2 several layers of clay and silt were encountered within the sandy sediments in the depth of the transition zone. The different layers are clearly visible in the resistivity measurements: In the transition from resistivities of about 80 Ωm around 45 m depth to 2 Ωm around 65 m depth the sand layers show slightly higher resistivities than the clay layers (fig. 1).

Since 2010 the depth of the transition zone as well as the kind of decrease of the apparent resistivity is very stable (fig. 1). Only within the first year large changes occurred, but these were caused by the readjustment of the local conditions (disturbed by drilling) to the undisturbed situation.

Temporal changes are only visible if single depths are considered. They are especially large in CLIWAT 2 (Ostland) in depths around 55 m (fig. 2). Here a sand layer confined by clay layers is found. In 2015 Miriam Ibenthal used a vertical 2D density-dependent groundwater flow model to explain the long-term resistivity measurements and showed within her master thesis at the University of Göttingen that these temporal changes could be explained by variations of the groundwater level, changing groundwater recharge rates and changing pumping rates of the nearby located drinking water supply wells (Ibenthal 2015). SAMOS will also be used in the projects TOPSOIL and GO-CAM (Wiederhold et al. 2018).

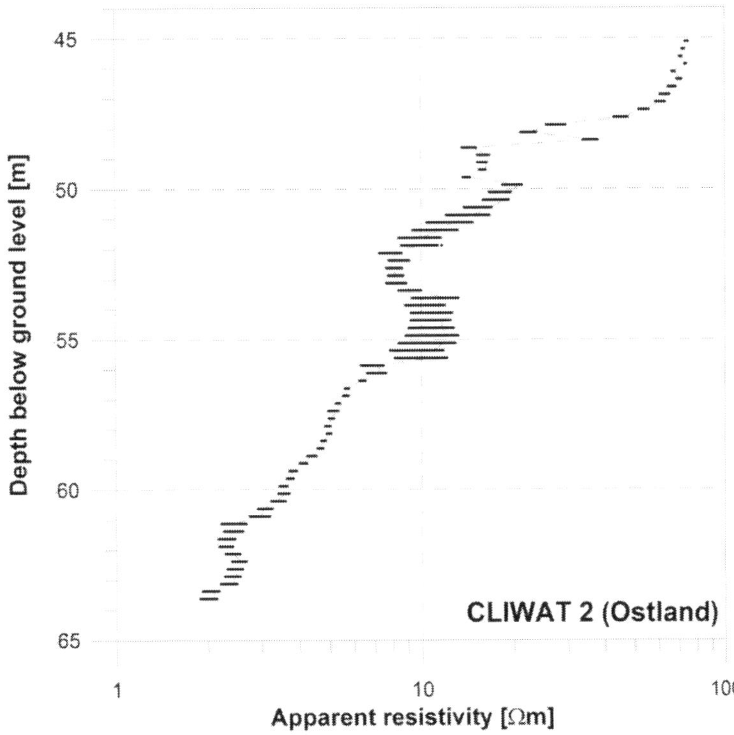

Figure 1. Time lapse of apparent resistivity in CLIWAT 2 between 2011 and 2017 (electrode spacing 0.25 m).

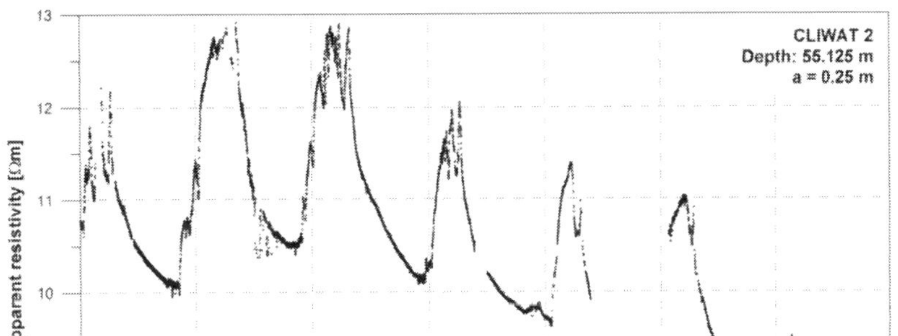

Figure 2. Time lapse of apparent resistivity in CLIWAT 2 in 55.1 m depth below ground level between 2011 and 2017 (electrode spacing a: 0.25 m).

REFERENCES

Ibenthal, M. 2015. A 2D density-dependent groundwater flow model to explain vertical distribution of electrical conductivity measurements and hydraulic heads of Borkum Island. Master Thesis, Georg August University Göttingen.

Ogilvy, R. D., P. I. Meldrum, O. Kuras, P. B. Wilkinson, J. E. Chambers, M. Sen, A. Pulido-Bosch, J. Gisbert, S. Jorreto, I. Frances, and P. Tsourlos. 2009. Automated monitoring of coastal aquifers with electrical resistivity tomography. Near Surface Geophysics 7: 367–375.

Poulsen, S. E., K. R. Rasmussen, N. B. Christensen, and S. Christensen. 2010. Evaluating the salinity distribution of a shallow coastal aquifer by vertical multielectrode profiling (Denmark). Hydrogeology Journal 18: 161–171.

Sulzbacher, H., H. Wiederhold, B. Siemon, M. Grinat, J. Igel, T. Burschil, T. Günther, and K. Hinsby. 2012. Numerical modelling of climate change impacts on freshwater lenses on the North Sea Island of Borkum using hydrological and geophysical methods. Hydrol. Earth Syst. Sci. 16: 3621-3643; doi: 10.5194/hess-16-3621-2012.

Wiederhold, H., H. Sulzbacher, M. Grinat, T. Günther, J. Igel, T. Burschil, and B. Siemon. 2013. Hydrogeophysical characterization of freshwater/saltwater systems – case study: Borkum Island, Germany. First Break 31: 109-117.

Wiederhold, H., W. Scheer, R. Kirsch, M. A. Rahman, and M. Ronczka. 2018. Saltwater intrusion under climate change in North-Western Germany – mapping, modelling and management approaches in the projects TOPSOIL and go-CAM. 25th Saltwater Intrusion Meeting, Gdansk, Poland, this volume.

Contact Information: Michael Grinat, Leibniz Institute for Applied Geophysics, Stilleweg 2, 30655 Hannover, Germany, Email: michael.grinat@liag-hannover.de

Location changes of "Wydrzany" groundwater intake in polish part of the Uznam Island aimed at groundwater state improvement

Ryszard Hoc[1], Andrzej Sadurski [2], Zenon Wiśniowski[1],
[1]Pomeranian Branch of Polish Geological Institut NRI, Szczecin;
[2]Marine Branch of Polish Geological Institut NRI, Gdańsk, and Nicholas Copernicus University, Tor, Poland

ABSTRACT

The water supply of the eastern part of the Uznam Island comes from two groundwater intakes: Zachód and Wydrzany. The threat to the resources part of the groundwater intake Wydrzany approach results from the progressive concentration of chloride ion, mainly in the southern part of the intake and locally in the central part of the water intake. After the launch of the intake at the end of the 1970s, the development of the water table depression, currently reaching the shore of the Lagoon, caused the frontal infiltration of the brackish waters of the Szczecin Lagoon into aquifers and a small amount of brine ascesion from the mesozoic strata. To determine the genesis of salinity, isotopic tests were carried out, including ^{37}Cl chlorine isotopes. Water for the Szczecin Lagoon and groundwater, including the aquifer of the Cretceous, have been sampled.

GEOLOGICAL AND HYDROGEOLOGICAL OUTLINE

The studied area of research is located in the western part of the Polish Baltic coast, in the the town of Świnoujście, West Pomeranian Voivodeship (fig. 1). There are two morphological forms in the landscape of this island. These are contrasting flat delta areas, being part of the so-called "Swina's Gates" and a varied ordinate of the upland of this island. The above mentioned forms of landscape differ from one to another not only in the genesis of the uprising, but also in different age. The decline is in Young Holocene, while the the upland area was shaped eventually during the deglaciation at the end of Pleistocene in this area. The "Świna Gate" it is the area between the Uznam and Wolin moraine uplands, which is a coastal lowland with an altitude in range of 0.3 ÷ 20.0 m above sea level, closing the passage between the Szczecin Lagoon and the Pomeranian Bay.

The outer waters for the island of Usedom are: poorly saline, with salinity lower than the open sea, due to the freshening influence of Odra River waters. The share of the "Wydrzany" intake in exploitation resources of groundwater in the Polish part of the Usedom island is Q = 300 m^3/h. To determine the cause of salinity and solve the problem associated with operational resources of the intake, cartographic, geophysical and laboratory analyses (chemical and isotopic) were made. The results of these works have been applied in the developed mathematical flow model of the considered area.

The multiaquifer formation in the eastern part of the Uznam Island, characterized in detail by numerous hydrogeological works (Matkowska, 1997) occurs in Quaternary sediments up to a depth of 30-50 m, depending on the shape of the Pre-Quaternary sediments. The intrusion of salt water into the Quaternary aquifer prevents natural inflow of rainwater.

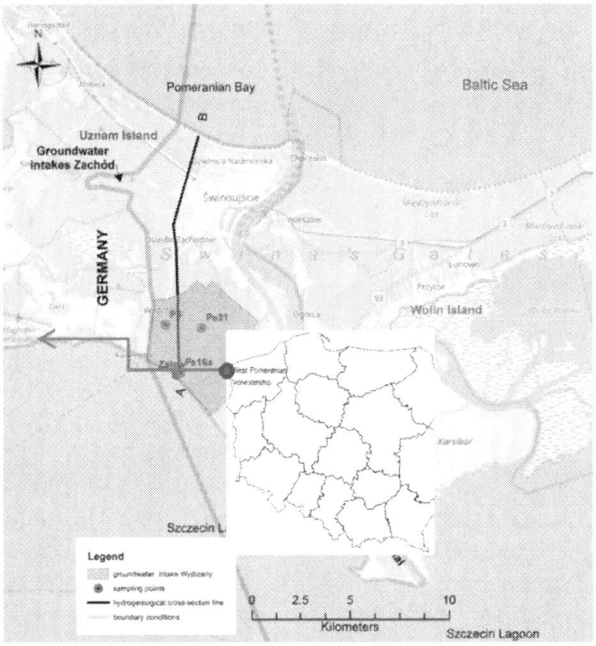

Figure 1. Location of the research area and groundwater intake "Wydrzany".

Hydrogeological conditions and flow systems of groundwater

The hydraulic head is higher in this aquifer than in the Cretaceous one. The infiltration conditions are generally very favorable. The 80% of the Uznam Island surface area is covered by permeable sediments. In the discussed area, there are also numerous cavities without outflow and inter-dune depressions which facilitate additionally infiltration. The thickness of the aeration zone depends mainly on the relief of the terrain surface and on the dune areas, it occurs at a depth of one to five m beyond the cone depression of the intake. Groundwater drainage takes place through the wells of the intakes and partial outflow to the surface waters. A small amount, these waters are drained by drainage ditches and pumping stations.

In the Cretaceous aquifer, there are mineralized waters and brines at depth, found in the sandy sediments of the lower Cretaceous at a depth of about 200 m. The groundwater circulation system in the discussed area can be reduced to the following scheme:
- under natural conditions, underground waters within the island are supplied only by infiltration of atmospheric precipitation,
- in the conditions of intensive exploitation of intakes, the lateral inflow of waters from the coastal and lake reservoirs is intensified.

Based on the analysis of the collected materials, three layers of the model were identified: Ist layer - groundwater level; IInd layer - weakly permeable sediments – muds and clays; IIIrd layer - the Quaternary aquifer, developed and fluvioglacial sands and gravels. The layer I and III remain in the hydraulic bond.

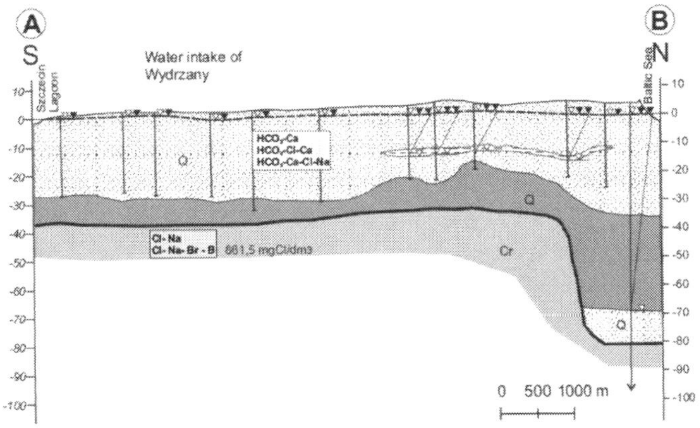

Figure 2. Hydrogeological cross-section through island of Uznam Island.

INTRUSION AND ASCENSION OF BRACKISH AND SALT WATER

During cartographic and geophysical works, the range and depth of occurrence of salt water in aquifer was determined. The occurrence of salinity in the first aquifer, which may be caused by infiltration of saline water from drainage ditches, has been found. The concentration of chloride ion in the years 2007 - 2015 on the "Wydrzany" intake varied from 15.0 mg / l to 447.0 mg / l and had an increasing trend.

METHOD AND SCOPE OF INVESTIGATION

The purpose of the research was to determine the genesis of the increase of chloride concentration in groundwater. Then, on the basis of the results, a solution was proposed in order to preserve the size of operational resources of the intake or even increase them, as well as to improve the status of groundwater in the Polish part of Uznam island. To solve the problem, geophysical and laboratory works were carried out, based on which a mathematical model of groundwater flow in the discussed area was developed. In order to explain the genesis of the salinity of the first aquifer in the southern part of the Uznam Island, it was made: geoelectrical logging, chemical analysis to determine the Cl^- ion for surface and groundwater samples, isotopic analysis of groundwater and surface water samples: $\delta^{34}S$ (SO_4^{2-}), $\delta^{13}C$ (dissolved inorganic carbon), $\delta^{37}Cl$, $\delta^{18}O$ (H_2O), $\delta^{2}H$ (H_2O).

The model of chloride ion transport in the MT3D program (MT3DMS version) has been developed for this task, which allows the calculation of mass transport in the advection-dispersion stream under steady state or transient flow conditions. Model calculations assume that chloride ions are transferred in the groundwater stream, like conservative tracer. The calculations did not take into account the dissolution and precipitation of mineral phases. The Cl^- ion ionization of the Cl^- ion was input by boundary conditions.

The basis for determination of external boundary conditions were the average concentrations of chloride ions observed in the waters of Szczecin Lagoon (800 mg/l), Świna R. (1000 mg/l) and Baltic Sea (2000 mg/l). The internal boundary conditions were applied to reflect the concentrations of chloride ions in the groundwater streams in the area of the "Wydrzany" intake. In the absence of data on the distribution of chloride ions in the vertical profiles of the aquifer, a constant initial value of concentrations in the entire profile of the aquifer was assumed.

RESULTS

The obtained results of isotopic research allow to see, on the one hand, the separateness of the P5 in relation to the other tested points, and on the other, the similarities between the composition of waters originating from the Szczecin Lagoon (indirectly the Baltic Sea) and groundwater included in the Ps 16a piezometer. The waters included in the piezometers Ps21 and Ps3a also show isotopic-hydrogeochemical similarity with the waters of the Szczecin Lagoon, however, the share of this "constituent" becomes smaller as the distance from the source of salinity increases.

As it results from the model calculations regarding water flow and chloride ions migration, in the conditions after the moved of the groundwater intake Wydrzany, saline water will not be supplied to it, or it will be irrelevant to water quality. In the figure below it can be seen that already in the initial conditions (t = 0 years) in the area of water runoff for chloride ion concentration, it is from 21 to 84 mg/l. After transferring the intake to the north, in the area with saline waters, chloride ion concentrations will decrease, and after 15 years in the "saline" area, the average Cl-ion concentrations fall below 250 mg/l. Changes in chloride ion concentrations in the area of the southern part of the island of Uznam Island, with the proposed exploitation of water on the "Wydrzany" intake, are presented in the figure below.

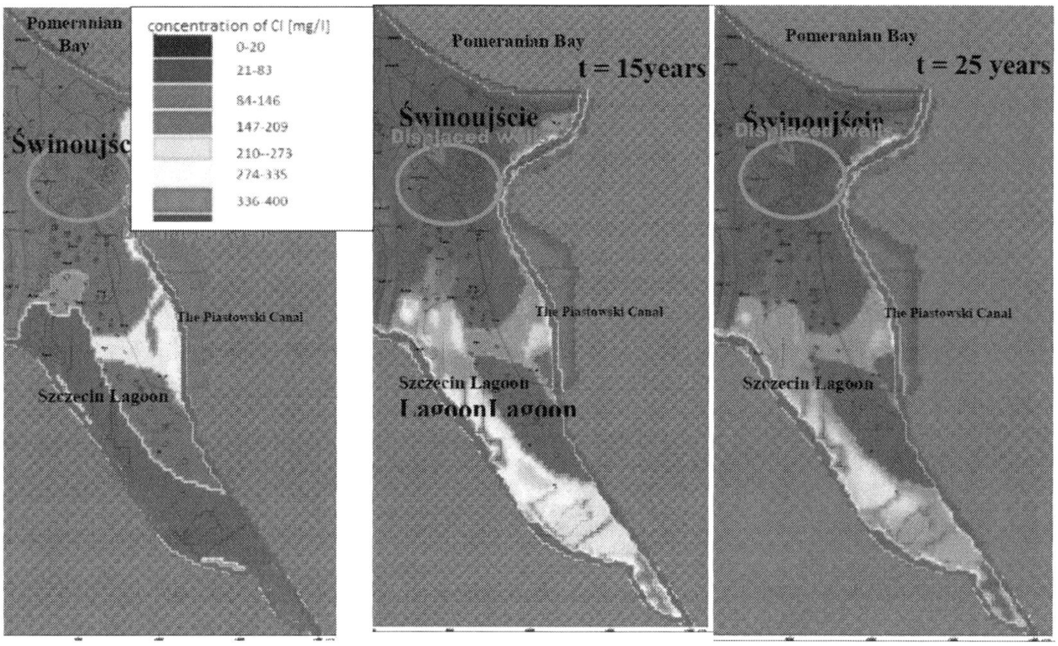

Figure 3. Changes in the salinity of main aquifer of usable level after dislocating water intake "Wydrzany", located on the Uznam Island.

PROPOSED SOLUTION FOR IMPROVING THE WATER CONDITION AND OPERATING RESOURCES

Based on analysis of water circulation conditions of water in the first and the second aquifer and also groundwater connection with surface waters, it was considered necessary to shift the resource area of the "Wydrzany" intake to the present area. In order to verify the validity of the proposed changes, a water balance was made for this intake in the first and second

aquifer. However, on the basis of simulations in the mathematical model, changes in the concentration of chloride ion in waters captured on this approach are depicted.
On the basis of model, field and laboratory works, it was confirmed that the lack of actions aimed at moving the groundwater intake Wydrzany to the north will cause a deepening of the salinity of the intake water. The positive scenario consisting in lowering the chloride concentration in the captured groundwater at the "Wydrzany" intake is related to the displacement of groundwater intake. This solution will allow maintaining disposable resources at the current level, and the salinity of water may decrease to approximately 30 mgCl/l. The confirmation of this conclusion is the result of modelling of chloride transport, which shows that in waters of the main aquifer there are concentrations in the range of 21 - 83 mg Cl/l in the range of impact of the "Wydrzany" intake. At the same time, in areas where groundwater salinity is currently observed, a steady decline in chloride ion concentrations is observed. The analysis shows that the proposed change, mainly in the area of the location of the resource area of the "Wydrzany" intake, will meet the basic environmental objective of improving the quality and quantity of groundwater on the Uznam Island.

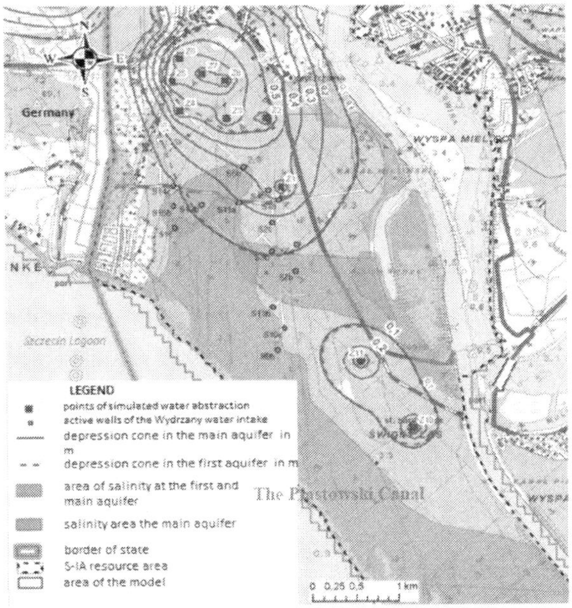

Figure 4. The range of the depression of the water intake after shifting the intake, on the background of documented saline groundwater.

LITERATURE

Matkowska Z., 1997 – Hydrogeological map of Poland in scale of 1:50 000, sheet Świnoujście with explanations [in Polish, English sum]. PGI-NRI. Warszawa.

Matkowska Z., Ruszała M., Wdowiak M., 1977 - Detailed geological map of Poland on a scale of 1: 50000 sheet Świnoujście i Międzyzdroje with explanations. [in Polish, English sum]. PGI-NRI. Warszawa.

Wiśniowski Z., Hoc R., Jezierski P., 2014 - Appendix 1 to the hydrogeological documentation setting the distribution resources of the groundwater of the Międzyodrza balance registration, the Szczecin region, the islands and the west part of the Wolin island. Unpublished. [in Polish, English sum]. PGI-NRI. arch. of Pomeranian Branch. Szczecin.

300 years of coastal salinization research in Germany – the Homann (1718) map of the Christmas Flood of 1717

Georg J. Houben[1]
[1]BGR, Hannover, Germany

ABSTRACT

The Christmas Flood of 1717 was one of the most destructive storm floods in the North Sea region and affected large parts of the shores of Germany, the Netherlands and Germany (e.g. Jakubowski-Tiessen 1992). It occurred in the night from the 24th to the 25th of December 1717, when a strong northwesterly storm front pushed massive volumes of water into the funnel-shaped German Bight. Adding to the astronomical high tide occurring this night, water levels rose up to 4 m higher than the mean tidal high water mark. This lead to widespread overtopping and breaching of the dikes, which had been neglected in the preceding years due to extended periods of war and unrest. Since the event happened at night, the population was unable to react. About 9,000 people in Germany lost their lives and around 2,500 in the Netherlands. The small German town of Jever alone lost 1,700 people. In the village of Stollhamm, located on the peninsula of Butjadingen, which was exposed to the flood from two sides, 582 out of a population of 1,200 perished and only a third of the houses were not destroyed. In Eastern Frisia, 922 houses were completely destroyed and 1,672 damaged. In all of the affected regions in Germany, at least 3,000 houses were completely destroyed. Agriculture was severely affected by the salinization of large tracts of agricultural land and the loss of 2,300 horses, 9,500 cows, 2,800 sheep and 1,800 pigs was recorded in Eastern Frisia. In the following years, famines and epidemic plagues took a further toll on the population. Many people emigrated. It took several decades to reconstruct the dikes and to restore the livelihoods of the population.

The 1717 flood was described in publications such as the "*The tearful Christmas joy of the Jever region. Or: Detailed news of the high water flood which inundated the governance of Jever during Christmas night 1717 and what damages it caused [...]* (Ummen 1718) and the rather dramatically titled „*New and improved war, murder, death, misery and hardship calendar for the year 1717 after the merciful holy birth of our Lord and Redeemer Jesus Christ. Which [...] also contains a detailed description of the terrible storms and the resulting almost supernatural water floods through which god's hand beset the countries at the North, Zuider and Baltic Sea on the holy Christmas Day 1717 [...], accompanied by a copper engraving clearly presenting this punishment by God.*" by Adelsheim (1719).

Floods in this region were (and still are) not uncommon and floods preceding the Christmas flood are well documented in historical documents. The 1717 flood, however, was the first that attracted the attention of cartographers. Only one year after the flood, a colored copperplate engraving map titled "*Geographical presentation of the miserable water flood in Lower Germany, which on the 25th of December 1717 AD, in the holy Christ-Night, with innumerable damages and losses of many thousands of humans, inundated a large part of the duchies of Holstein and Bremen, the shires of Oldenburg, Frisia, Groningen and North Holland*" was published in Nuremberg (Homann 1718). It shows an outline of the flooded area (green shading). This is probably the first largely accurate map of a coastal inundation. The salinization of soil and groundwater is, however, not explicitly mentioned. The flooded zones in the Northern Netherlands are also shown in an insert map.

The cartographer of the map was Johann Baptist Homann (March 20, 1664 in Oberkammlach – July 1, 1724 in Nuremberg). In 1702, he opened a publishing house in Nuremberg, Germany, specializing in the production and sale of maps. His company became a very influential publisher of maps, which were also renowned for their artistic quality. Consequently, Homann became a member of the Royal Academy of Sciences in Berlin in 1715 and cartographer at the imperial court in Vienna. The publishing house existed until 1848.

The size of the map is 58 cm by 47 cm. The base map is probably much older and contains several errors. For example, the island of Juist is missing and the island of Nordstrand is shown in its shape prior to the 1634 Burchardi flood, which destroyed much of the island. Prints of the Homann map must have been sold in relatively high quantities during the 18th century, since copies can still be found in several European archives and some are even up for sale on online platforms. An internet search easily yields access to digital copies.

Homann and his coworkers probably neither mapped the flood extent themselves in the field nor inspected the installations shown in the insert images. They probably relied on descriptions by third parties, such as the reports by Ummen (1718) and Adelsheim (1719), which contain sufficient details about which cities were struck. Homann then probably only connected these dots to arrive at his outline of the flooding. Although this resulted in some errors, the general outline of the flooded marsh area is largely correct.

The map is adorned by depictions of antique deities, such as Neptune with his trident, goddesses of wind (Boreas?) and allegorical mythical beings. The crowned female figure to the right could be Hammonia, a Latin personification of the city Hamburg (the figure is placed close to the location of the city on the map). The map contains two text frames, one of which describes the flood and some of its consequences (in German). There, Homann puts the number of casualties in Germany at 18,140. The other contains a more general description of the problems of coastal zones (in German), followed by some excerpts from the "Metamorphoses" of the Roman poet Ovid (Publius Ovidius Naso) in Latin. The latter describe floods caused by the wrath of the gods and cities that have sunk below sea level.

Some inserts show pictures of a breached dam, a bucket elevator used for the removal of salt water through sluices and a sluice door in a dam that closes itself when water comes from the seaward side but opens automatically when water drains from the landward side. The latter images are, however, not realistic, the elevator would not work in the form shown and the depicted dam is excessively steep. The extent of the flooding is also not exactly reproduced in all details. For example, the German Frisian Islands are shown in light colors, indicating that they were not flooded. However, it is a known fact that they were strongly affected by the Christmas flood. Homann himself mentions the flooding of some islands in the first text frame. The map contains a few curious details, e.g. the description of a ship, which, on its way to France, was washed onto land through a breached dam with its complete cargo, near the present-day town of Cuxhaven.

Up to this day, the region covered in the Homann (1718) map has to deal with the threat of inundations. Securing a sustainable water supply for the population is an ongoing challenge, since saline groundwater often occurs kilometers away from the coast. Although the Christmas flood was not the only one that inundated the North Sea coast, the Homann map of 1718 provides an invaluable resource for hydrogeologists trying to delineate the boundary between fresh, brackish and saline water in this coastal zone. It also helps to explain how and when the salinization of coastal groundwater took place.

REFERENCES

Adelsheim, P. 1719. Neuer und Verbesserter Kriegs-(,) Mord- und Tod-(,) Jammer- und Noth-Calender/ Auf das Jahr nach der gnadenreichen heiligen Geburt unsers Herrn und Heilands Jesu Christi M DCC XIX. In welchem Nebenst der Beschreibung des Gewitters/Erwehlungen/der Planeten Lauff und Gang/samt deroselben natürlichen Zuneigungen/ auch zu finden ist Eine ausführliche Beschreibung Der entsetzlichen Stürme/ und daher verursachten fast über natürlich-hohen Wasser-Fluten/ womit GOttes Hand am H.(eiligen) Christ-Tag 1717, und den 25. Hornung (= Februar) 1718 die Länder an der Nord-(,) Süder- und Ost-See heimgesuchet, Nebst einem Kupfer /welcher diese Strafe Gottes deutlich vorstellet. Verlag Johann Andrea Endters sel. Sohn und Erben, Nuremberg.

Homann, J.B. 1718. Geographische Vorstellung der jämerlichen Wasser-Flutt in Nieder-Teutschland, welche den 25. Dec(ember) A(nn)o 1717, in der heiligen Christ-Nacht, mit unzählichen Schaden, und Verlust vieler tausend Menschen, einen großen theil derer Herzogth(ümer) Holstein und Bremen, die Grafsch(aft) Oldenburg, Frislandt, Gröningen und Nort-Holland überschwemmet hat. Nuremberg.

Jakubowski-Tiessen, M. 1992. Sturmflut 1717. Die Bewältigung einer Naturkatastrophe in der Frühen Neuzeit. Munich, Oldenbourg.

Ummen, C.J. 1718. Die Mit Thränen verknüpffte Weynachts-Freude Jeverlandes. Oder Eine ausführliche Nachricht der hohen Wasser-Fluht/ Wodurch die Herrschafft Jever in der Christ-Nacht 1717. überschwemmet/ und was dadurch für Schaden verursachet worden In gebundener Rede entworffen/ und mit weitläufftigen Anmerckungen erläutert von Conrad Joachim Ummen. 32 p., Bremen. (http://digital.lb-oldenburg.de/ihd/content/thumbview/242095)

Contact Information: Georg J. Houben, Federal Institute for Geosciences and Natural Resources (BGR), Stilleweg 2, 30655 Hannover, Germany, Phone: +49-511-6432373, Email: georg.houben@bgr.de

Tectonically conditioned brine leakage into usable freshwater aquifers – implications for the quality of groundwater exploited in central Poland

Dorota Kaczor-Kurzawa[1]
[1]Polish Geological Institute – National Research Institute (PGI-NRI), Kielce, Poland

ABSTRACT

Brine leakage areas, which are identical with zones of chloride ion content anomalies (Cl^- >60 mg/dm^3) in usable aquifers, were mapped and examined on the basis of chemical and isotopic analyses. These zones are predominantly developed in tectonic conditions enabling the inflow of Mesozoic saline waters and brines into freshwater aquifers: 1) fault zones, 2) hydrogeological windows above salt anticlines and elevated tectonic blocks, 3) salt diapirs. The natural process of brine migration has been accelerated in some areas due to groundwater exploitation. Consequently, the decline of groundwater quality on many intakes has been reported, which is a result of the elevated content of chloride, Natrium and Ammonium ions.

INTRODUCTION

This paper describes the phenomenon of brine leakage into freshwater usable aquifers and is associated with groundwater quality decline affecting intakes operating in central Poland. The studied area (28,000 km^2) corresponds to the region of well developed salt tectonics forms within the Permian-Mesozoic complex (Fig.1). In this area, groundwater is extracted from aquifers developed within Mesozoic calcareous and sandstones series as well as Paleogene, Neogene and Quaternary sand series.

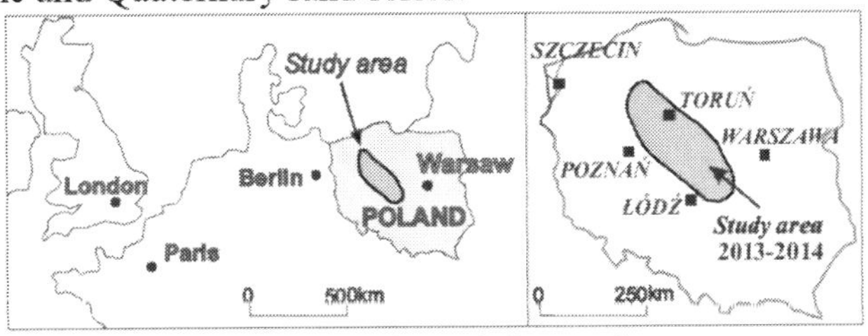

Figure 1. Location of the study area.

METHODS

The preliminary stage of the study consisted in the mapping of the areas in which the brine leakage into freshwater aquifers is detectable. Such areas were named chloride anomaly zones, and were defined by groups of wells with the concentration of Cl^- ion in groundwater exceeding the value of 60 mg/dm^3 – the upper limit of hydrogeochemical background. Spatial differentiation of water salinity in usable aquifers was documented with over 12,000 archival chemical analyses from the data bank of the Polish Geological Institute. Chemical data were verified for each intake. The results of analyses indicating that the elevated chlorides concentration in water originated due to anthropogenic pollution were rejected. The next stage of research was focused on specific chloride anomaly zones and on intakes affected by groundwater salinization. It included detailed chemical and isotopic (δ^2H, $\delta^{18}O$) analyses of groundwater samples as well as groundwater table measurements.

RESULTS & DISCUSSION

The hydrochemical and isotopic data allow for detecting at least 23 chloride anomaly zones in usable freshwater aquifers. They cover about 20% of the investigated area.

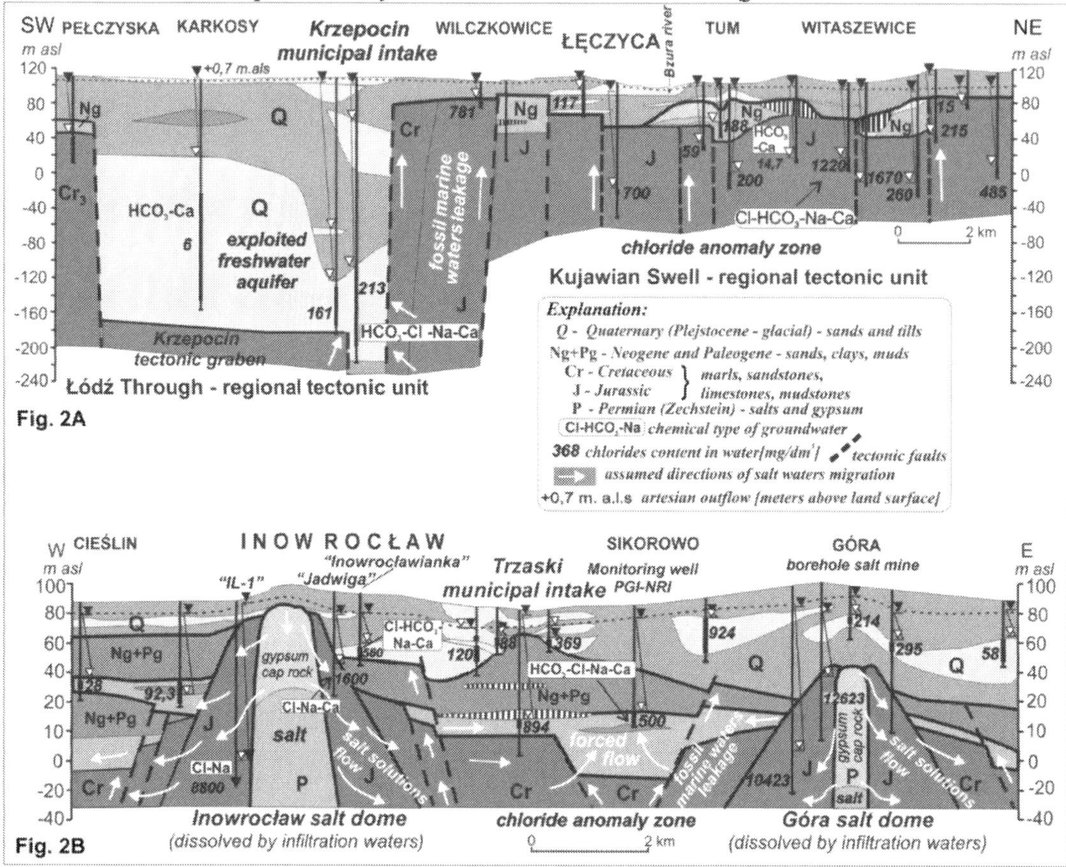

Figure 2. Hydrogeological cross-sections through A) Łęczyca-Żychlin and B) Inowrocław chloride anomaly zones.

Tectonic determinants of chloride anomaly zones development within freshwater aquifers

The correlation of hydrochemical, geophysical and drilling data confirmed that chloride anomalies have been predominantly developed in geologic situations, enabling the inflow of salt waters from deeper parts of Mesozoic complex into the useful aquifers. These are: 1) tectonic fault and fracture zones, especially those running along the borders of tectonic units (Fig.2A); 2) hydrogeological windows above uplifted crests of salt anticlines and elevated tectonic blocks; 3) mature salt diapirs currently being dissolved by infiltrating waters (Fig.2B). The conditions enabling the upward migration of salt waters through the tectonically produced pathways were confirmed with the archival results of pressure tests executed for Mesozoic brine-bearing aquifers in deep research boreholes (Kaczor 2006; Kaczor-Kurzawa 2017).

https://doi.org/10.1051/e3sconf/20185400012

Chemical and isotopic record of brine leakage into freshwater aquifers

The upward migrating saline waters and brines have been mixing with fresh, drinking waters, which are exploited on groundwater intakes (Fig.3A). As a result of the mixing process, the groundwaters extracted on intakes situated within the detected chloride anomaly zones are converted to chloride type (Fig.3B). Their TDS value (up to 3.6 g/dm^3) as well as ion relations are differentiated because of the various proportions between the mixed components of fresh and salt waters. In cation group, Natrium and Potassium ions prevail over Calcium. The composition of the anion group is dominated by bicarbonates ions, but the content of chloride ions changes from 20% to 60%. These waters are often characterized also by elevated K, Sr, B, Mg, Ba ion concentrations.

The leakage of saline waters, 'older' than modern infiltration waters (supposedly diluted Mesozoic brines), into freshwater aquifers, was proved by chemical and isotopic tests (δH and δO) executed for groundwater samples from 9 working intakes (Fig.4) situated in 5 major chloride anomaly zones (Kaczor-Kurzawa 2017).

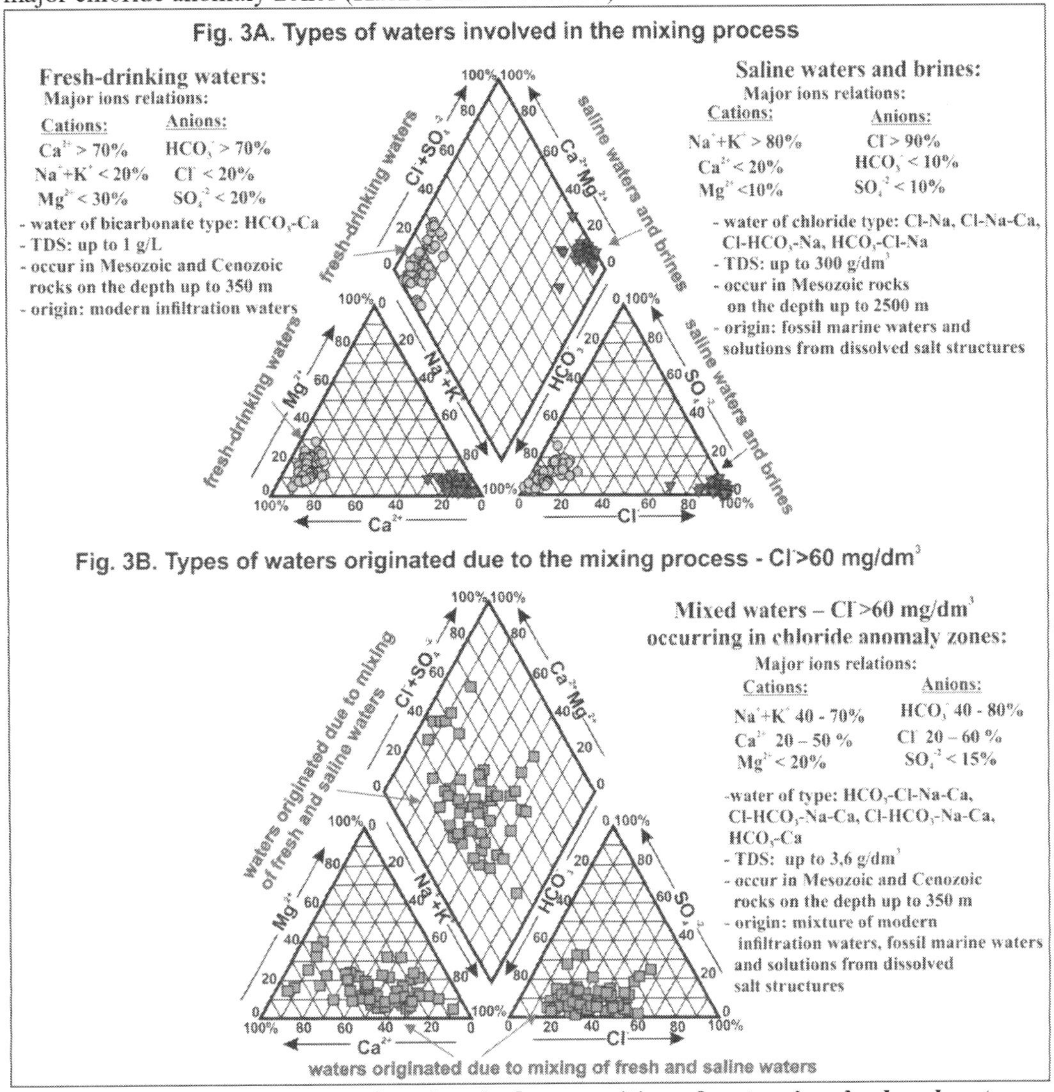

Figure 3. Pipper plots presenting chemical composition of waters involved and waters originated due to mixing process.

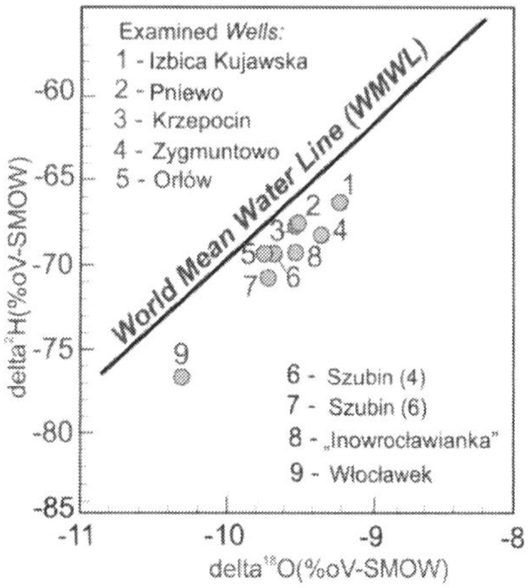

Figure 4. Graph of dependency between δ²H and δ¹⁸O in the analyzed waters compared to WMWL

Brine leakage implications for the quality of exploited groundwater

Brine leakage is responsible for a significant decline in groundwater quality, which has been reported from numerous intakes working in the indicated chloride anomaly zones, due to the high concentration of chlorides (>250 mg/dm^3), Natrium (>200 mg/dm^3), Ammonium (> 0.5 mg/dm^3) ions, in some cases associated with elevated contents of sulfate (> 250 mg/dm^3) and Boron (> 1.0 mg/dm^3) ions. In general, chloride contents in groundwater increase with the depth, which reflects the inflow of saline waters from the deeper rock complexes. However, we can also observe the coexistence of fresh and salt waters with highly differentiated Cl^- ion contents, occurring at similar depths in neighboring wells (Fig.2A). This means that high chloride concentrations are related to a close presence of salt water migration routes, such as tectonic faults and fracture groupings in Mesozoic rocks. However, being located at a long distance from them, chloride concentrations decrease due to the dilution effected by infiltrating fresh waters. Therefore, only part of groundwater well populations within indicated anomaly zones display elevated chlorides concentrations. An attempt to estimate the scale of the geogenic salinization phenomenon included about 12,000 wells with documented chloride concentrations in groundwater. Only in about 300 (3%) of those wells did Cl^- ion content exceed the standard value (250 mg/dm^3) for drinking water. However, these calculations are primarily based on the data coming from the period of the wells being constructed and documented, usually many years ago.

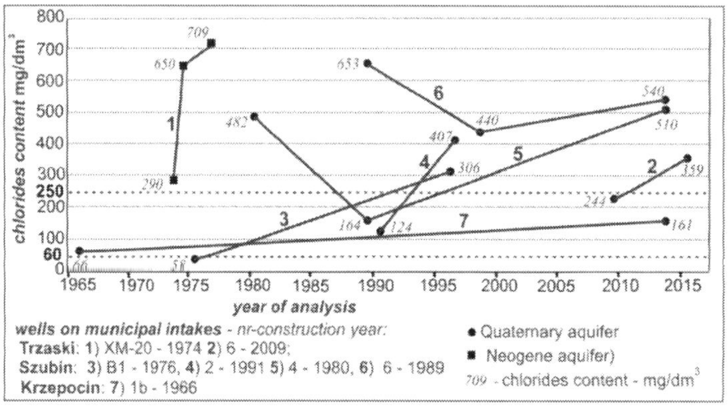

Figure 5. Graph of the temporal variability of chloride concentration in waters from selected intakes.

In many cases, intake activity has been inducing and/or accelerating salt water flow towards the exploited aquifers, which caused a degradation of groundwater quality (Szubin and Trzaski intakes - Fig.2B, 5) and/or can lead to it in the near future (Krzepocin intake - Fig.5).

CONCLUSIONS

The origin and location of the brine leakage areas, identical with chloride anomaly zones, are dependent on tectonic elements within the Permian-Mesozoic complex and on variations of their Cenozoic cover, the factors conditioning the inflow of Mesozoic salt waters into useful aquifers. In many cases, the Cl^- ion anomalies detected in groundwater of usable aquifers can be correlated with the spontaneous outflows of diluted brines, which created the natural Cl^- ion anomalies on the land surface. Their existence has been ascertained through soil research, and especially through botanical examinations describing the ecosystems of halophilous plants. The natural process of brine migration and leakage, occurring in the chloride anomaly zones, has been intensified by a long-term exploitation on groundwater municipal and industrial intakes. This can lead to a progressing salinization and degradation of groundwater quality, due to the elevated content of chlorides, Natrium, Ammonium and sulfates ions, exceeding the regulatory limits. That phenomenon affects dozens of currently working intakes, increasing water supply costs for numerous villages and towns.

ACKNOWLEDGEMENTS

This work was supported by the Ministry of Science, Poland, under Project no. 61.8301.1301.00.0

REFERENCES

Kaczor, D. 2006. The salinity of groundwater in Mesozoic and Cenozoic aquifers of NW Poland - origin and evolution. Studia Geologica Polonica, 126:5-76.

Kaczor-Kurzawa, D. 2014 – The salinization of useful Cenozoic aquifers by ascending Mesozoic brines – characterization on the basis of hydrochemical data from northern and central Poland. Proceedings, 23 SWIM, Husum: 200–203.

Kaczor-Kurzawa, D. 2017 - Geogenic chloride anomalies in groundwater of useful aquifers in Central Poland. Przegląd Geologiczny, 65: 1282–1289 (in Polish with English summary).

Contact Information: Dorota Kaczor–Kurzawa, Polish Geological Institute - National Research Institute, Holy Cross Branch, ul. Zgoda 21, 25-953 Kielce; e-mail:dorota.kaczor-kurzawa@pgi.gov.pl

Hydrochemical characterization of various groundwater and seepage water resources located in the Bay of Puck, Southern Baltic Sea

Żaneta Kłostowska[1,2], Beata Szymczycha[1], Karol Kuliński[1], Monika Lengier[1] and Leszek Łęczyński [2]
[1] Institute of Oceanology, Polish Academy of Sciences, Sopot, Poland
[2] University of Gdańsk, Department of Marine Geology, Gdynia, Poland

INTRODUCTION

Submarine groundwater discharge (SGD) has been recognized as a significant source of water and chemical substances such as trace metals, nutrients, organic and inorganic carbon to the coastal zone (Oberdorfer et al. 1990; Charette et al. 2001; Slomp et al. 2004; Burnett et al. 2006; Moore 2010). In some regions SGD has been responsible for a deterioration of the coastal environment affecting the local economy and management of the area (Valiela et al. 2002; Slomp et al. 2004; Andersen et al. 2007). Particularly sensitive to the SGD are enclosed or semi-enclosed reservoirs of a limited water exchange with the open sea (Cyberski 1993). The Bay of Puck, southern Baltic Sea, due to its morphological and hydrological features, is a good example of a basin partially separated from the open sea waters (Nowacki 1993; Nowacki 1993a). Moreover, the bay itself is divided into two parts by the Rybitwa Shallow (the average depths: 21m in the outer and 3m in the inner part, that additionally impedes highly the water renewal in the inner part) (Korzeniewski et al. 1994; Urbanski et al. 2007). The Bay of Puck borders with the continental land in the west and the thin Hel Peninsula in the north and north-east. Both, the continental part and the peninsula differ in terms of land use and management. Interestingly, the first studies on the SGD in the Baltic Sea region were also performed at the Hel Peninsula (Sadurski 1987). They were related to the seawater intrusion due to the intense exploitation of aquifers. As a consequence several groundwater wells were closed along the Hel Peninsula. Later a reverse phenomenon has started to be observed and scientists became interested in the identification and quantification of the groundwater seepage to the marine environment (Piekarek-Jankowska et al. 1992; Piekarek-Jankowska 1994; Piekarek-Jankowska et al. 1994). The anomalies in the chloride distribution due to the groundwater discharge were identified in the pore waters of the Bay of Puck sediments (Bolałek 1992). Recently, the fluxes of different chemical substances via SGD to the Bay of Puck have been estimated based on the experimental studies performed at the Hel Peninsula, off Hel (Szymczycha et al. 2012; Szymczycha et al. 2014; Szymczycha et al. 2016). In fact, the loads of P, Mn and DIC via SGD to the Bay of Puck reported by the authors are comparable to the loads supplied by the local rivers entering the basin. This suggests that the SGD can also be an important source of other chemical constituents and thus has a potential to shape the functioning of the unique ecosystem of the Bay of Puck (Szymczycha et al. 2012; Szymczycha et al. 2014; Szymczycha et al. 2016). Those estimations and conclusions were based on the assumption that the composition of the groundwater coming from the side of the Hel Peninsula and the continental part are similar. Our present research hypothesis assumes that the different aquifers exploitation, land use and management, diversified geological structure and differences in land-sea interactions along the Hel Peninsula and at the continental part differentiate the chemical composition of SGD coming from both these sides. The main aim of this study was to identify the provenance of the groundwater discharged to the bay and processes influencing its composition. This was done based on the chemical composition of SGD.

METHODS

The research was carried out in years 2016-2017. In order to identify SGD sites salinity was used as a groundwater tracer (Szymczycha et al. 2012). Three active areas of SGD off Hel Peninsula (Hel, Jurata, Chałupy), and three off continental part of the bay (Puck, Swarzewo and Osłonino) were identified based on salinity and chloride measurements (Figure 1). Water samples were collected along the salinity transition zone: groundwater from piezometers located in the coastal zone, groundwater and seepage water from the subterranean estuary (groundwater – seawater mixing zone) and seawater. From each point, 10 mL sample for ions content analysis (Mg^{2+}, Ca^{2+}, K^+, Na^+) were collected and filtered using syringe filters (0.45 μm pore size) to the PTE vials, which were previously digested for 24 h in nitric acid. The sample for metals analysis were conserved with 3M HNO_3. 180 ml of water were collected for HCO_3^-, SO_4^{2-} and Cl^- analysis. Samples for HCO_3^- were conserved with $HgCl_2$. Samples for SO_4^{2-} and Cl^- were kept in refrigerator till the time of the analysis. The major ions (Ca^{2+}, Mg^{2+}, Na^+, K^+, Cl^-, SO_4^{2-}, HCO_3^-) content was analyzed in 198 samples. The measurements of Ca^{2+}, Mg^{2+}, Na^+, K^+ were conducted by means of AAS (SHIMADZU 6800). Concentrations of HCO_3^- and Cl^- were determined by potentiometric titration using ion-selective electrodes (ISE) Cerko Lab System Potentiometry for precise titration. Chloride ions were titrated with silver nitrate (0.1 M $AgNO_3$) and the bicarbonate ions were titrated with 0.1 molar hydrochloric acid (0.1 M HCl). The titration equivalent was determined by the first derivative (Dojlido 1987, Gajkowska-Stefańska et al. 2007). SO_4^{2-} was quantified by the precipitation of a sparingly soluble $BaSO_4$ in an acidic medium (Szczepaniak 2011). In addition, in the samples were measured parameters: the ORP, pH and oxygen content, with a multiparametric meter. Separation of individual types of water was made based on Cl^- content (Bolałek 1992). The seepage samples with chloride ions content smaller than 1 $gCl^-\cdot dm^{-3}$ were identified as groundwater samples.

RESULTS

The major ions ratios in all groundwater samples fluctuated in the range: from 0.11 to 0.94 for Ca^{2+}/Cl^-, from 0.08 to 1.23 for Mg^{2+}/Cl^-, from 2.23 to 93.50 for Na^+/Cl^-, from 0.03 to 0.66 for K^+/Cl^-, from 0.01 to 0.32 for SO_4^{2-}/Cl^-, from 0.07 to 0.39 for HCO_3^-/Cl^-, from 0.57 to 3.42 for Mg^{2+}/Ca^{2+}, while for seawater samples were in the range from 0.01 to 0.02 for Ca^{2+}/Cl, from 0.01 to 0.06 for Mg^{2+}/Cl^-, from 0.58 to 0.75 for Na^+/Cl^-, from 0.01 to 0.02 for K^+/Cl^-, from 2.98 to 4.79 for Mg^{2+}/Ca^{2+}. The ionic ratios in water samples from piezometers ranged from 0.04 to 1.97 for Ca^{2+}/Cl^-, from 0.01 to 0.32 for Mg^{2+}/Cl^-, from 1.11 to 88.68 for Na^+/Cl^-, from 0.01 to 0.44 for K^+/Cl^-, from 0.01 to 0.06 for SO_4^{2-}/Cl^-, from 0.01 to 1.16 for HCO_3^-/Cl^- and from 0.20 to 2.20 for Mg^{2+}/Ca^{2+}. Water samples collected in the groundwater wells were characterized by the range: from 0.45 to 5.67 for Ca^{2+}/Cl^-, from 0.09 to 0.62 for Mg^{2+}/Cl^-, from 0.78 to 0.85 for Na^+/Cl^-, from 0.08 to 0.34 for K^+/Cl^-, from 0.03 to 4.49 for SO_4^{2-}/Cl^-, from 2.90 to 20.23 for HCO_3^-/Cl^- and from 0.18 to 0.35 for Mg^{2+}/Ca^{2+}. The maximum ratios in groundwater samples collected in the subterranean estuary were observed in Chałupy region and equal to 0.79 for Ca^{2+}/Cl^-, 1.23 for Mg^{2+}/Cl^-, 25.17 for Na^+/Cl^- and 0.66 for K^+/Cl^-, respectively. Comparable results were detected also in Swarzewo. In case of water samples coming from piezometers, the maximum ratios were observed in Hel and equal to 0.72 for Ca^{2+}/Cl^-, 0.27 for Mg^{2+}/Cl^-, 34.87 for Na^+/Cl^-, 0.08 for K^+/Cl^-, 0.06 for SO_4^{2-}/Cl^-, 0.46 for HCO_3^-/Cl^-, while the lowest ratios were recorded in Jurata and equal to 0.04, 0.01, 0.01, 0.0048 and 0.01, respectively. The ratios for groundwater samples collected

 https://doi.org/10.1051/e3sconf/20185400013

in wells were highest in Swarzewo and equal to 5.67 for Ca^{2+}/Cl^-, 0.62 for Mg^{2+}/Cl^-, 0.85 for Na^+/Cl^-, 0.34 for K^+/Cl^-, 4.49 for SO_4^{2-}/Cl^-, 20.24 for HCO_3^-/Cl^-, except Mg^{2+}/Ca^{2+} (0.35) that was identified in the Jurata.

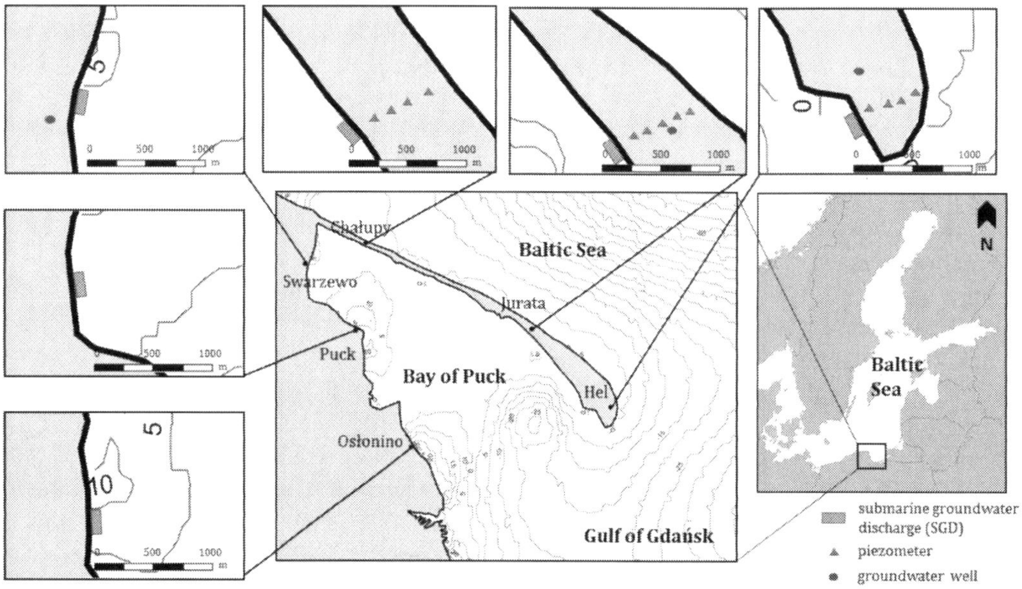

Figure 1. Map of the Bay of Puck showing the active areas of submarine groundwater discharge (SGD), locations of piezometers and groundwater well, along the Hel Peninsula (Hel, Jurata, Chalupy), and in the continental part of the bay (Puck, Swarzewo and Osłonino).

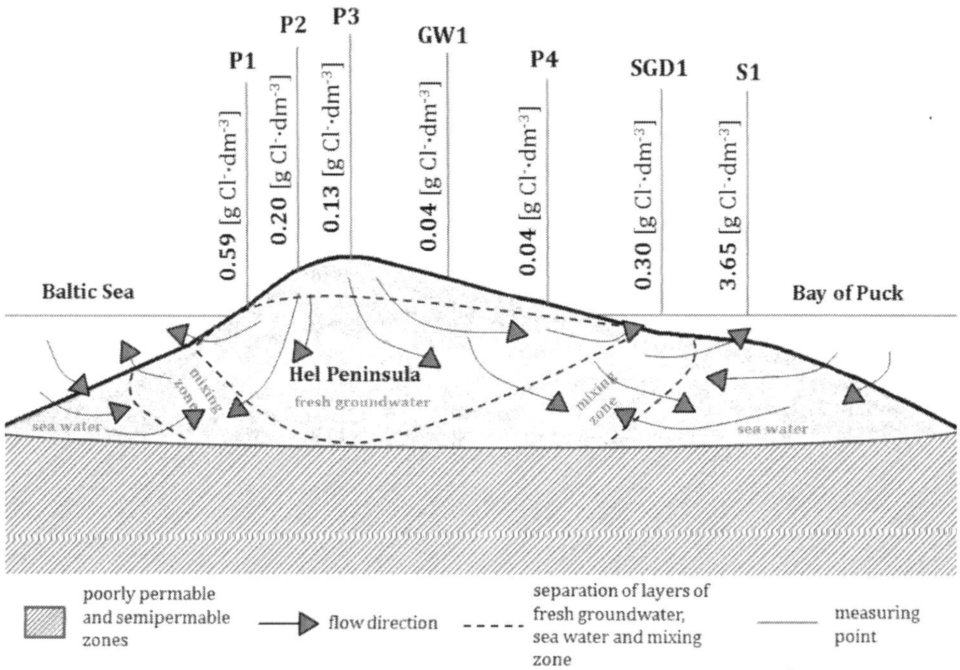

Figure 2. Schematic distribution of chloride concentration [gCl⁻·dm⁻³] at submarine groundwater discharge (SGD), piezometers (P), groundwater well (GW) and sea water (S) in the Hel Peninsula, of Hel. Sampling points are marked schematically.

DISCUSSION AND CONCLUSIONS

The preliminary results indicate that SGD composition in Chałupy and Swarzewo are comparable. This suggests that groundwater discharged from both these sites located in the Inner Bay of Puck can have similar source. This hypothesis will be further verified by the analysis of the oxygen stable isotopes. For Hel and Jurata the ionic composition of the groundwater samples is mainly influenced by the land-sea interactions. Thus the increase of seawater level and atmospheric conditions such as storms will result in higher seawater intrusion (Sadurski, 1987). The vertical cross-section of groundwater composition located in Hel (Figure 2) is conditioned by the processes associated with the movement of seawater masses. This is observed at both sites of the peninsula. Osłonino and Swarzewo characterized with totally different ratios in comparison to other areas which can be explained by possible different source of groundwater relating to different land use and management. The results show that there is no clear separation between continental part and the peninsula in terms of groundwater composition. Further studies are necessary to understand the groundwater provenance and factors influencing SGD in the Bay of Puck.

ACKNOWLEDGMENTS

Acknowledgments The study reports results obtained within the framework of the following projects: UMO-2016/21/B/ST10/01213 sponsored by National Science Center and WaterPUCK financed by the National Centre for Research and Development (NCBR) within BIOSTRATEG program.

REFERENCES

Andersen S., Baron L., Gudbjerg J., Gregersen J., Chapellier D., Jakobsen R., Postma D., 2007. Discharge of nitrate-containing groundwater into a coastal marine environment. J. Hydrol., 336, 98–114

Bolałek J. 1990, Ionic macrocomponets of the interstitial waters of Puck Bay. Oceanologia, 33, 131-159

Burnett W.C., Aggarwal P.K., Aureli A., Bokuniewicz H.J., Cable J.E., Charette M.A., Kontar E., Krupa S., Kulkarni K.M., Loveless A., Moore W.S., Oberdorfer J.A., Oliveira J., Ozyurt N., Povinec P., Privitera A.M.G., Rajar R., Ramessur R.T., Scholten J., Stieglitz T., Taniguchi M., Turner J.V., 2006. Quantifying submarine groundwater discharge in the coastal zone via multiple methods. Sci. Total Environ., 367, 498-543

Charette M.A., Buesseler K.O., Andrews J.E., 2001. Utility of radium isotopes for evaluating the input and transport of groundwater-derived nitrogen to a Cape Cod estuary. Limnol. Oceanogr., 46, 465–470

Cyberski J. 1993. Hydrologia zlewiska i morfometria zatoki. Korzeniewski K., Zatoka Pucka. Gdańsk: Instytut Oceanografii Uniwersytetu Gdańskiego. 40

Dojlido J., 1987. Chemia wody, Arkady, Warszawa

Gajkowska-Stefańska L., Guberski S., Gutowski W., Mamak Z., Szperliński Z. 2007, Laboratoryjne badania wody, ścieków i osadów ściekowych, Oficyna Wydawnicza Politechniki Gdańskiej, 92-96, 113-117, 181-187

Korzeniewski K. 1994. Zatoka Pucka. Inst. Ocean. UG.

Krall L., Trezzi G., Garcia-Orellana J., Rodellas V., Mörth C. M., Andersson, P. 2017. Submarine groundwater discharge at Forsmark, Gulf of Bothnia, provided by Ra isotopes. Mar. Chem., 196, 162–172.

Kryza J., Kryza H. 2006. The analytic and model estimation of the direct groundwater flow to Baltic Sea on the territory of Poland. Geologos 10, 154-165.

Moore WS, 2010. The Effect of Submarine Groundwater Discharge on the Ocean. Annu. Rev. Mar. Sci., 2, 59-88.

Nowacki J.. 1993a. *Morphometric characteristic of Puck Bay*, [in:] *Puck Bay*, Eds. Korzeniewski K., Fundacja Rozwoju Uniwersytetu Gdańskiego, Gdańsk, 71–78.

Nowacki J. 1993. Thermics, salinity and density of water, [in:] Puck Bay, K.Korzeniewski (ed.), Fundacja Rozwoju Uniwersytetu Gdańskiego, Gdańsk, 79–111.

Oberdorfer JA, Valentino MA, Smith SV, 1990. Groundwater contribution to the nutrient budget of Tomas Bay, California. Biogeochemistry 10, 199-216.

Pempkowiak J., Szymczycha B., Kotwicki L. 2011. Podwodny dopływ podziemny do morza Bałtyckiego. Rocznik Ochrona Środowiska, 12, 1, 17–32.

Piekarek-Jankowska H., Bolałek J. 1992. Jon chlorkowy w wodach porowych osadów dennych Zatoki Puckiej. Wyd. UG, Gdańsk.

Piekarek-Jankowska H. 1994. Zatoka Pucka jako obszar drenażu wód podziemnych. Wyd. UG, Gdańsk.

Piekarek-Jankowska H., Matciak M., Nowacki J. 1994. Salinity variations as an effect of groundwater seepage through the seabed (Puck Bay. Poland).Oceanologia, 36,1, 33-46.

Piekarek-Jankowska H. 2007. Podmorski drenaż wód podziemnych gdańskiego system wodonośnego: Gdański system wodonośny. Edited by Kozerski B, 2010. Wydawnictwo Politechniki Gdańskiej, Gdańsk, Poland, 34-49.

Peltonen K., 2002. Direct Groundwater Inflow to the Baltic Sea. TemaNord, Nordic Councils of Ministers, Copenhagen, Holand, 79.

Sadurski A. 1987. Warunki hydrogeologiczne i hydrochemiczne Mierzei Helskiej, Geol. Q., 1987, 31 (4), 767-782.

Slomp C.P., Van Cappellen P. 2004. Nutrient inputs to the coastal ocean through submarine groundwater discharge: controls and potential impact. J. Hydrol., 295, 64-86.

Szczepaniak W. 2011. Metody instrumentalne w analizie chemicznej. Wydawnictwo Naukowe PWN, Warszawa.

Szymczycha B., Vogler S., Pempkowiak J. 2012. Nutrient fluxes via submarine groundwater discharge to the Bay of Puck, Southern Baltic, **Sci**. Total Environ., 438, 2012, 86-93.

Szymczycha B., Maciejewska A., Winogradow A., Pempkowiak J. 2014. Could submarine groundwater discharge be a significant carbon source to the southern Baltic Sea?, Oceanologia, 56, 327–347.

Szymczycha B., Kroeger K. D., Pempkowiak J. 2016. Significance of groundwater discharge along the coast of Poland as a source of dissolved metals to the southern Baltic Sea, Mar. Pollut. Bull., 109, 1, 151-162.

Urbański J., Grusza G., Chlebus N. 2007. Fizyczna typologia dna Zatoki Gdańskiej. Gdynia: Pracownia Geoinformacji Zakładu Oceanografii Fizycznej, Instytut Oceanografii UG. 8.

Uścinowicz Sz., Miotk-Szpiganowicz G. 2011. The Baltic Sea: Location, Division and Catchment Area: Geochemistry of Baltic Sea Surface and Sediments. (ed) Uścinowicz Sz, 2011. Polish Geological Institute-National Research Institute, Warsaw, Poland, 13-17.

Valiela I., Bowen J.L., Kroeger K. D. 2002. Assesment of models for estimation of land-derived nitrogen loads to shallow estuaries. Appl. Geochem., 17, 935-953.

Contact Information: Żaneta Kłostowska, Institute of Oceanology, Polish Academy of Sciences, Powstańców Warszawy 55, 81–712 Sopot, Poland Email: klost@iopan.pl

Fourier series solution for an anisotropic and layered configuration of the dispersive Henry Problem

B. Koohbor[1], M. Fahs[1], B. Belfort[1], B. Ataie-Ashtiani[2,3], C. T. Simmons[3]
[1] Laboratoire d'Hydrologie et Géochimie de Strasbourg, University of Strasbourg/EOST/ENGEES - CNRS, Strasbourg, France
[2] Department of Civil Engineering, Sharif University of Technology, Tehran, Iran
[3] National Centre for Groundwater Research & Training and College of Science & Engineering, Flinders University, Adelaide, Australia

ABSTRACT

Henry Problem (HP) still plays an important role in benchmarking numerical models of seawater intrusion (SWI) as well as being applied to practical and managerial purposes. The popularity of this problem is due to having a closed-form semi-analytical (SA) solution. The early SA solutions obtained for HP were limited to extensive assumptions that restrict its application in practical works. Several further studies expended the generality of the solution by assuming lower diffusion coefficients or including velocity-dependent dispersion in the results. However, all these studies are limited to homogeneous and isotropic domains. The present work made an attempt to improve the reality of the SA solution obtained for dispersive HP by considering anisotropic and stratified heterogeneous coastal aquifers. The solution is obtained by defining Fourier series for both stream function and salt concentration, applying a Galerkin treatment using the Fourier modes as trial functions and solving the flow and the salt transport equations simultaneously in the spectral space. In order to include stratified heterogeneity, a special depth-hydraulic conductivity model is applied that can be solved analytically without significant mathematical complexity. Several examples are proposed and studied. The results show excellent agreement between the SA and numerical solutions obtained with an in-house advanced finite element code.
Keywords: Henry problem; Semi-analytical solution; Velocity-dependent dispersion; Stratified heterogeneity; Fourier series

INTRODUCTION

Henry Problem has always been important in theoretical and practical investigation of general concepts and characteristics of seawater intrusion (SWI). It is an abstraction of SWI in a vertical cross-section of a confined coastal aquifer perpendicular to the shoreline. A schematic representation of the HP is given in Figure 1. The most important aspect of HP stems from the existence of a semi-analytical (SA) solution (Henry 1964). This solution is obtained using the Fourier series method applied to the stream function form of the flow and salt concentration equations. This method combines the exactness of the analytical methods with an important extent of generality in describing the geometry and boundary conditions of the numerical methods.

In the scope of previous works concerning the SA solution of the HP, several simplifying assumptions were applied to make the calculation of SA solution feasible. While all the studies deal with uniform diffusion coefficient, recently, Fahs et al. (2016) developed a new SA solution with velocity dependent dispersion. Anisotropy and heterogeneity are primary characteristics of real aquifers that the SA solutions of the HP do not account for. Thus the aim of this work is to address this gap by extending the SA of the dispersive Henry problem to heterogeneous and anisotropic domain. In order to address the heterogeneity and to model it analytically, the Gardner exponential permeability function (Gardner 1958) was considered. This function is widely used for layered aquifers. It makes the handling of the space derivatives of permeability emerging in the process, much more applicable. The approach developed to get the SA solution is described in the next section. Then, the results of the SA solution are compared against an in-house finite element code throughout different test cases investigating the effect of anisotropy and heterogeneity.

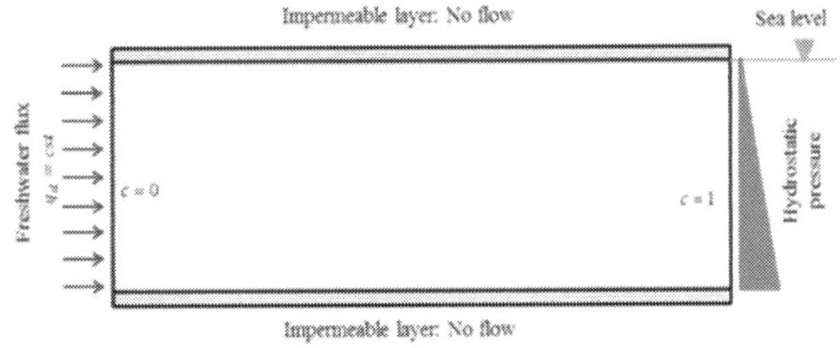

Figure 1. Schematic configuration of Henry Problem.

The semi analytical solution

The mathematical model of SWI is based on coupled variable-density flow and salt transport partial differential equations. Under Oberbeck-Boussinesq approximation and steady-state conditions the flow system is as followed:

$$\nabla . q = 0 \tag{1}$$

$$q = -K\left(\nabla h + \frac{\rho - \rho_0}{\rho_0} \nabla z\right) \tag{2}$$

where **q** is the Darcy's velocity $[LT^{-1}]$; ρ_0 the freshwater density $[ML^{-3}]$; **K** the freshwater hydraulic conductivity tensor $[LT^{-1}]$ (diagonal with components K_x and K_z); h the equivalent freshwater head $[L]$; ρ the density of mixture fluid $[ML^{-3}]$ and z is the elevation $[L]$.
The salt transport process can be described by the advection-dispersion equation:

$$\varepsilon \frac{\partial c}{\partial t} + q \nabla c - \nabla . (\varepsilon D_m I + D) \nabla c = 0 \tag{3}$$

Where c is the dimensionless solute concentration [-]; D_m the molecular diffusion coefficient $[L^2T^{-1}]$; ε the porosity [-], **I** the identity matrix And the dispersion tensor **D** is defined by:

$$D = (\alpha_L - \alpha_T)\frac{q \times q}{|q|} + \alpha_T |q| I \tag{4}$$

where α_L and α_T are the longitudinal and transverse dispersion coefficient respectively $[L]$.
Flow and transport equations are coupled via the linear mixture density equation:

$$\rho = \rho_0 + \Delta\rho . c \tag{5}$$

where $\Delta\rho = \rho_1 - \rho_0$ is the difference between the freshwater (ρ_0) and saltwater (ρ_1) density. The geometry of the aquifer is simplified to a rectangle of length (ℓ) and depth (d). The freshwater recharge flux per unit of width imposed on the inland side is noted q_d $[L^2T^1]$.
By differentiating Darcy's law with respect to x and z direction and eliminating pressure and applying stream function theory the system is results as followed:

$$r_k \frac{\partial^2 \psi}{\partial z^2} + \frac{\partial^2 \psi}{\partial x^2} - r_k \frac{\partial \psi}{\partial z} \frac{1}{K_x} \frac{\partial K_x}{\partial z} - \frac{\partial \psi}{\partial x} \frac{1}{K_z} \frac{\partial K_z}{\partial x} - \frac{K_z \Delta\rho}{\rho_0} \frac{\partial c}{\partial x} = 0 \tag{6}$$

where $q_x = -\frac{\partial \Psi}{\partial z}$ and $q_z = -\frac{\partial \Psi}{\partial x}$ and $r_K = K_z/K_x$ is the anisotropy coefficient that corresponds to the anisotropic nature of the domain that is being studied in this work. The non-dimensional form of the Eq(6) is defined as follows:

$$r_k \frac{\partial^2 \Psi}{\partial Z^2} + \frac{\partial^2 \Psi}{\partial X^2} - r_k \frac{1}{K_x} \frac{\partial K_x}{\partial Z} \left(\frac{\partial \Psi}{\partial Z} + 1 \right) + \frac{1}{K_z} \frac{\partial K_z}{\partial X} \frac{\partial \Psi}{\partial X} - NG \left(\frac{\partial C}{\partial X} + \frac{1}{\xi} \right) = 0 \tag{7}$$

where:

$$X = \frac{x}{d} \; ; \; Z = \frac{z}{d} \; ; \; \Psi = \frac{\psi}{q_d} - Z \; ; \; C = c - \frac{X}{\xi} \; ; \; \xi = \frac{\ell}{d} \tag{8}$$

$NG = \frac{K_z d \Delta\rho}{\rho_0 q_d}$ is the local gravity number which compares the buoyancy flux to the inland freshwater flux (q_d).

In order to address the layered heterogeneity of the aquifer, the depth-hydraulic conductivity model is defined by the following equation:

$$K_x(Z) = K_{x,0} e^{\Upsilon Z} \quad ; \quad K_z(Z) = K_{z,0} e^{\Upsilon Z} \tag{9}$$

where $K_{x,0}$ and $K_{z,0}$ are the hydraulic conductivity at the bottom of the aquifer in x- and z-directions respectively. Υ is the rate of hydraulic conductivity change with depth.

The non-dimensional form of the salt transport equation is presented as follows:

$$b_m \left(\frac{\partial^2 C}{\partial X^2} + \frac{\partial^2 C}{\partial Z^2} \right) - \frac{\partial \Psi}{\partial Z} \frac{\partial C}{\partial X} + \frac{\partial \Psi}{\partial X} \frac{\partial C}{\partial Z} - \frac{1}{\xi} \frac{\partial \Psi}{\partial Z} - \frac{\partial C}{\partial X} - \frac{1}{\xi} + \Delta_{1,1} \frac{\partial^2 C}{\partial X^2} + 2\Delta_{1,2} \frac{\partial^2 C}{\partial X \partial Z} + \Delta_{2,2} \frac{\partial^2 C}{\partial Z^2} + \left(\frac{\partial C}{\partial X} + \frac{1}{\xi} \right) \left(\frac{\partial \Delta_{1,1}}{\partial X} + \frac{\partial \Delta_{1,2}}{\partial Z} \right) + \frac{\partial C}{\partial Z} \left(\frac{\partial \Delta_{1,2}}{\partial X} + \frac{\partial \Delta_{2,2}}{\partial Z} \right) = 0 \tag{10}$$

where $b_m = {}^{\varepsilon D_m}/_{q_d}$ and $\Delta_{i,j} = {}^{D_{i,j}}/_{q_d}$.

The unknowns are expanded into infinite Fourier series that satisfy the boundary conditions:

$$\Psi = \sum_{m=1}^{Nm} \sum_{n=0}^{Nn} A_{m,n} \sin(m\pi Z) \cos(n\pi X/\xi) \tag{11}$$

$$C = \sum_{r=0}^{Nr} \sum_{s=1}^{Ns} B_{r,s} \cos(r\pi Z) \sin(s\pi X/\xi) \tag{12}$$

where $A_{m,n}$ (resp. $B_{r,s}$) are the Fourier series coefficients for the stream function (resp. concentration). *Nm* and *Nn* are the truncation orders for the stream function in the X- and Z-directions, respectively. *Nr* and *Ns* are the ones for salt concentration. The Fourier series expansions are then appropriately substituted in Eqs. (7) and (10). Then, a Galerkin treatment is applied with the Fourier modes as trial functions. This leads to the following system of equations (for flow and salt transport respectively):

$$R_{g,h}^F = \varpi_h^1 \pi^2 \xi A_{g,h} \left(r_K g^2 + \frac{h^2}{\xi^2} \right) + \varpi_h^1 r_k \xi \Upsilon \sum_{m=1}^{Nm} m A_{m,n} \Pi_{g,m} + \pi^2 NG_0 h \sum_{r=0}^{Nr} \tilde{B}_{r,h} \Gamma_{g,r} + 2\pi \, NG_0 \delta_{h,0} \Gamma_{g,0} + \frac{2 r_k \xi \Upsilon}{\pi} \Pi_{g,0} \delta_{h,0} = 0 \qquad (g = 1..Nm, h = 0..Nn) \tag{13}$$

$$R^T_{g,h} = \varpi^1_h b_m \pi^2 \xi B_{g,h}\left(g^2 + \frac{h^2}{\xi^2}\right) - \frac{\pi}{4}\sum_{m=1}^{Nm}\sum_{n=0}^{Nn}\sum_{r=0}^{Nr}\sum_{s=1}^{Ns} B_{r,h}A_{m,n}\left(s.m.\eta_{g,m,r}\theta_{h,n,s} - r.n.\kappa_{g,m,r}\lambda_{h,n,s}\right)$$
$$- g\sum_{n=0}^{Nn}\tilde{A}_{g,n}\Lambda_{h,n} - \varpi^2_g\sum_{s=1}^{Ns}B_{g,s}\Lambda_{h,s} - \frac{2}{\pi}\Lambda_{h,0}\delta_{g,0}$$
$$- 4\sum_{1}^{Np} W_{p_i}F^{Disp}(X_{p_i}, Z_{p_i})\cos(g\pi Z_{p_i})\sin(h\pi X_{p_i}/\xi) = 0$$
$$(g = 0..Nr, h = 1..Ns) \qquad (14)$$

where R^F (resp. R^T) is the residual vector for the flow equation (resp. transport equation) and $\delta_{i,j}$ is the Kronecker delta function. Np refers to the number of integration points used to evaluate the dispersion terms. W_P, X_P and Z_P are the integration weight and the coordinates of the integration points, respectively. NG_0 $\left(K_{Z,0}d\Delta\rho/\rho_0 q_d\right)$ is the local gravity number at the aquifer bottom surface.

In this work a new technique is developed to solve the system of equations in the spectral space (Eqs 13 and 14). With this technique the Fourier coefficients of the stream function are analytically calculated in terms of the salt concentration and the spectral transport equation is solved the Newton's method, with only the Fourier series coefficients of the salt concentration as primary unknowns.

RESULTS

A FORTRAN code has been developed to solve the final system of nonlinear equations resulting from the Fourier series method. To examine the correctness of this code, we compared it against a full numerical solution obtained using an advanced in-house model (Younes et al. 2009). All the simulations of the numerical model are developed under the transient regime for a long duration to reach the steady-state solution.

Two dispersive cases are considered using the same parameters as in Abarca et al. (2007). The first case deals with a homogenous domain ($\Upsilon = 0$) while in the second one, the aquifer is assumed to be stratified ($\Upsilon = 1.5$). Anisotropy is acknowledged with ($r_k = 0.66$). The non-dimensional parameters and the corresponding physical parameters used for these cases are given in Table 1. For the homogenous cases, oscillation-free solution has been obtained using 4,725 Fourier modes (*Nm=15; Nn=90; Nr=20 and Ns=160*). This is equal to the number of Fourier modes used in Fahs et al. [2016] for the isotropic domain. This indicates that the anisotropy does not affect the stability of the Fourier series solution. For the heterogeneous case, the same truncation modes. For the heterogeneous case, the same number of Fourier modes (4,725) leads to unstable solution with some unphysical oscillations at the aquifer top. In fact, in this upper zone, the local permeability is 5 times more important than at the aquifer bottom leading to stronger buoyancy effects. An oscillation-free solution has been obtained with 6,405 coefficients (*Nm*=15, *Nn*=90, *Nr*=20 and *Ns*=240). Several runs confirm that heterogeneity affects only the truncation order of the concentration Fourier series in the x-direction. The SA results for both cases (homogenous and heterogeneous) are depicted in Figures 2 and 3, respectively.

As mention previously, both test cases are simulated using an in-house numerical code. The physical parameters used in the numerical simulations are given in Table 1. The numerical isochlors for both test cases (homogenous and heterogeneous) are depicted in Figure 2 and 3, respectively. These Figures highlight the excellent agreement between the analytical and numerical solutions. It should be noted that, with the new technique developed in this work for solving the HP in the spectral space, the solution can be obtained with a reduced number of unknowns. For instance, in the heterogeneous case, the final system is solved using 4,800 unknowns instead of 6,405.

Table 1. Non-dimensional and physical parameters used in the semi-analytical solution and the numerical model for the verification test cases.

Numerical Model			Semi-analytical Solution		
Parameters	**Value**	**Cases**	**Parameters**	**Value**	**Cases**
$\Delta\rho$ $[kg/m^3]$	25	Both cases	$\overline{NG}$	3.11	Both cases
$\rho_0 [kg/m^3]$	1000	Both cases	b_m	5×10^{-4}	Both cases
$q_d [m^2/s]$	6.6×10^{-5}	Both cases	$b_L = \frac{\alpha_L}{d}$	0.1	Both cases
d $[m]$	1	Both cases	$r_\alpha = \frac{\alpha_T}{\alpha_L}$	0.1	Both cases
$\ell [m]$	4	Both cases	r_K $[-]$	0.66	Both cases
$\overline{K}_z$ $[m/s]$	8.213×10^{-3}	Both cases	Υ $[-]$	0 1.5	Homogeneous cases Heterogeneous cases
$r_K [-]$	0.66	Both cases			
$\varepsilon [-]$	0.35	Both cases			
D_m $[m^2/s]$	3.300×10^{-8}	Both cases			
α_L $[m]$	0.1	Both cases			
α_T $[m]$	0.01	Both cases			
Υ $[-]$	0 1.5	Homogeneous cases Heterogeneous cases			

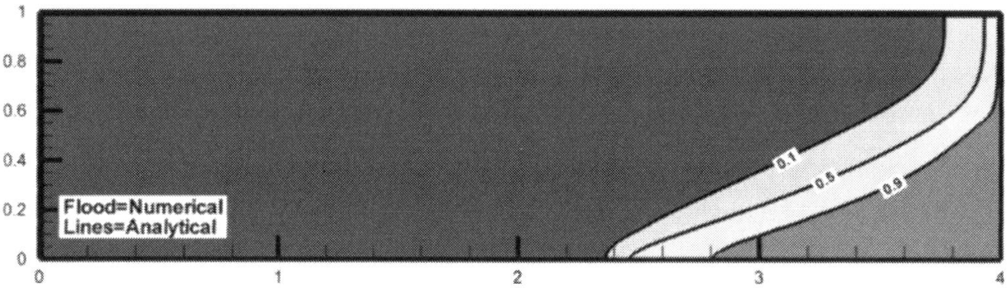

Figure 2. Semi-analytical and numerical isochlors (10%, 50% and 90%) for the homogeneous case ($\Upsilon = 0$).

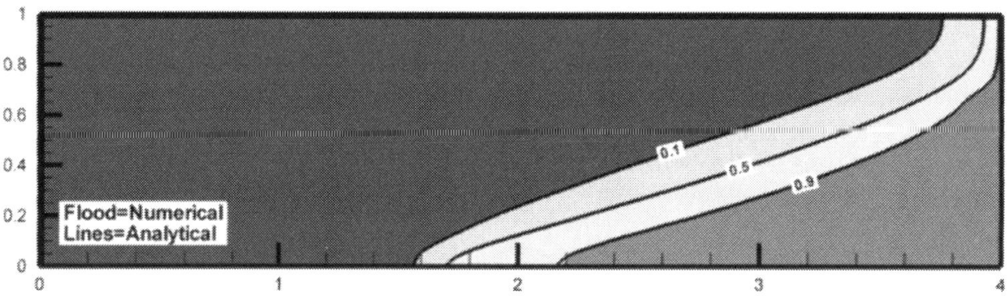

Figure 3. Semi-analytical and numerical isochlors (10%, 50% and 90%) for the heterogeneous case ($\Upsilon = 1.5$).

CONCLUSION

In this paper, the SA solution of the dispersive Henry problem is generalized to anisotropic and stratified heterogonous porous media. The stratified heterogeneity described with Gardner's model is considered to derive the SA solution. A major contribution of this work is to derive the first SA solution of SWI with DDF model in an anisotropic and heterogeneous domain with velocity-dependent dispersion. The SA solution is useful for testing and validating DDF numerical models in realistic configurations of anisotropy and stratification. The different test cases in homogeneous and heterogeneous domain with diffusive or dispersive configurations have demonstrated excellent agreement between SA solution and numerical results. While in the most existing studies, the effect of anisotropy and heterogeneity is mainly discussed on the position of the saltwater wedge, the SA solution will be investigated (in future works) to provide a deeper understanding of these effects on several metrics characterizing SWI.

REFERENCES

Abarca, E., J. Carrera, X. Sanchez-Vila, and M. Dentz. 2007. Anisotropic dispersive Henry problem. *Advances in Water Resources*, *30*, 913–926, doi:10.1016/j.advwatres.2006.08.005.

Fahs, M., B. Ataie-Ashtiani, A. Younes, C.T. Simmons, and P. Ackerer. 2016. The Henry problem: New semianalytical solution for velocity-dependent dispersion. *Water Resources Research*, *52*, 7382–7407, doi:10.1002/2016WR019288.

Gardner, W. R. 1958. Some Steady-state solutions of the unsaturated moisture flow equation with application to evaporation from water table. *Soil Science*, *85*, 228.

Henry, H. 1964. Effects of dispersion on salt encroachment in coastal aquifer. *U.S. Geol. Surv. Water Supply Pap.*, *1613*, C70–C84.

Younes, A., M. Fahs, and S. Ahmed. 2009. Solving density driven flow problems with efficient spatial discretizations and higher-order time integration methods. *Advances in Water Resources*, *32*, 340–352, doi:10.1016/j.advwatres.2008.11.003.

Contact Information: Marwan FAHS, Laboratoire d'Hydrologie et Géochimie de Strasbourg, University of Strasbourg/EOST/ENGEES, CNRS, 1 rue Blessig 67084 Strasbourg, France, Phone : +33-3-68 85 04 48, Email : fahs@unistra.fr

E3S Web of Conferences 54, 00015 (2018) https://doi.org/10.1051/e3sconf/20185400015
SWIM 2018

Major groundwater reservoir nr 112 in the coast of Gdańsk Bay

Zbigniew Kordalski[1], **Andrzej Sadurski**[2]
Marine Branch of Polish Geological Institute NRI, 5 Kościerska str., 80-328 Gdańsk, Poland

ABSTRACT

The main groundwater reservoirs (MGR) in Poland have been established during last 9 years and finished in 2017 yr. (Mikołajków, Sadurski edit.). The most reach in groundwater resources hydrogeological structures in the country where deliminated and the scope of limited and forbidden human activities for their protection where established. Area of main groundwater reservoir nr 112 , named - Żuławy Gdańskie - comprises predominantly City of Gdańsk and area of Gdańsk and Pruszcz Gdański Counts and slightly exceeds 100 km^2. The area undertaken for the investigation and modelling was 363.8 km^2. The aim of MGR delimitation and hydrogeological documentation was the water resources protection based on physical planning proper, responsible water management (Herbich et al., 2009). The methods of hydrogeological researches, groundwater resources calculation and delimitation of MGR is presented in the paper.

LOCATION AND NATURAL CONDITIONS

133 major groundwater reservoirs, named later MGR, were established in the country during last 9 years (*www.psh.gov.pl*). Some of them are in the coastal zone of the Southern Baltic Sea, as shown on fig. 1. MGR nr 112 is in the area of Gdańsk region and is presented in the paper. Investigation area is situated on the boundary area of moraine plateau of Cashubian Lakeland, western part of Vistula delta plain and Gdańsk Bay, and is given on fig. 1. The MGR nr 112 comprises two parts, one on the Vistula delta in the South and second one on the seaside terrace in Gdansk in the North. The Mesozoic strata that belong to the young cover of East European Platform are finished in Upper Cretaceous by the marine sediments.

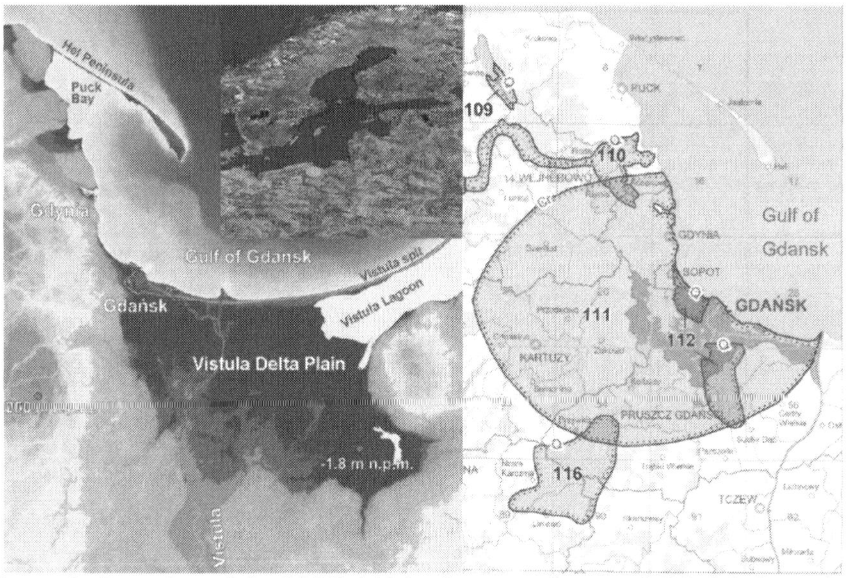

Figure 1. Location of the MGR 112 in the coastal area of Gdańsk Bay after Mikołajków et al. (2017).

Lithology of Cretaceous strata cause the limitation of regional circulation system and depth of fresh groundwater occurrence (Sadurski, 1989). In the middle of Cretaceous sediments extends sandy series with glauconite. Thickness of this series is up to 200 m in the central part of the MGR nr 111, named Gdańsk Upper Cretaceous reservoir. Above this series, in the roof of Cretaceous occur marls, sandy limestones and gaizes. The roof of Cretaceous on the area of MGR nr 112 was stated on the ordinates from 90 to 100 m b.s.l. The Cainozoic and Cretaceous aquifers create the multilayer water bearing system in the regional, transition and local groundwater circulations recognized (Kozerski edit. 2007).

The regional base of drainage of MGR nr 111 and 112 is the Gdańsk Bay, Vistula delta plane and Seaside terrace in Gdańsk. This is also the limit of two hydrogeochemical environments; salty and brackish sea water and fresh inland groundwater.

The groundwater of Quaternary aquifer are fresh of HCO_3 – Ca chemical type, locally needs only simple treatment in the water works station due to iron and manganese natural enrichment.

GROUNDWATER RESOURCES

Groundwater resources of MGR nr 112, that could be used for water supply, have been calculated separately for distinguished aquifers being hydraulic connection as the regional system. The total value of groundwater resources of quaternary aquifer is 2900 m^3/h, but 2700 m^3/h (it means 23,652 million m^3/a) came from the area of MGR nr 112.

The biggest groundwater extraction in the area of Gdańsk and Sopot cities has been registered during the second part of 80 yr of XX century and the total exploitation of water intakes reached 8000 m^3/h. This total value of exploitation exceeded natural, renewable resources and caused bad chemical state - salt water intrusion and dynamic – hydraulic head and flows in many places of the area. Since 90. of XX century is observed decline of groundwater exploitation up to 3000 m^3/h at present (Szelewicka, Kordalski 2013). Groundwater intakes are only source for water supply in the area of MGR nr 112 and also MGR nr 111. The groundwater are dominant for municipal and industrial water works.

CONCEPTUAL MODEL

Hydrogeological, conceptual model comprises layout of the strata extends, aquifers geometry and properties and also semi permeable strata and map of hydraulic heads of the groundwater of different aquifers, that is the initial boundary condition. Scheme of hydrogeological profiles 2 according to description given in table 2.is presented on the on figure below.

Delimitation of the MGR nr 112 limits and its protection area was done using the criteria elaborated for the whole country (Herbich et al., 2009):

- safe yield of the Wells $Q>70\ m^3/h$,
- transmissivity of the aquifer $T>10\ m^2/h$,
- good quality state of groundwater,
- possibility of new construction of water intake of high exploitation ratio $Q>10\ 000\ m^3/24$ h.

Table. 1. Profile of modelled strata (I, III, V – semipermeable, separation strata; II, IV, VI – aquifers).

Numer and profile of modelled strata	Characteristics of modelled strata
I Convertible	Impermeable sediments near terrain surface, locally with sands.
II Convertible	Quaternary – Miocene porous aquifer, mainly sands, loams and gravel.
III Confined	Miocene and lower Pleistocene aquifer – fine sands, sands and also silts and glacial tills.
IV Confined	Oligocene sands of Oligocene and sand and gravel in buried structure of Pleistocene, mainly on the Seaside Terrace and in the Vistula delta plane.
V Confined	Impermeable and semi permeable sediments – silts and loams of Paleogene and roof of Cretaceous – gaizes, marls and sandy limestones.
VI Confined	Main aquifer of Cretaceous – sands with glauconite (MGR 111).

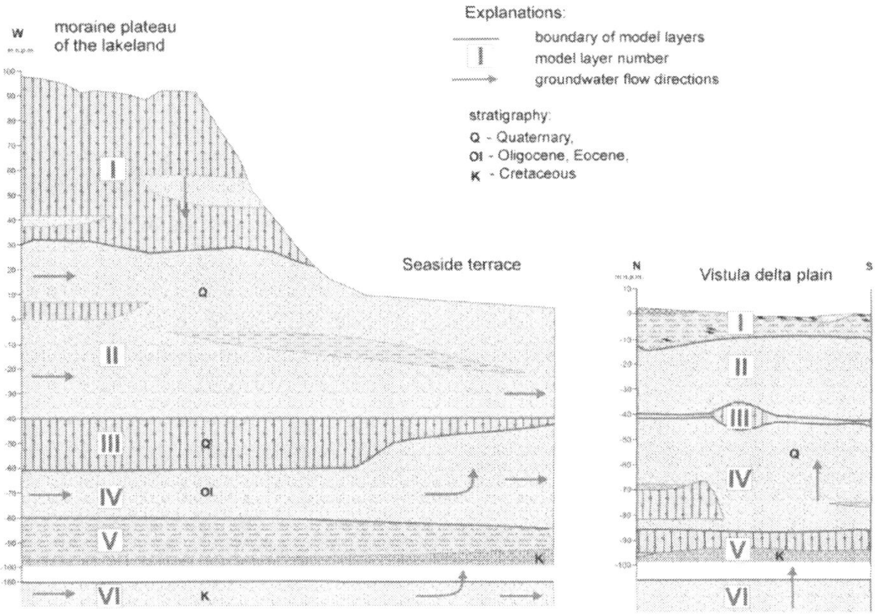

Figure 2. Hydrogeological diagram of the conceptual model of groundwater flows.

MOTHODS OF INVESTIGATION

Realization of these objectives needed many investigations and solutions, as follow:

a) hydrogeological (conceptual) model of MGR construction, that include the 3D distribution of hydraulic and geological parameters to establish the boundary condition of the aquifers,
b) verification of groundwater reservoir parameters for water resources and intensity of groundwater flow,

 https://doi.org/10.1051/e3sconf/20185400015

c) optimum location of new water intakes and their protection area estimation.

Chemical analysis of groundwater samples comprises mainly: TDS, Temperature, electrolytic conductivity, colour, pH, basicity, NH_4, NO_3, NO_2, Cl, SO_3, Na, K, Ca, Mg, phosphates, fluorine, iron compounds, manganese, bicarbonates, silica and also Br, B, Ba, P, Ti, Li, Be, V, Co, Ni, As, Se, Mo, Ag, Cd, Sn, Sb, Tl, U, BTX, pesticides, PAH and mercury.

NUMERICAL MODEL

Investigation area was divided into grid $\Delta x = \Delta y = 200$ m, in 124 columns and 153 rows. The total area cover 363.8 km^2. Distribution of calculation grid was sufficient for analysis of groundwater flow area and for imaging of geological structures reflection together with hydrogeological and hydraulic properties in frames of the MGR 112.

The code MODFLOW–2000 (MODular FLOW, 1996) as a part of Groundwater Modeling System (GMS) was used for the simulation of groundwater flow in the water reservoir considered. For the water balance estimation the code ZONE BUDGET was used.

Identification and test of model has been carried by the difference value between measured hydraulic head and calculated on the model. The next goal of model investigation was the protection area of proposed two parts of MGR nr 112. It results from the 25th years flow time of groundwater, including the time of vertical percolation from the surface.

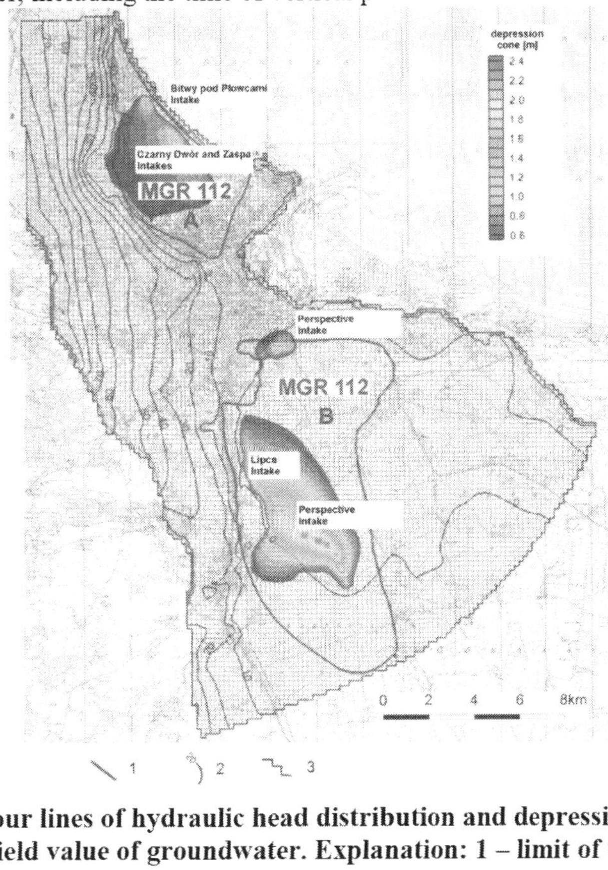

Figure 3. Contour lines of hydraulic head distribution and depression cone of water intakes at safe yield value of groundwater. Explanation: 1 – limit of the reservoir; 2 – hydraulic head of Quaternary aquifer of MGR nr 112; 3 – boundary of the investigation area.

CONCLUSIONS

There are two MGR in the area of Gdańsk region; one nr 112 was established as Holocene–Pleistocene aquifers and the second one is nr 111 recognized in the Cretaceous strata, below. Potential reason of the groundwater resorces pollution are serious for the first MGR nr 112 and can be mentioned as overexploitationm of water intakes, dumping site and sewage and industrial pollutants also from the past centuries. Percolation and flow time of groundwater is a good criterion for estimation of vulnerability of the area on the surface of MGR nr 112. The protection area of MGR was determine for isochrone of 5, 15 and 25 years of percolation and flow time of water. Very high and high vulnerability of the quaternary aquifer range 66.9 km^2 in part A in the North and 52.4 km^2 in part B in the South (see fig. 3). Total protection area comprises 119.3 km^2.

Forecast of the pollutants migration time to the wells was possible by groundwater flow modelling for each of MGR also in the coastal area of the Baltic Sea. Calculation of safe yield was done by modelling too, especially according to the risk of salt water intrusion. The safe, total exploitation of the wells of water intakes of this reservoir has been calculated and referes to the both parts of the MGR under consideration. Restrictions and limitation on surface within the protection area implemented to the developing plans of the Gdańsk City.

REFERENCES

McDonald M. G., Harbaugh A. W., 1996 – User's Documentation for MODFLOW-96, an update to the U.S. Geological Survey Modular Finite-Difference Ground-Water Flow Model, U.S. Geological Survey.

Herbich P., Kapuściński J., Nowicki K., Prażak J., Skrzypczyk L., 2009 – Merthod of the protection area of MGR delimitation for developing planes and water management plan sof the river basins. Ministra of Environment Publ. Warsaw.

Kozerski B., edit., 2007 – Gdańsk's Hydrogeological Flow System. Techn. Univ. of Gdańsk Publ. Gdańsk.

Mikołajków J., Sadurski A. edit., 2017 – Major Groundwate Reservoirs in Poland, [in polish]. PGI-NRI Publ. Warsaw.

Mikołajków J. et al., 2017 – Map of Major Groundwater Reservoirs in Poland, [in polish]. PGI-NRI Publ. Warsaw.

Sadurski A., 1989 – Upper Cretaceous Groundwater Flow system of Eastern Pomerania. MMinning and Metal. Academy Pobl. Serie Geology, vol. 46. Cracow.

Szelewicka A., Kordaloki Z., 2013 Hydrodynamics changes in the drainage area of Hydrogeological System on the basis of recent research. Bull. of PGI-NRI, vol. 456: 595-600. Warsaw.

Szelewicka A., Karwik A., Kordalski Z., 2013 – Hydrogeological documentation of MGR nr 112 – Żuławy Gdańskie for protection area establishment of this major groundwater reservoir. Unpublished. Arch. of Marine Branch of PGI-NRI. Gdańsk.

Groundwater chemistry and origin of the visutla delta plain

Arkadiusz Krawiec[1], Andrzej Sadurski[1, 2]
[1] Nicolaus Copernicus University, Faculty of Earth Sciences, Toruń, Poland
[2] Polish Polish Geological Institute NRI, Gdańsk, Poland

ABSTRACT

River estuaries are peculiar areas from the point of view of their groundwater origins, frequently connected with the diversification of groundwater chemistry. In the area of the Vistula delta in a number of places, the increased concentration of chloride and fluoride ions has occurred (Kozerski et al. 1987). These issues have already been considered in the studies of Kozerski (1981), Kozerski and Kwaterkiewicz (1984, 1988). The origin of salt and brackish water in the area of Żuławy was already the subject of scientific research in the 1930s (Schroeder 1931), and subsequently continued by Sadurski (1985, 1989), Burzyński and Sadurski (1989), Zuber at al. (1990, 2000) and Krawiec (2013).

INTRODUCTION

The Vistula delta plain constitues a flat plain formed by the accumulation of alluvial deposits. The area of Vistla delta can be dividied into: Żuławy Gdańskie, Wielkie and Żuławy Elbląskie in its eastern part. Around 30% of this region is below the sea level and this depression achieves the level of -1.8 metres above the sea level in the vicinity of Drużno Lake. The highland of the Kashubian Lake District forms the natural western border of the Vistula delta whereas the eastern border is formed by the Elblag Hills and the highland of the Iława Lake District (Fig.1). These highlands reach the altitude of over 200 metres above the sea level and were formed by the sediments of subsequent glaciations.

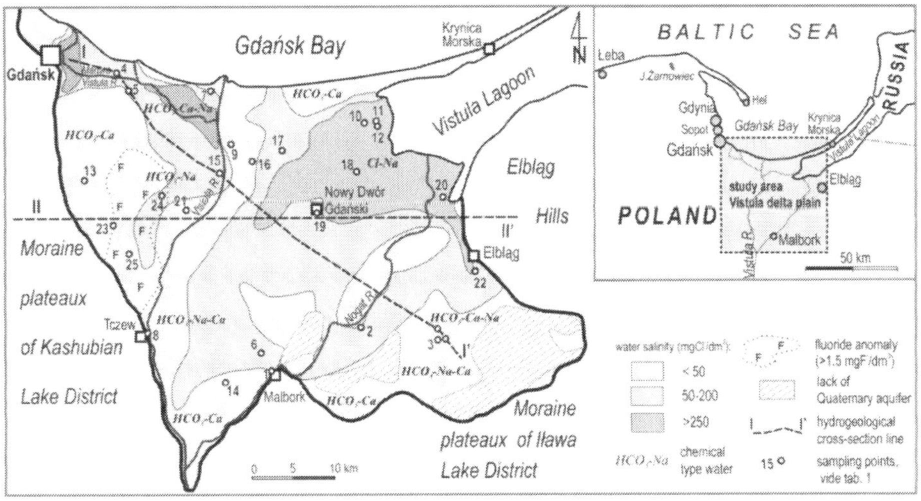

Figure 1. Location of study area and chloride and fluoride anomaly of the Vistula delta plane.

GEOLOGY AND HYDROGEOLOGY

In the area of the Vistula delta there are the Pleistocene and Holocene sediments, which mainly consist of glacial till and interglacial limnic-fluvial sediments in the form of sands, loams and silts. The Holocene alluvia of the Vistula delta were formed as sands, loams, sludge and peats of fluvial and marine origin in their northern part (Fig. 2).

Under the deposits of the Cainozoic and Palaeozoic sedimentary cover of the Precambrian East European Platform can be found. On the bedrocks of the Quaternary strata there occur discontinuous sediments of the Neogene and the Palaeogene formed as glauconitic fine-grained sands, occasionally with loam intercalations. In a large part of the Żuławy area these sediments were destroyed and on the surface of the sub-Quaternary period sand and marl sediments of the Upper Cretaceous. The sub-Quaternary surface bedrock is located at the elevation from -80 to -120 m sea level (Fig. 2).

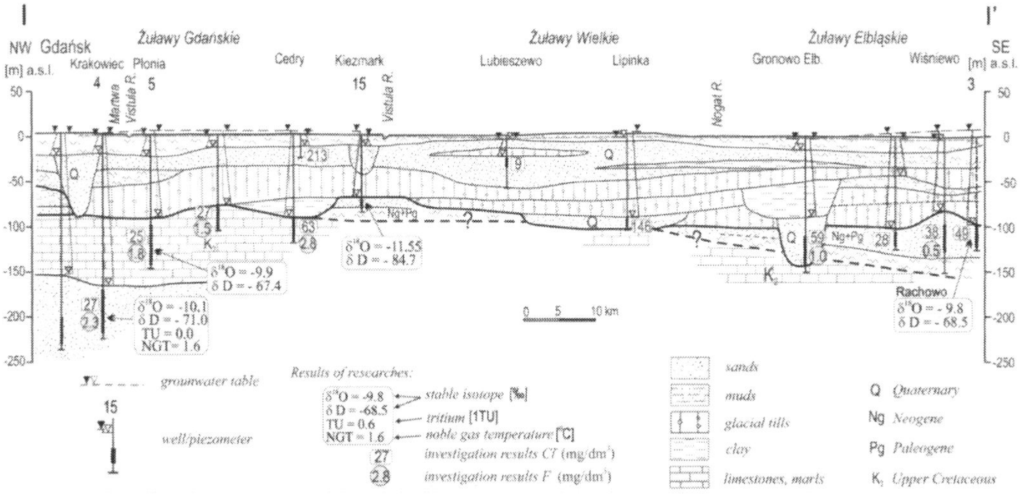

Figure 2. Hydrogeological cross-section I – I' in the Żuławy Wiślane area.

In the area under consideration there occurs multi-level aquifer, but the main utility significance can be ascribed to the Pleistocene aquifer, the so-called uneven-aged and Cretaceous sediments (Kozerski, Kwaterkiewicz 1988). The Pleistocene and Holocene aquifer commonly occurs in the area of Żuławy. The thickness of the deltaic Holocene aquifer hardly ever exceeds 20 m. This aquifer is supplied by waters from the surrounding highlands as well as the ascension of waters from the deepest aquifer (Burzyński and Sadurski 1989; Zuber at al. 2000).The aquifers that are the most abundant in ground waters occur in the marginal zone of the western part of the Vistula delta – the drinking water intake for the city of Gdansk called „Lipce" and in the eastern part, in the vicinity of the Nogat River, where the municipal water intake called „Letniki" is situated.

The uneven-aged aquifers are formed of the earliest Quaternary entities, sand sediments of the Palaeogene and Neogene periods, and occasionally of the parts of surface bedrock of the Cretaceous (Sadurski 1989; Kozerski, Kwaterkiewicz 1988).This level is separated from the Pleistocene and Holocene level by a series of glacial till, and occasionally also loams and silts. It occurs in a large part of the Żuławy area, but its absence was reported in the northern part of Żuławy Gdańskie and Wielkie. The thickness of sand sediments of the uneven-aged complex most commonly ranges from 10 to 20 m, and in the structures of buried valleys it can reach even 40 metres.

The Cretaceous aquifer is connected with a series of carbonate formations at the roof and lower-lying sands with glauconite (Sadurski 1989). The waters connected with a series of carbonate and silica occur in the eastern parts of Żuławy Gdańskie and the southern part of Żuławy Wielkie. Żuławy Wiślane and the Gulf of Gdansk constitute the regional drainage base for groundwater flowing down from the surrounding lakeland elevations as well as from the Elbląg Hills.

Detailed model tests have demonstrated that the lateral inflow from the highlands ranges from around 60% of the balance total in the recreated "natural" conditions to 65% with the mandatory amount of water abstraction (Kulma et. al. 2002). The average rate of infiltration water recharge in the area of Żuławy does not exceed 1% of annual precipitation level.

Anomalously high concentration of chloride and fluoride ions, as well as iron compounds are recorded both within the Quaternary and Cretaceous aquifers. In the marginal zones of Żuławy Wiślane, the groundwater of the Pleistocene and Holocene aquifers as well as that of the deeper uneven-aged level are both most frequently of the HCO_3-Ca type. This is mainly freshwater. In the more distant areas from the upland edges, within the Quaternary aquifer of Vistula delta occur groundwater of the HCO_3-Ca-Na type can also be found. The admixtures of the so-called young relict waters of marine origin can also occur here, which results in changes in water chemistry to the Cl- HCO_3 - Na type.

RESULTS AND DISCUSSION

The highest concentration of Cl^- ions in the Pleistocene and Holocene aquifers has been recorded in the north-eastern part of Żuławy Wielkie and in the northern part of Żuławy Elbląskie (Fig. 1). The high concentration of chlorides, connected with brackish water ingression from surface watercourses, have also been recorded in the vicinity of Gdańsk (Kozerski, Kwaterkiewicz 1988),where the natural groundwater cycle was disrupted due to the launch of large municipal water intakes in the north-western part of Żuławy. On the Cretaceous aquifer in the area of Żuławy there occurs water of HCO_3-Na and Cl-HCO_3-Na types (Kozerski and Kwaterkiewicz 1984; 1988; Sadurski 1985; 1989). In some parts of this area, for example in the north-eastern part of Żuławy, for instance, Gozdawa - tab.1, no 18, the water of Cl-Na type can also be found. In the area of Żuławy Gdańskie and the eastern and southern marginal zones of the Vistula delta there occurs freshwater (tab.1, no 4-9).

Table 1. Isotope and noble gas data from the Żuławy Wiślane area.

No	Sampling strat./depth	Data	Cl^- [mg/l]	Tryt (TU)	$\delta^{18}O$ (‰) V-SMOW	$\delta^{2}H$ (‰) V-SMOW	$\delta^{13}C$ (‰) V-PDB	^{14}C (pMC)	NGT [^0C]	$^4He_{excess}$ $10^{-6}cm^3g^{-1}$	"age" water
1	Malbork K/96-121	1986 1998	107	0.0±1.0	-8.8 -9.0	-63.0 -64.0	-5.2	4.5	0.2	1.3	^{14}C = 12.7 ka
2	Letniki K-1 K/137-250	2000^A			-9.9	-71.4	-9.0	14.6			^{14}C =7.5 ka
3	Rachowo Tr/106-128	2000^A	49		-9.8	-68.5	-12.0	18.7			^{14}C = 7.8 ka
4	Krakowiec K/171-230	1986 1998	27	0.0±1.0	10.0 -10.1	-67.0 -71.0	-7.0	2.1	1.6	0.681	^{14}C = 21.5 ka
5	Płonia M. K/112-147	2000^A			-9.9	-67.4	-5.6	0.7			^{14}C =28.6 ka
6	Kościelecz. Q/64-76	1987	81	0.0±1.0	-9.7	-68.0	-6.8	13.1			^{14}C = 7.1 ka
7	Świbno K-2 K/94-120	1998	27		-12.0	-85.0			2.2	1.56	infiltr. in a cool climate

8	Tczew. K-1 K/160-180	1998	152	0.0±1.0	-9.4	-66.0	-8.0	14.0	4.2	0.941	^{14}C = 6.9 ka
9	Drewnica K/96-130	1986	24	0.0±1.0	-12.1	-85.0	-6.0	1.7			^{14}C = 22.9 ka
10	Grochowo K/120-150	2010			-12.90	-97.0					infiltr. in a cool climate
11	Łaszka Q /2-5	2010	453	3.2±0.3	-8.89	-68.5	-7.6	77.3			modern;
12	Łaszka Q / 2.6-22	1986	2600	0.7±1.0	-8.8	-64.0	+2.2	28.0			holocene
13	Roszkowo Q/87-106	1987	3	0.0±1.0	-10.0	-66.0	-6.6	6.7			^{14}C = 12.4 ka
14	Stara Koś. Q/1.7-15	2010	86		-8.9	-65.3					holocene
15	Kiezmark Tr /71-83	2000^{A}	60		-11.55	-84.7	-6.6	2.0			^{14}C = 21.3 ka
16	Niedzwiedz Q-K	2010	202		-10.6	-77.6					infiltr. in a cool climate
17	Wiśniówka Q-K	2010	408	0.5±0.3	-10.14	-76.1					infiltr. in a cool climate
18	Gozdawa K/94-115	1986 1998 2010	1287 1830 1554	0.0±1.0 0.4±0.3	-10.8 -11.0 -10.9	-79.0 -80.0 -82.0	-3.4 -4.5	3.8 1.1	3.5	8.24	^{14}C = 10.5 ka ^{14}C= 24.2 ka
19	N. Dw. Gd. Q/21-36	2010	156		-8.47	-62.7					holocene and admix. of salt water (older)
20	Nowakowo K2/99-136	1987	632	0.3±1.0	-12.5	-88.0	-4.0	3.0			^{14}C = 14.9 ka
21	Długie Pole K/188-235	1986 1998	525 485	0.0±1.0	-10.0 -10.1	-70.0 -72.0	-4.9	1.8	0	2.18	^{14}C = 19.7 ka
22	Elbląg. K-4 ok. 120 m	1986 1998	 278	0.0±1.0	-11.1 -11.2	-78.0 -81.0	-7.4	8.0	3.8	10.4	^{14}C = 10.8 ka
23	Ostrowite Q/10-40	2010	16.4	1.9±0.4	-9.76	-69.5	-11.1	41.2	7.5	<0.14	^{14}C = 1.6 ka holocene
24	Trutnowy K/85-98	1987 2010	25.8 25.7		-10.1 -9.99	-67.0 -72.6	-6.4	7.0	2.4	0.70	^{14}C = 11.8 ka H+G
25	Krzywe K. Q/10-26	2010	7.2	0.3±0.3	-9.75	-69.3			7.5	0.21	holocene

Uncertainty (standard deviation): $\delta^{18}O$ – ±0.05 ‰, δD – ±1 ‰, NGT (noble gas temperature) – ±0.7 °C, $^{4}He_{excess}$ – ±10 %; Sampling points are shown in Fig. 1; Q - Quaternary, K - Cretaceous

The origin of the anomalously high concentrations of fluorine is connected with the leaching of fluoroferous minerals in the upper part of the Cretaceous strata and ion exchange and high pH > 8. The analysis of satellite pictures conducted by Daniel-Danielewska and others (1986) demonstrated the presence of cracking and tectonic discontinuities in the roof of the Cretaceous in the area of Żuławy and corroborated the existence of vertical geotectonic deformations occurring in the Holocene. The possibility is indicated both by the chemical composition of water at this level similar to water from the Cretaceous aquifer and the isotopic composition. Fig. 1 presents the location of salinity zones and places where fluoride anomalies have been found in the area of Żuławy Wiślane (Lidzbarski et al. 2007).

The results of the isotopic research as well as determinations of noble gases have confirmed that the oldest waters occur in the northern and central parts of Żuławyat the depth of just below 100 metres (Gozdawa, Kiezmark, Drewnica, Grochowo). These are mainly infiltrating waters in the cool climate of the late Pleistocene (the Vistula glaciation). The flow temperatures specified from the concentrations of noble gases (NGT) range from 0 to 3^{0}C.

The values of stable isotopes' deltas are around $\delta^{18}O$ = -12 ‰ and$\delta^{2}H$ = -85 ‰, but there are here also waters the age of which based on ^{14}C parameters exceeds 12,000 years (tab. 1), and their isotopic compositions are similar to the waters of today's infiltration in this area ($\delta^{18}O$ = -9.8 ‰, $\delta^{2}H$ = -69 ‰). Zuber et al. (2000) claim that the isotopic compositions of these waters have undergone changes as a result of evaporation of surface water reservoirs before the process of infiltration and call them evaporated „glacial" waters from the end of the last glaciation. They could also constitute infiltrating waters from the water-logged reservoirs on the pre-frontier of the ice sheet.

Slightly younger waters, the age of which ranges from 7,000 to 13,000 years, occur in water intakes at similar depths, but located closer to the edges of Lakeland plateaux (Roszkowo, Rachowo, Elbląg, Letniki, Malbork, Kościelniczki and Tczew). In shallower water intakes at depths 20–30 m, there occur waters which mainly originate from the Holocene infiltration.

In the central and northern parts of Żuławy, where hydraulic gradients are very low, ($i \rightarrow 0$), zones of hindered water exchange as well as zones of stagnant water can be found. In such places the so-called young relict waters could survive until nowadays (Kozerski, Kwaterkiewicz 1984; Zuber et al. 2000) or at least the components of such waters mixed with waters of different origin, for example, contemporary or possibly earlier waters permeating to shallower aquifer levels as a result of ascension from the Mesozoic ground.

The decomposition of the Holocene organic matter in the deltaic series causes the emission of CH_4 and CO_2, observed in the wells in the village of Łaszka, which prevents the correct interpretation of the age of water by means of the radiocarbon dating.The composition of stable isotopes in the water samples from Łaszki no. 11 and 12, which have revealed heavier isotopic compositions than waters of today's infiltration, indicates also the possibility of water admixtures from lakes, bayou or bays which no longer exist. The diagram of water circulation in this area is shown in fig. 3.

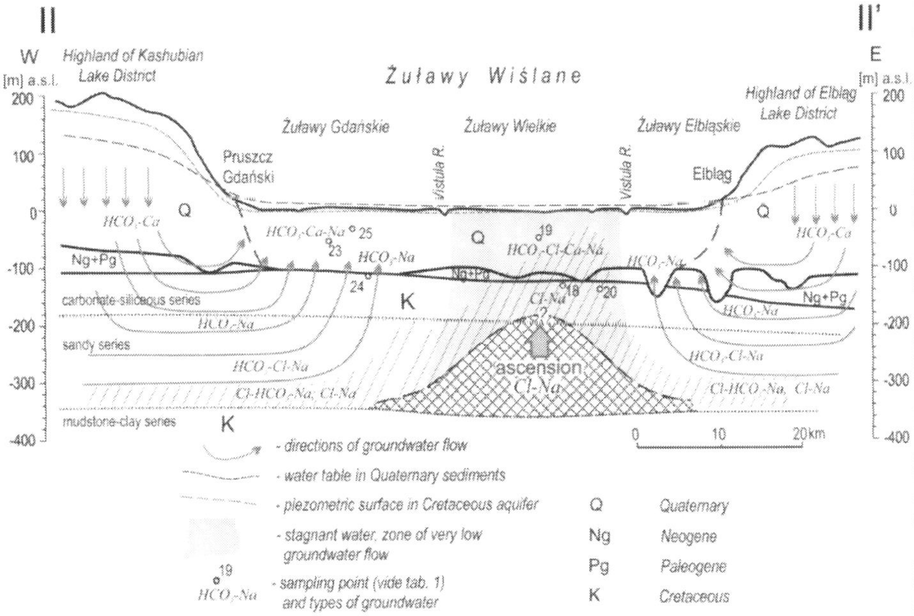

Figure 3. Schematic hydrochemical cross-section through the Żuławy Wiślane area (by Burzyński, and Sadurski 1989 modified by Krawiec 2013).

CONCLUSIONS

The origin of the brackish waters of the deltaic series in the north-eastern part of Żuławy can be explained by the remains of the former Littorina Sea in the poorly permeable loamy and sandy sediments (Kozerski, Kwaterkiewicz 1984, 1988). The results of the most recent isotopic determinations and archive research findings (Zuber et al. 2000; Krawiec 2013) indicate the law content of ^{14}C in these waters, which would corroborate this hypothesis, simultaneously excluding the possibility of sea water intrusion.

REFERENCES

Burzyński, K. and A. Sadurski. 1989. Influence of near slop drainage of Lakeland plateau on the Cretaceous aquifer on the Vistula delta plane. [in polish, eng. Sum]. Kwart. Geol. t. 33 (2): 301–312.

Daniel-Danielska, B., S. Kibitlewski and A. Sadurski. 1986 – Geology of the Vistula delta plane and the satellite images. [in Polish, eng. sum]. Geol., t.30, 3/4: 643–658.

Kozerski, B. 1981. Salt water intrusions in to coastal aquifers of Gdansk region. Proc. 7th Salt Water Intrusion Meeting, Uppsala, (27): 83–89.

Kozerski, B. and A. Kwaterkiewicz. 1984. Zones of salinity of groundwater and its dynamics on the Vistula delta plane. [in polish, eng.sum]. Arch. Hydrotech., 31 (3): 231–255.

Kozerski, B. and Kwaterkiewicz A., 1988. The origin of groundwater salinity of Quaternary aquifer in the Gdańsk region. Proc. „Aktualne Problemy Hydrogeologii". [in polish, eng. Sum]. Technical University of Gdańsk Publ. Gdańsk, vol. IV, part I: 93–104.

Kozerski, B., Maciaszczyk A., Pazdro Z. and Sadurski A., 1987. Fluorine anomaly in the groundwater in the Gdańsk Region. [in polish, eng. Sum.]. Ann. Soc. Geol. Pol., 57 (3–4): 349–374.

Krawiec, A. 2013. The origin of chloride anomalies in the groundwaters of the Polish Baltic coast. [in polish, eng.sum]. Nicholas Copernicus Univ. Publ.Toruń. pp. 1-143

Kulma, R., Haładus A., and Lidzbarski M,. 2002. Groundwater resources of the Vistula delta plane. [in polish, eng.sum.]. Polish Academy of Sciences. Gospodarka Surowcami Mineralnymi. 18, z. 2: 117–135.

Lidzbarski, M. 2007. Region of lower Vistula. In: Regional hydrogeology of Poland. [in polish]. Paczyński B., Sadurski A., Edit. P.G.I. Publ., Warsaw.

Sadurski, A. 1985. Hydrogeochemical conditions of the Cretaceous aquifer in the Gdańsk area. [in polish, english sum.], Kwart. Geol., 29, 2: 405–418.

Sadurski, A. 1989. Upper Cretaceous groundwater system of Eastern Pomerania. [in polish, english sum.] Bull. of Geology Academy of Mining and Metallurgy Publ., Cracov : nr 46: 1–137.

Schroedter M., 1931. *Die salzhaltigen grundwasser an der Weichselmundung*. Geol. u. chem. Untersuchungen Danzig. Statist. Mitt.: 167.

 https://doi.org/10.1051/e3sconf/20185400016

Zuber, A., Kozerski B., Sadurski A., Kwaterkiewicz A. and Grabczak J., 1990. Origin of brackish waters in the Quatemary aquifer of the Vistula delta. In: Proceed. of 11th Salt Water Intrusion Meeting. Gdańsk: 249–262.

Zuber, A., Sadurski A., Weise S.M., Rübel A.,. Osenbrück K. and Grabczak J,. 2000. Isotope and noble gas data of the Gdańsk Cretaceous Aquifer, Northern Poland. In: Hydrogeology of the Coastal Aquifers. University of Nicholas Copernicus University Publ. pp. 181–186.

Salt water intrusion in the breakthrough valley of the river Aa between the Flemish coastal plain and the Saint Omer basin (France)

Luc Lebbe[1] , Devlin Depret[2], Jasper Claus[3] and Gert Jan Devriese[4]
[1]Department of Geology Ghent University, Ghent, Belgium
[2] G. Smeyers N.V. , Zandhoven, Belgium
[3]International Marine & Dredging Consultants (IMDC), Antwerp, Belgium
[4] Pidpa, Antwerp, Belgium

ABSTRACT

The salinity distribution in a breakthrough valley has been studied based on field observations and using a numerical model. The studied valley is located in Northern France between the Saint Omer basin and the Flemish coastal plain. The field campaign starts with the installation of observation wells. In these observations wells the electrical conductivity is logged versus depth with an electro-magnetic device. These wells also allow the necessary measurements to deduce the fresh water heads. Some of the wells are used for the performance of a pumping test. The lateral variation of the electrical conductivity of the water in the drainage canals is measured. With a numerical model the evolution of the flow and distribution of the fresh and salt water are simulated. Two simulations are made with the same schematization of the groundwater reservoir consisting of quaternary marine sediments, silt and clay and with the same hydraulic and solute parameters and boundary conditions but with two different initial salinity distributions. The results of both simulations show that the salinity distribution in the quaternary marine sediments depends not only on the lateral variation of the watertable but also on the shape of the incision of the palaeo-valley into the clay of Ypresian age. The salinity distribution in the lower part of the quaternary marine sediments depends first and foremost on the shape of the incision of the palaeo-valley into the clay of Ypresian age whereas the salinity distribution in the upper part of the quaternary marine sediments depends first and foremost on the lateral variation of the watertable.

INTRODUCTION

On the edges of coastal plains, and even beyond, salt water intrusions can occur in valleys which are surrounded or flanked by hills. An example of such an area is the Saint Omer basin and its connection to the Flemish coastal plain (Figure 1). The Saint Omer basin is a small subsiding basin where the river Aa rises at its southern edge and flows in the north-western direction through the basin and in northern direction through the breakthrough valley into the Flemish coastal plain. The breakthrough valley is incised in the clay of Ypresian age (YC) and is filled with quaternary marine sediments (QMS). The valley is flanked by hills where the YC is covered with quaternary continental silt (QCS). The QMS are filled with salt, brackish and fresh water. The flow and the salinity distribution are studied by the aid of a numerical model which is verified and supported by field observations and tests.

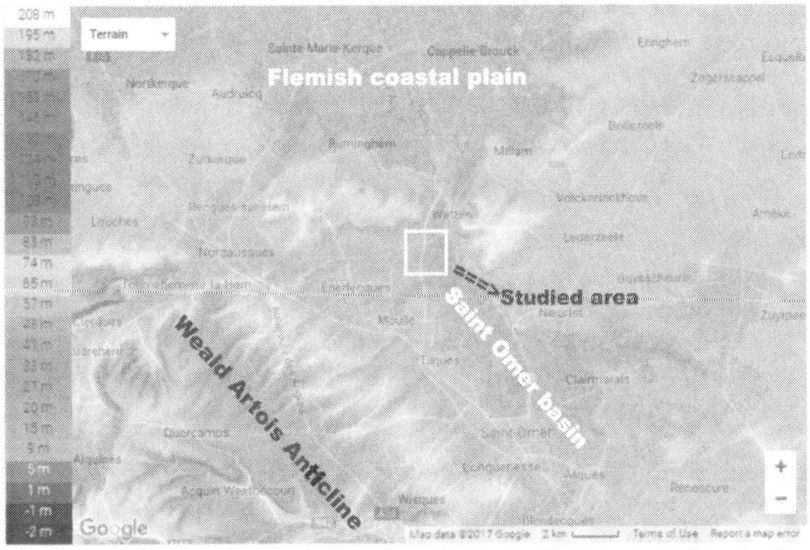

Figure 1. Topographic map with indication of the studied area and geological areas.

3D MODELLING

The field work starts with the installation of a number of observation wells. The observations wells allow the measurements of fresh water head, the loggings of electrical conductivities of the depositions around the wells with an electro-magnetic device and the performance of a pumping test. In Figure 2 the lithostratigraphic cross section is shown along with electrical conductivity (EC) measured with the EM39-device. At the pumping test site a 3m thick peat layer occurs. Above the peat there is a sequence from top to bottom of clay, silt, peat, sand and silt in which the electrical conductivity gradually increases with depth. The EC of the sediments reaches a maximum value of 380 mS/m in the lower two meter of the peat. Below the peat occurs a thick sand layer in which the screen of the pumped well PP is placed. The EC of the sand varies around 280 mS/m. With the pumping test three hydraulic parameters are identified: the horizontal conductivity and specific elastic storage of the sand and the hydraulic resistance of the peat layer.

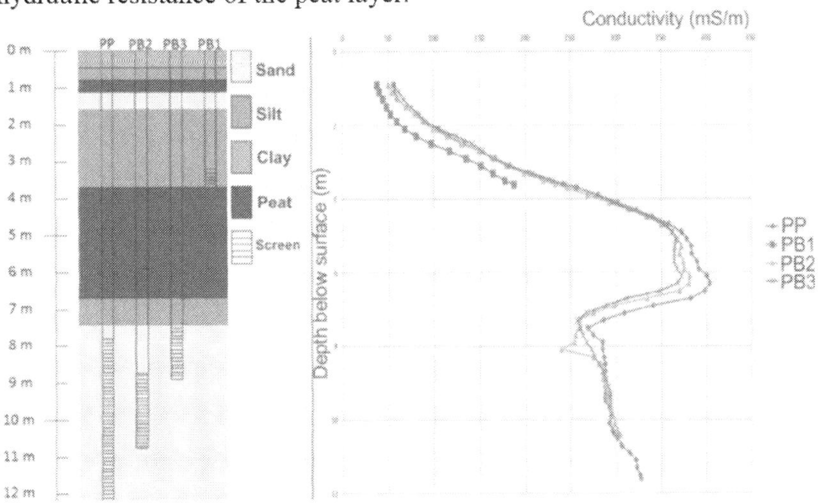

Figure 2. Lithostratigraphic cross section at the site of the pumping test along with the EM39 measurements in the wells of the pumping test PP, PB2, PB3 and PB1.

In the central part of the studied area the lateral EC variation of the water in the drainage canals is measured on 22 October 2014. From these measurements the total dissolved solids (TDS) are derived. The TDS values vary between 300 mg/l and 5 g/l. The water in the drainage canals in the north western corner of the measured area has a TDS larger than 2g/l.

3D MODELLING

The MOCDENS3D-code (Lebbe & Oude Essinck, 1999) is applied to simulate the 3D density dependent groundwater flow. The modeled area is located in the breakthrough valley of the river Aa between the Flemish coastal plain and the Saint Omer basin. It is a rectangular area of 2000 m on 1620 m. The finite-difference grid consists of 200 columns, 162 rows and 24 layers. All finite-difference cells have the same size. These cells have a thickness of 1.5 m and have a squared base plane with a side of 10 m. The base of the uppermost layer is 15 mNGI. The uppermost nine layers consist of active and inactive cells. Due to the relative high topographic difference the number of the active cells in the uppermost layer is limited but these number increases in the downward direction. All cells of layers 10 to 24 are active. The northern, southern, eastern and western boundaries are impervious as well as the base of lowermost layer. In all uppermost active cells there is constant inflow of fresh water (Ps =0.1 %, 280 mm/year). All uppermost active cells are drained. The estimated drainage level is based on the topographic data. The dense network of drainage canals are considered as rivers in the model as well as the canalized Aa (le canal de Grand Gabarite).

Three hydrogeological units are considered: the quaternary marine sediments (QMS), the silt or the quaternary continental sediments (QCS) and the clay of Ypresian age (YC). Due to the lack of detailed maps of the heterogeneities of the QMS these sediments are simplified to a homogeneous anisotropic layer with a horizontal and vertical conductivity of respectively 2.5 and 0.04 m/d. The QCS and YC are less heterogeneous than the QMS. The assumed horizontal and vertical conductivity of the QCS are respectively equal to 0.04 and 0.002 m/d whereas the assumed values for the YC are respectively 0.002 and 0.0001 m/d. The spatial variation of the transmissivities and hydraulic resistances between the finite-difference cells are derived with the aid of the topographic map and the map of the top of the YC. Two different simulations of the salinity distribution are made.

Simulation 1

The initial salinity distribution in the QMS is based on the estimated areal variation of the level of the interface between fresh and salt water. This areal variation is based on the application of the Ghyben-Herzberg relation (Todd, D.K., 1959) and the estimated average level of the watertable. The estimated watertable level is in its turn based on the topographic data. When the finite-difference cells with QMS are situated completely above the interface the initial salt water percentage (Ps) is set equal to zero. When these cells are situated completely below the interface the initial Ps = 100%. When the salt fresh water interface is located in a QMS cell the Ps is set equal to the percentage of the volume of the cell below this interface. The fresh water has a Ps = 0% and a TDS $\leq$ 800 mg/l. The salt water has a Ps = 100% and a TDS $\geq$ 28 g/l. An initial Ps of 1% is attributed to the finite-difference cells which are situated in the QCS. The cells situated in the YC obtain an initial Ps of 5%. The applied value for the water-conducting porosity is 0.38. The longitudinal dispersivity is set equal to 0.3 m and the transversal dispersivity for horizontal and vertical flow are respectively equal to 0.05 and 0.03 m.

First, the evolution of the flow and salinity distribution is discussed by the aid of horizontal and vertical cross-sections. In these cross sections the fresh water head configurations are first shown between 1.75 and 17 mNGI, respectively the minimum and maximum value in the modeled area. With this representation it was not possible to show the small but important variation of the fresh water head in the valleys. Because these small variations in the fresh water head are important in the framework of the flow and salinity distribution in the valleys, the fresh water heads are represented between the interval of 1.75 and 2.35 mNGI (see Figure 3).

The flow and the salinity distribution in the central part of the valley and in the upper part of the QMS are similar as in the major part of the Flemish coastal plain (e. g. Vandenbohede & Lebbe, 2000; Van Meir & Lebbe, 2002 and Vandenbohede et al., 2004). There, the depth of the transition zone between fresh and salt water has a strong negative correlation with the average level of the watertable. A slightly higher watertable corresponds with a larger depth of the fresh/salt transition zone and *vice versa.* On the edges of the valley there is an inflow of fresh water from the elevated areas where QCS covers the YC. This inflow and the watertable which is slightly higher than in the central part of the valley, result in a deeper transition zone. In the upper part of the QMS the flow and the evolution of the salinity distribution are slow.

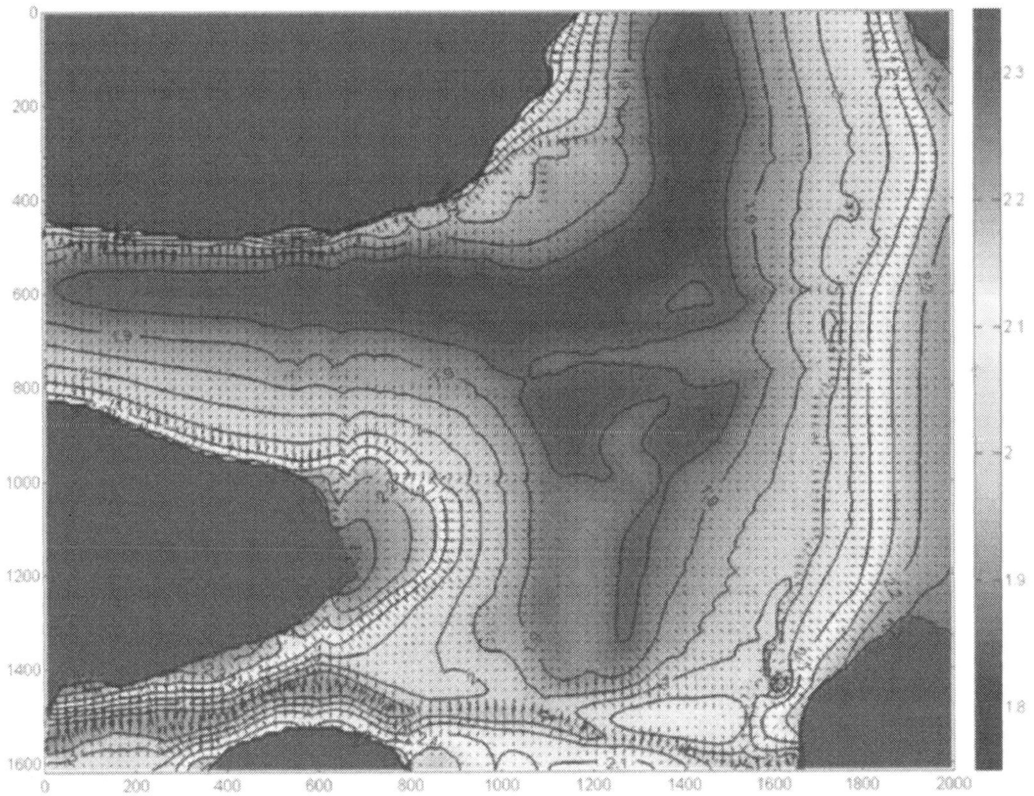

Figure 3. Fresh water head between the interval 1.75 -2.35 mNGI of layer 10 which is located between 1.5 and 3 mNGI and horizontal groundwater flow vectors after 50 years (simulation 1).

 https://doi.org/10.1051/e3sconf/20185400017

In the lower part of the QMS the flow after 50 years of simulation is considerably larger than in the upper part of the QMS. This flow is strongly dependent from the gradient of the top of the YC and of the initial salinity distribution which was estimated with the Ghyben-Herzberg relation. This flow results in a faster evolution of the salinity distribution in the lower part of the QMS. There occurs a flow of salt water to the QMS located in the deeper incision in the YC. This is in contrast with former simulations of salinity distribution made in other parts of the Flemish coastal plain where, however, the top of the YC is rather flat and where the fresh/salt water transition zone is located far above this top. In these last cases the flow of salt water above the top of the YC is very limited.

Simulation 2

In the second simulation the QMS are initially completely filled with salt water (Ps=100%). The initial salt water percentage of the QCS and the YC is the same as in the first simulation, respectively equal to 1 and 5%. The evolution of the salinity distribution is simulated for the same hydraulic and solute transport parameters and with the boundary conditions for a period of 800 years, the estimated time since the last marine influence in the studied area. The evolution of the flow and salinity distribution is discussed and compared with the first simulation. At the end of both simulations the salinity distribution in the lower part of the QMS are similar. In the upper part of the QMS the salt water content is smaller at the end of simulation 1 in comparison to simulation 2. This is due to the fact that the freshening process of the QMS lasted over a longer time period during simulation 1.

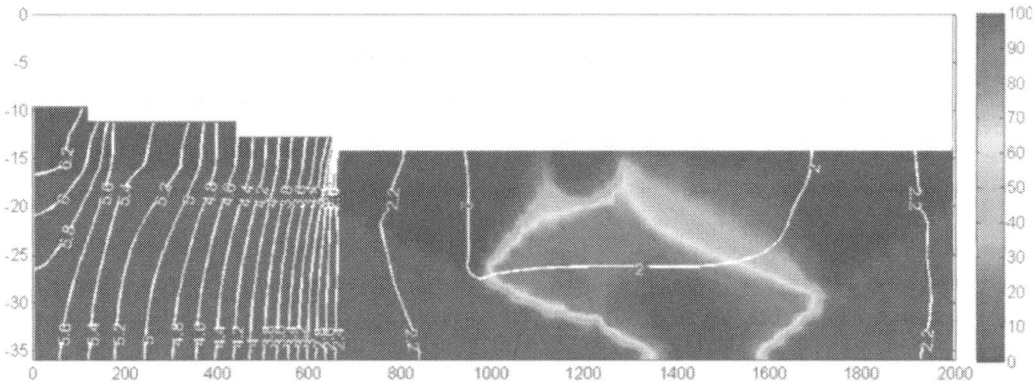

Figure 4. Salt water percentage in row 116 after 800 years of recharge with fresh water (simulation 2 with an initial Ps of 100% in the QMS) the white contour lines correspond with the fresh water head.

As a conclusion one can state that the salinity distribution in the QMS depends not only on the lateral variation of watertable but also on the shape of the incision of the palaeo-valley into the YC. The salinity distribution in the upper part of the QMS depends first and foremost on the lateral variation of the watertable whereas the salinity distribution in the lower part of the QMS depends first and foremost on the shape of the incision of the palaeo-valley into the YC. In this studied case the Ghyben-Herzberg relation was not appropriate for the estimation of the initial salinity distribution in the lower part of the QMS.

DISCUSSION AND CONCLUSIONS

In the breakthrough valley of the river Aa between the Saint Omer Basin and the Flemish coastal plain occurs a salt water intrusion. This salt water intrusion was studied by the aid of field observations and numerical model simulations. A limited number of observations wells allow the collection of diverse data to compare with the results of the numerical model such as the fresh water head, the variation of electrical conductivity of the deposits around the wells and the execution of a pumping test. The measured variation of the electrical conductivity of the water in the drainage canals furnishes valuable data to assess the evolution of the salinity distribution. The results of the simulations show that the evolution of the salinity distribution in the studied area depends also on the shape of the incision of the palaeo-valley into the clay of Ypresian age. This shape influences this evolution in addition to other characteristics of which the foremost is the lateral variation of the watertable. In the studied case the present day salinity distribution was not obtained after a relative short simulation period (e.g. 50 years) with an initial salinity distribution which was obtained by the aid of the Ghyben-Herzberg relation and the lateral variation of the watertable. This last mentioned method was, however, successful in other areas of the Flemish coastal plain where the top of the clay of Ypresian age is rather flat and where the fresh/salt water transition zone is located above this top. In the studied case where the top of the clay of Ypresian age has steep slopes there is a better method which, however, requested more computer time. In this method the simulation starts with the quaternary marine sediments completely filled with salt water and for a simulation time since the last marine influence in the area.

REFERENCES

Lebbe L. and Oude Essink, G. 1999. MOC DENSITY / MOCDENS3D-code. p. 434-439, in Chapter 12. Survey of Computer codes and Case Histories. Eds. Sorek, S. & Pinder, G.F. in: *Seawater Intrusion in Coastal Aquifers, Concepts, Methods and Practices*. Eds. Bear, J., Cheng, H-D, Herrera, I., Sorek, S. and Ouazar D. Kluwers Academic Publishers.

Todd, D. K., 1959. Ground Water Hydrology, New York, John Wiley & Sons. 336 p.

Vandenbohede, A. & Lebbe, L. (2000). Water quality distribution in the eastern coastal plain of Belgium: Influence of intercalated peat beds. *Proc. 16th Salt Water Intrusion Meeting*, Miedzyzdroye, Poland, p.133-140.

Vandenbohede, A., Linster, T. & Lebbe, L. (2004). Modelling of density dependent groundwater flow in the south-western Belgian coastal plain. Proc. of the 18th Salt Water Intrusion Meeting, 31 May-03 June 2004, Cartagena, Spain, 207-214.

Van Meir, N. & Lebbe, L. (2002). 3D-density-dependent modeling of sea-level rise scenarios around De Haan (Belgium). Proc. 17th Salt Water Intrusion Meeting, Delft, The Netherlands, p.73-81.

Contact Information: L. Lebbe, Ghent University, Faculty of Science, Geology Department, Krijgslaan 281 (S8), B 9000 Gent , Belgium, Phone: +32-9-2644653, Email: luc.lebbe@ugent.be

Hydrogeological researches in the 4D cartography program in the coastal zone of the southern Baltic

Mirosław Lidzbarski[1], Ewa Tarnawska[1]
[1]Marine Branch of Polish Geological Insitute - NRI, 5 Kościerska str., 80-328 Gdańsk

ABSTRACT

The aim of the pilot programme of the 4D cartography in the coastal zone of the southern Baltic is to develop appropriate forecasts concerning the southern Baltic coastline alterations with reference to the geological structure of the coast and anticipated changes of climate. The effects of these changes on the infrastructure, aquatic plant habitats as well as major fresh water reservoirs responsible for water supply for both municipal and industrial purposes will also be thoroughly analysed.

The project is executed as part of the tasks of the Polish Geological Survey, aiming to monitor and recognize potential geological and hydrogeological hazards (Kramarska at al., 2014).

GOALS OF THE PROJECT

The basis for the development of the programme is the geological and hydrological research conducted on the pilot section of the seashore between Władysławowo and Lubiatowo. More meticulous research was carried out in the vicinity of Jastrzębia Góra, where the coastal erosion in the active part of the cliff coast is increased and the conservation activities do not seem to produce the expected outcome. The research, which has been conducted since July, 2012, aims to create accurate spatial geological-morphometric models of the coastline at a specific time record (4D). It is also essential that the hydrogeological conditions affecting the erosion of the coastline should be recognized. Groundwater circulation systems have been identified and the correlation between the pace of the geodynamic changes occurring on the cliff coast and the geological structure and hydrogeological conditions has been established.

Erosive processes occurring on the seashore are shaped not only by outer factors, but also by inner ones. The abrasion of cliff coast seems to be particularly dangerous as the occurrence of landslides is affected by hydrogeological conditions, land-use planning as well as human pressure.

INVESTIGATION AREA

Conservation measures which are supposed to prevent, or at least reduce, the adverse phenomena have been taken on selected sections of the cliff coast. The planning of remedial and maintenance work appears difficult and is highly vulnerable due to the lack of reliable data concerning the occurrence of the groundwater circulation system in the area of the cliff.

METHODS AND RANGE OF RESEARCH

As part of fieldwork, detailed hydrogeological and hydrochemical mapping was conducted. Within the research area, all occurrences of groundwater and surface water, particularly salt waters, were documented. The preliminary assessment of the origin and chemical composition of groundwater and surface waters from springs was also carried out. The

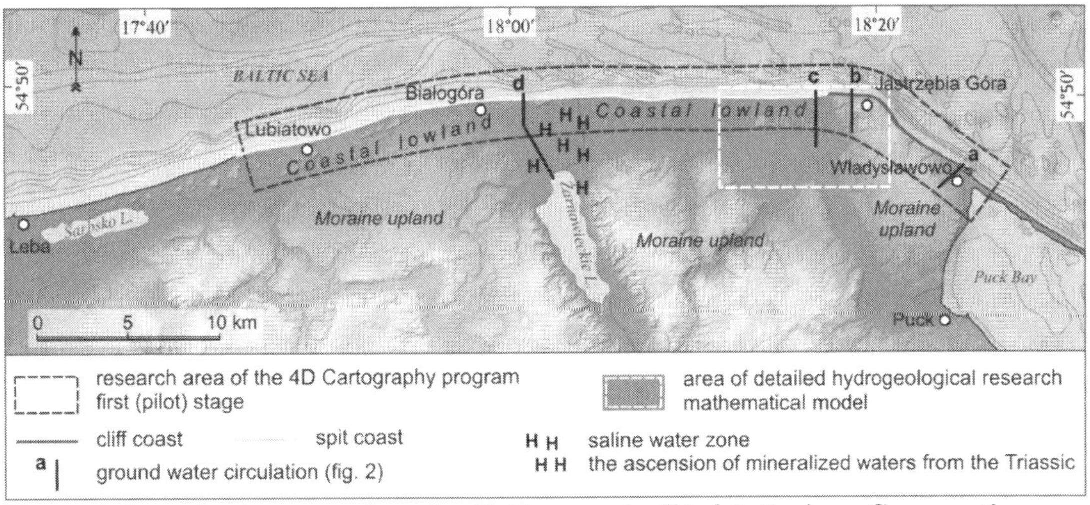

Figure 1. Investigation area along the Baltic coast in Gdańsk Region. Cross-sections are given on fig. 2.

obtained information along with the hydrogeological archive data and the results of geological research have provided the basis for the creation of a three-dimensional mathematical model of hydrogeological processes. The internal structure, boundaries as well as spatial geometry of the aquifer were determined along with its full hydrogeological description. The groundwater dynamics, the intensity of horizontal and vertical flows, as well as salt water and brackish water zones were also recognized.

Geophysical methods and shallow drilling (probing) both in the cross-border land or sea area were applied. The terrestrial laser scanner was used to register all the changes in the morphology of the seashore (Kamiński at al., 2012).

HYDROGEOLOGICAL MODEL

For the purpose of numerical modelling of hydrogeological processes, the programme ModFlow version 2000 from the software package GMS 8.3 has been used. The research area of 7.2 km^2 has been digitized by means of the square grid 10x10 m (fig. 3). The total number of computational blocks amounted to 408480. On the basis of the adopted conceptual model, the system of aquifers occurring in the area of research has been subjected to certain schematization. During the process of aquifer aggregation, the directions of groundwater flows as well as piesometric pressure differences were taken into consideration. As a result four model layers have been distinguished. The identification of directions of groundwater flows and the groundwater flow balance have been conducted in three basic variants simulating different levels of underground water consumption:

Variant 1 – the current state of groundwater abstraction,

Variant 2 – the lack water intakes exploitation, displaying the restoration of natural piesometric surface in the area of research;

Variant 3 – the simulation of hydrogeological conditions, taking into consideration the projected climate changes (rise in the sea level – drainage base by 1 metre).

The hydrogeological conditions of the coastal zone include two aquifer levels: the Quaternary level and the Neogene and Palaeogene level, which remain in a close hydraulic connection with each other. The coastal highlands also include intermoraine aquifer levels, and in the area of the coastal lowland the Pleistocene and Holocene aquifer connected with fluvioglacial, aeolian and marine sediments can be found. In the sandbars fresh groundwater commonly occurs in the dune sands.

The groundwater recharge occurs as a result of rainwater infiltration. The groundwater flow in the direction of the Baltic Sea predominates in the horizontal flows. Local groundwater circulations of limited range are generated within the coastal dune zone as well as in first subsurface aquifers, from which groundwater flows towards the seashore, local watercourses and drainage ditches. The system of perched water circulation occurring in the vadose zone is more complex. In the immediate surroundings of the active part of the cliff coastline in Jastrzębia Góra four levels of perched water have been identified.

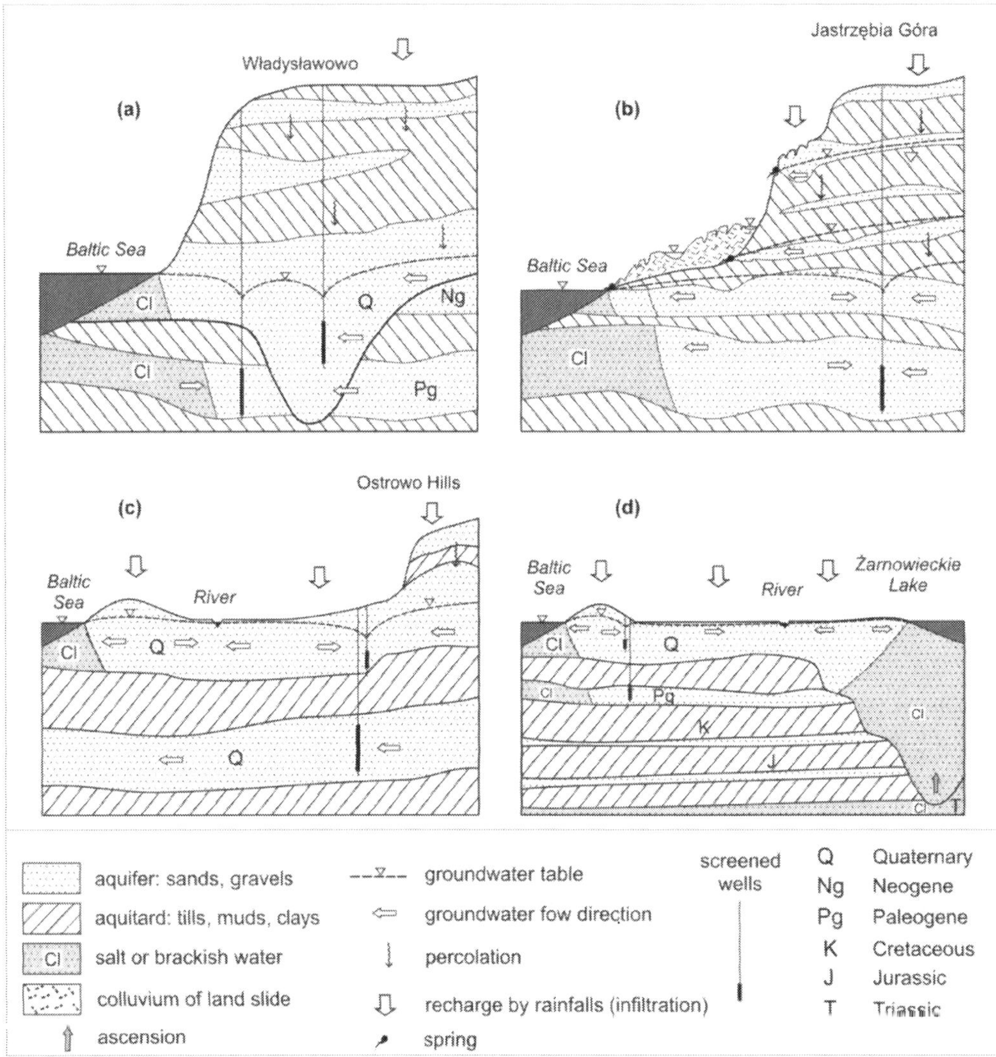

Figure 2. Different groundwater circulation systems along the investigated coasts.

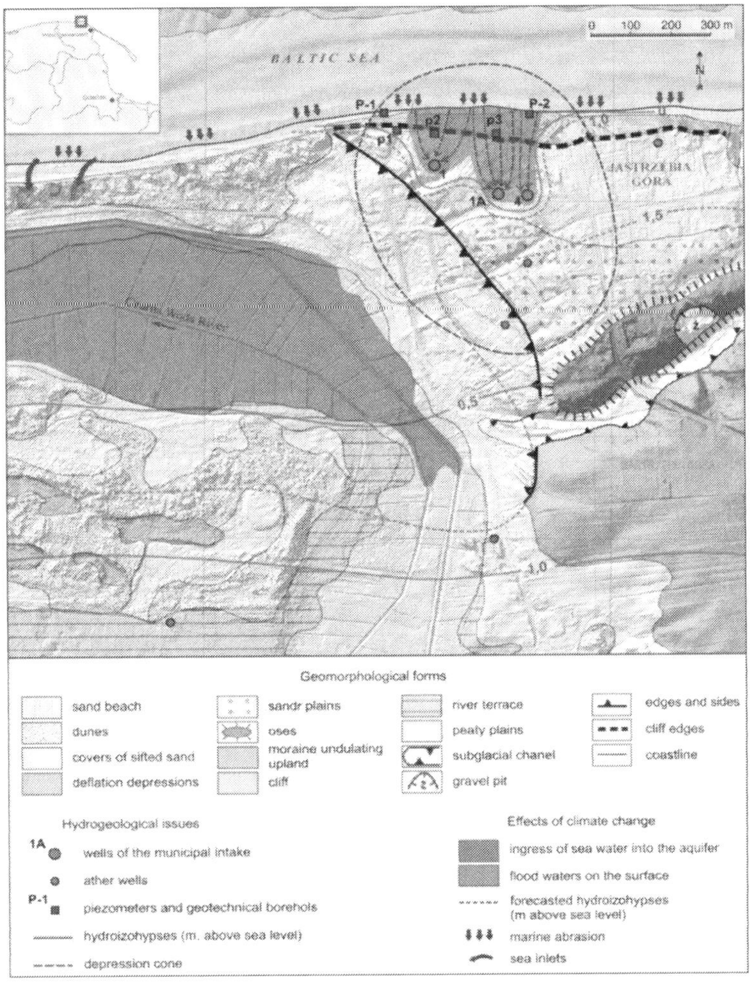

Figure 1. Predicted effects of climate change in the coastal zone based on model tests.

In the coastal zone of the Baltic Sea in question two base drainage zones have been recognized – the offshore zone in the cliff section and the coastline zone in the dune section of the shore. In Jastrzębia Góra and other areas of excessive consumption of groundwater, the offshore drainage through poorly permeable seabed sediments is reduced, and in the case of excessive exploitation can even become temporarily reversed, which in turn contributes to marine waters ingression (Prussak, 1987). To the north of Żarnowiec Lake the groundwater flow extends to the Triassic layers, where chloride concentrations exceed 11 g/dm^3.

The results of flow modelling allowed for the analysis of hydrodynamic alterations coming from the projected climatic changes. The computational simulation was based on an extreme variant which assumes the rise of the Baltic Sea level by 1 metre. The scenario analysing the change of hydrodynamic conditions with such a high drainage base demonstrated far-reaching changes in the hydrographic system as well as in the system of groundwater circulation. The total quantity of groundwater participating in the circulation will decrease by approximately 15%. The flow rate of water draining away to the Baltic Sea will increase at the expense of the drainage through surface water (Lidzbarski, Tarnawska, 2015).

More extensive impacts of groundwater at the foot of the cliff as well as in the colluvia zone by approximately 10% should also be expected. As a result of the rise in the drainage base, the geotechnical conditions of the soil at the foot of the cliff will become weakened. They will become more prone to potential landslides. Damming up of the marine waters which frequently occurs in this area will also contribute to the loss of stability of cliff slopes. As a result of the hydrodynamic changes, the seashore will be more susceptible to abrasion, not only in the cliff zone, but on the sandbar stretches of the seacoast as well. In the coastal zone, the terrain surface will be flooded by groundwater and seawater, which will in effect create considerable lagoon-like pools. Furthermore, as a result of marine water intrusion, groundwater in the coastal zone will be affected with significant salinity. In the vicinity of Żarnowiec Lake, the zone of ascending water will blend with saline water as a result of marine water intrusion.

CONCLUSIONS

Groundwaters occurring in the coastal zone have a detrimental effect on the stability of the Cliff coast. They weaken the strength properties of cohesive soil. The hydrodynamic change forecast has been formulated by means of the three-dimensional model of hydrogeological conditions of the coastal aquifer system. Taking into consideration the possible scenario of climatic changes, it is predicted that the abrasion threat to the seashore will rise in the cliff coast zone. The sandbar coastline will be exposed to destructive processes from two sides: the increased abrasion from the sea and the flooding of coastal plains from the land. The information obtained during the research work will be helpful in forecasting mass wasting (geohazards) within the area of the Cliff coast.

LITERATURE

KAMIŃSKI M., KRAWCZYK M., ZIENTARA P., 2012 — Recognition of geological structures of Jastrzębia Góra cliff by geoelectrical imaging method for landslide geohazards estimation. [in polish, eng. sum]. PGI-NRI Biul., vol. 452: 119–130.

LIDZBARSKI M., TARNAWSKA E., 2015 – Hydrogeological investigations of the cliff seashore for geohazards estimation and forecast. [in polish, eng. sum]. Przegl. Geol., vol. 63, nr 10.

KRAMARSKA R., UŚCINOWICZ G., JURYS L., JEGLIŃSKI W., PRZEŹDZIECKI P., FRYDEL J., TARNAWSKA E., LIDZBARSKI M., DAMRAT M., WOŹNIAK M., 2014 — Pilot project of 4D mapping on the seashore of southern Baltic. [in polish]. Unpublished. Arch. Of PGI-NRI. Marine Branch. Gdańsk.

PRUSSAK W., 1987 — Hydrogeological documentation of groundwater resources of Quaternary-Tertiary aquifer. [in polish]. Unpublished. POLGEOL Com. Warsaw.

Ferruginous groundwaters as a source of P, Fe, and DIC for coastal waters of the southern Baltic Sea: (Isotope) hydrobiogeochemistry and the role of an iron curtain

Marko Lipka[1], **Michael E. Böttcher**[1], Zijun Wu[1,4], Jürgen Sültenfuß[2], Anna-K. Jenner[1], Julia Westphal[1], Olaf Dellwig[1], Peter Escher[1,5], Iris Schmiedinger[1], Vera Winde[1,6], Ulrich Struck[3]

[1]Geochemistry & Isotope BioGeoChemistry Group, Leibniz Institute for Baltic Sea Research (IOW), Warnemünde, FRG; [2]Environmental Physics Department, Bremen University, Bremen, FRG; [3]Natural History Museum, Berlin, FRG; [4]State Key Laboratory of Marine Geology, Tongji University, Shanghai, PRC; [5]current address: Ecoandmore, Freiburg, FRG [6]current address: LUBW, Lake Research Institute (ISF), Langenargen, FRG

ABSTRACT

We report first results from a study on water and element exchange across the land-ocean boundary at the southern Baltic Sea. The focus is set on ferruginous fresh ground waters escaping at the shore line, flowing in air contact before entering a subterranean mixing zone with brackish Baltic Sea water. The present study combines the results from multiple sampling campaigns that investigated the composition of several springs as well as the surface and subsurface development of fresh waters on the way to the Baltic Sea. This is achieved by a combination of hydro- and solid-phase geochemical and stable isotope measurements with ground water dating. Results are compared to the composition of groundwaters recovered from wells in the catchment area and the local isotope meteoric water line developed for Warnemünde. The spring water is shown to be impacted by the dissolution of biogenic carbon dioxide and marine carbonate as well as the oxidation of pyrite in glacial sediments. Dating yields a surprisingly high diversity between some closely associated springs with average ages of about 25 to 32 years, but different mixing proportions with older tritium-free ground-water.

INTRODUCTION

Fresh waters, like rivers, streams or submarine ground water discharge (SGD), entering the estuarine and coastal areas, provide important pathways for element and water flow at the land-ocean transition. Since these waters may be enriched in (micro) nutrients, metals, and dissolved inorganic carbon (DIC) (Knee & Paytan 2010; Winde et al. 2014; Donis et al. 2017; Böttcher et al. 2018) the input is linked to potential societal issues, like eutrophication, the development of algal blooms and hypoxia, the increase or decrease of the buffer capacity of coastal waters (associated to green-house effect, as well as ocean acidification), to mention a few.

In the present study we focused on ferruginous fresh groundwaters emerging at the shore line of the southern Baltic Sea (close to Kühlungsborn) and their role as sources for P, DIC, SO_4^{2-} and metals and the factors impacting the fate in the transition zone. We apply a multi-tracer combination of hydrogeochemical with stable isotope (H, C, O, S) analysis and the dating of the springs (tritium and noble gases) to define the sources of fresh water and dissolved elements, processes affecting the surface water in contact with the atmosphere (absorption of oxygen, desorption of carbon dioxide) and upon subterranean mixing with brackish Baltic

 https://doi.org/10.1051/e3sconf/20185400019

Sea water. The precipitation of FeOOH ('iron curtain') at the sediment surface and in the sediments acts as a temporal trap for dissolved phosphate and the (temporal?) impact by the activity of iron-oxidizing bacteria is indicated. The results are further compared to the composition of groundwaters sampled from drinking and observation wells near the Baltic Sea coastline as well as the local isotopic meteoric water line established for Warnemünde.

METHODS

Water samples from the springs and the stream, underlying pore waters, as well as the water column of the Baltic Sea were taken and analyzed following established procedures (Kowalski et al. 2012; Winde et al. 2014). Water aliquots were immediately filtered (0.45 µm membrane filters) for further analyses by ICP-OES (Thermo iCAP 6300 Duo, Thermo Fisher Scientific) and a QuAAtro nutrient analyzer (SEAL Analytical). ICP-OES and SEM-EDX were used to analyze the composition of acidic sediment extracts and the microstructure of FeOOH precipitates, respectively. Measurements of tritium, helium isotopes and neon followed Sueltenfuß et al. (2009). Stable C and S isotopes were analyzed with a Thermo Finnigan MAT253 mass spectrometer attached to a Gasbench II or and Thermo elemental Flash analyzer. Stable water (H, O) isotope measurements were conducted by means of Picarro CRDS systems. All stable isotope results are given in the conventional δ-notation (equivalent to milli Urey) versus respective international standards. Hydrogeochemical results were additionally subjected to a thermodynamic analysis via PHREEQC.

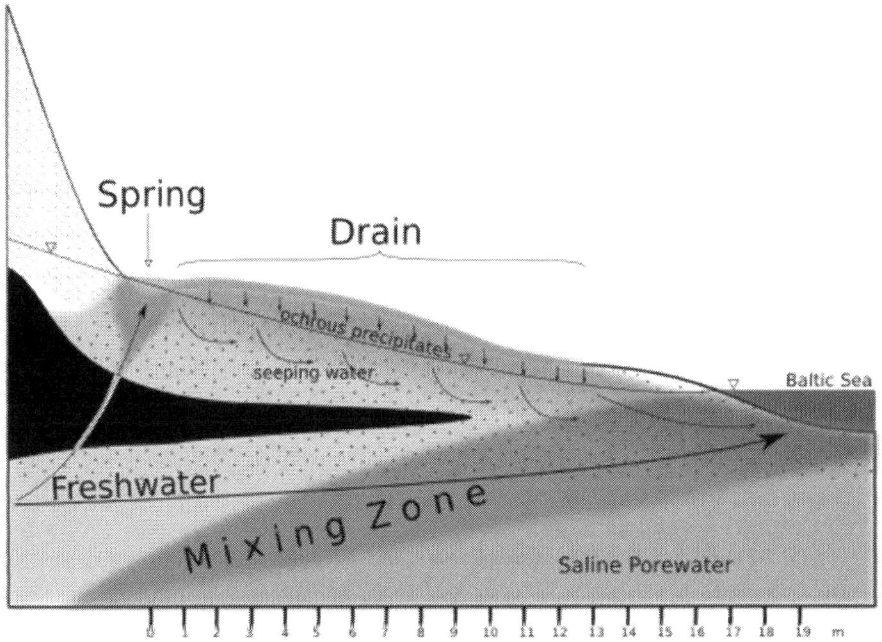

Figure 1. Sketch showing the generation of ferruginous ground water springs and a the development of a subterrestrial mixing zone for SGD with Baltic Sea water at the shore line. Black: Fissured clayish layer.

https://doi.org/10.1051/e3sconf/20185400019

RESULTS AND DISCUSSION

Spatial and seasonal dynamics of spring waters

Ferruginous fresh groundwaters springs occuring at the shore line of the southern Baltic Sea were investigated on a seasonal base for a period of more than 7 years. The geostationary springs emerge in small pits of up to about 70 cm diameter with up to 15 cm water depth. The shape of the pits and the stream beds, in particular closer to the Baltic Sea water line varied considerably due to regular sediment re-structuring driven by wind and wave activity. The outflowing ground waters are oxygen-free, but rich in dissolved iron, sulfate, phosphate and dissolved inorganic carbon (DIC; Fig.2). They showed constantly minor relative differences between springs through the years, with an exception for redox-sensitive elements like Fe^{2+} that are further impacted by storm-related redistribution of surface sediments.

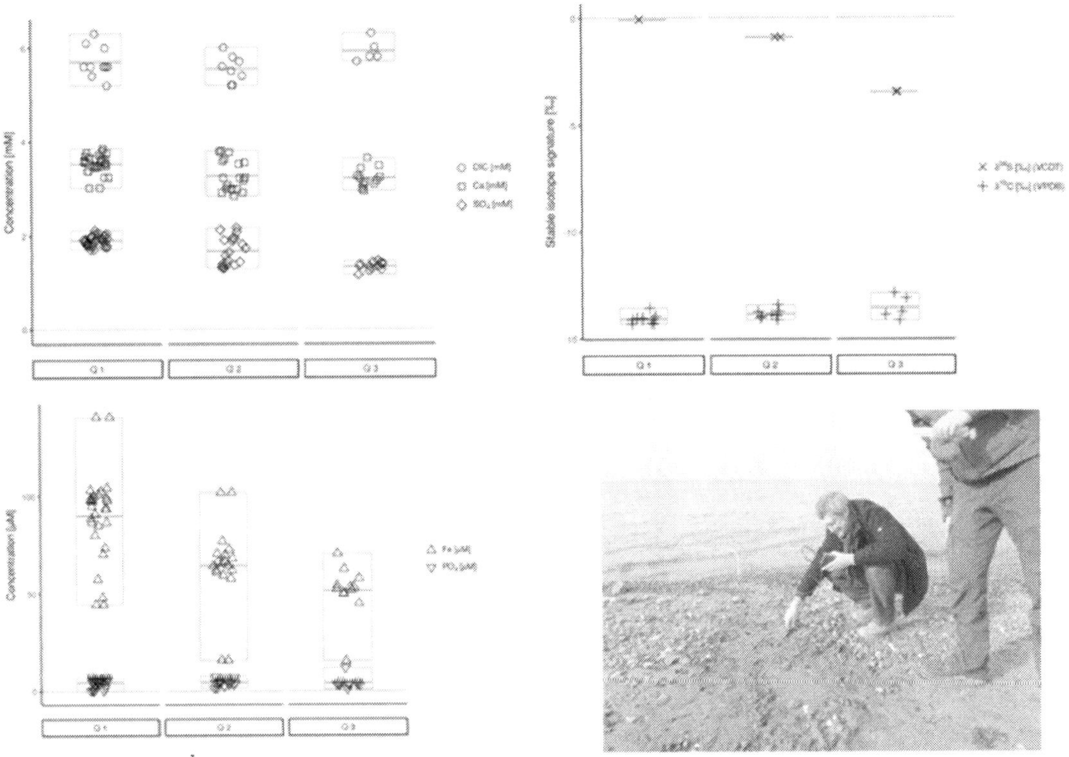

Figure 2. A-C Temporal dynamics of selected dissolved major and minor components and the stable isotope composition of DIC and SO_4 for three springs waters. D: In-situ measurements at spring Q1 during winter time.

Tritium-noble gas dating indicates ages between 25 and 35 years, with surprisingly high different mixing contributions from older tritium-free waters. This indicates a high complexity of the hydrogeology of the coastal aquifers with aquifers essentially build of glacial deposits. The observed differences in ground water composition reflect heterogeneous flow path conditions (including residence times) before reaching the Baltic Sea shore line, highlighting the importance of local and regional investigations in the evaluation of the impact SGD discharge on the transport of substances (DIC, nutrients, metals) into coastal ecosystems. The hydrogeochemical and stable isotope composition indicates that the spring water composition is controlled by the dissolution of biogenic carbon dioxide in the unsaturated soil zone of the catchment area followed by the dissolution of marine carbonate and biogenic pyrite in the sediments building up the aquifers (e.g., Zhang et al. 2012; Donis et al. 2017).

Processes impacting stream water composition in air contact

Each springs leads to a surface run-off towards the Baltic Sea. After flowing for several meters in contact with the atmosphere, the fresh water is lost to underground drainage through the beach sands reaching a mixing zone with brackish Baltic Sea waters and finally ending as submarine ground water discharge (SGD). Surface flow is associated with the uptake of gaseous oxygen, the microbially promoted precipitation of iron oxi(hydroxi)des, adsorption of phosphate (Fig.3), the loss of isotopically light carbon dioxide into the atmosphere, and probably minor calcite precipitation. Spring waters are saturated with respect to calcite, but undersaturated with respect to other rock-forming minerals like dolomite and gypsum, turning into calcite supersaturated solutions along the surface flow-path. According to our present conceptional understanding (Fig.1) an essentially anaerobic flow path directs fresh ferruginous ground water into a mixing zone with aerobic Baltic Sea water. This leads to physico-chemical and geomicrobiological changes in the composition of the aqueous solution and the surrounding sediments. This process is superimposed by the spring waters that are modified by the precipitation of FeOOH and associated adsorption of phosphate and arsenic, before seeping away into the mixing zone. This iron curtain acts as a temporal sink for Fe and P in the coastal sediments that may be transported into the Baltic Sea ecosystem during storm events.

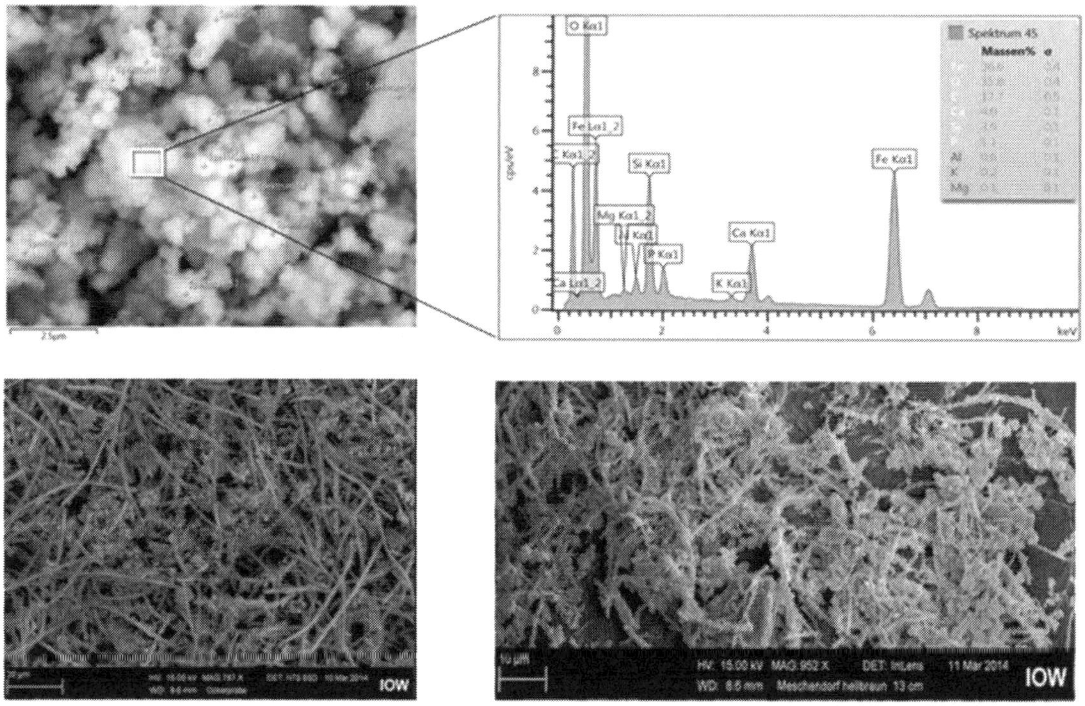

Figure 3. A and B: SEM-EDX analysis of a FeOOH sample from the surface sediments of the stream. C and D: Residual leftovers of Fe oxidizing bacteria (e.g., Leptothrix) in the stream bed.

Comparison with ground waters in the catchment and other SGD sites

The hydrogeochemical and stable isotopic composition of the ferruginous springs is close to those found in some ground waters of Mecklenburg-Western Pomerania. Coastal pore waters in front of the Hütelmoor area are impacted by subterrestril freshening due to SGD originating from the rewetted peatland. The estimated composition of this fresh water component differs substantially from the springs found close to Kühlungsborn, likely due to a substantial overprint by early diagenetic reactions taking place in the marine sediments and a hydrological cycle in the peatland that may differ from the one in the catchment of the ferruginous springs.

ACKNOWLEDGEMENTS

This study was partly conducted within the framework of the Research Training Group 'Baltic TRANSCOAST' funded by the DFG (Deutsche Forschungsgemeinschaft) under grant number GRK 2000 (www.baltic-transcoast.uni-rostock.de). The research was further supported by BMBF within the BONUS^{+} project AMBER, and Leibniz IOW. Thanks are due to R. Bahlo, A. Köhler, and I. Scherff for technical support. This is Baltic TRANSCOAST publication no. GRK2000/0009.

REFERENCES

Böttcher M.E., U. Mallast, G. Massmann, N. Moosdorf, M. Müller-Petke, and H. Waska. 2018. Coastal-Groundwater interfaces (submarine groundwater discharge). In (Krause S, ed.) Ecohydrological Interfaces, Wiley Science, in press.

Donis D., F. Janssen, B. Liu, F. Wenzhöfer, O. Dellwig, P. Escher, A. Spitzy, and M.E. Böttcher. 2017. Biogeochemical impact of submarine ground water discharge on coastal surface sands of the southern Baltic Sea. Est. Coast. Shelf Sci., 189: 131-142.

Knee K.L., and A. Paytan. 2011. Submarine groundwater discharge: A source of nutrients, metals, and pollutants to the coastal ocean. In: Wolanski E., D.S. McLusky. (eds.) Treatise on Estuarine and Coastal Science, 4: 205-233.

Kowalski N., O. Dellwig, M. Beck, M. Grunwald, T. Badewien, H.J. Brumsack, J.E.E. van Beusekom, and M.E. Böttcher .2012. A comparative study of manganese dynamics in pelagic and benthic parts of two tidal systems of the North Sea. Estuar. Coast. Shelf Sci., 100: 3-17.

Sültenfuß, J., M. Rhein, and W. Roether. 2009. The Bremen Mass Spectrometric Facility for the measurement of helium isotopes, neon, and tritium in water. *Isotopes in Environmental and Health Studies*, 45(2), 1-13.

Winde, V., M.E. Böttcher, P. Escher, P. Böning, M. Beck, G. Liebezeit, and B. Schneider. 2014. Tidal and spatial variations of DI^{13}C and aquatic chemistry in a temperate tidal basin during winter time. Journal of Marine Systems, 129: 394-402.

Zhang Y.-C., C.P. Slomp, H.P. Broers, H.F. Passier, M.E. Böttcher, E.O. Omoregie, J.R. Lloyd, D.A. Polya, and P. van Cappellen. 2012. Geochemical and microbial signatures of denitrification linked to pyrite oxidation in a sandy aquifer. Chem. Geol., 300-301: 123-132.

Contact Information: Michael E. Böttcher, Geochemistry & Isotope Biogeochemistry Group, Marine Geology Department, Leibniz IOW, D-18119 Warnemünde, FRG, email: michael.boettcher@io-warnemuende.de.

On the use of COMSOL Multiphysics for seawater intrusion in fractured coastal aquifers

B. Mozafari[1], M. Fahs[2], B. Ataie-Ashtiani[1,3], C. T. Simmons[3], Rafic Younes[4]
[1]Department of Civil Engineering, Sharif University of Technology, Tehran, Iran
[2]Laboratoire d'Hydrologie et Geochemie de Strasbourg, University of Strasbourg/ EOST/ENGEES, CNRS, Strasbourg, France
[3]National Centre for Groundwater Research & Training and College of Science & Engineering, Flinders University, Adelaide, Australia
[4]Faculty of Engineering, Lebanese University, Rafic Harriri Campus, Hadath, Beirut, Lebanon

ABSTRACT

COMSOL Multiphysics is a comprehensive simulation software environment for a wide range of applications. COMSOL has an interactive interface that facilitates the modeling procedure and allows an easy coupling of different physical processes. The Subsurface Flow module extends the COMSOL modeling environment to applications related to fluid flow in saturated and variably saturated porous media. COMSOL is increasingly used in the investigation of geophysical, hydrogeological and environmental phenomena. The main goal of this work is to explore the ability of COMSOL for simulating seawater intrusion (SWI) in fractured coastal aquifers. Numerical modeling of such a problem is of high interest as fractured/karstic coastal aquifers are widespread and processes of SWI in the presence of fractures remains poorly understood. We set up a COSMOL model for the popular Henry problem. The accuracy of COMSOL is highlighted by comparison against the semianalytical solutions for simple homogeneous aquifers. For fractured aquifers, the performance of COSMOL is evaluated by comparison against an in-house finite element model based on the discrete fracture model and against the results of existing works. Given its versatility and flexibility, COMSOL shows promise as a tool for SWI in coastal aquifers.

INTRODUCTION

Numerical models are considered as irreplaceable tools for the simulation of SWI in coastal aquifers *[Werner et al., 2013]*. Although several numerical codes are available, modeling SWI remains a vibrant research discipline of high interest of the scientific community. Modeling of SWI reveals several multifaceted challenges that are not common in other hydrogeological applications. These challenges are related to the physical processes that require coupling between variable-density flow and transport equations, the specific type of seaward boundary conditions and the variable velocity field induced by the salinity buoyant effects. The modeling challenges become more complex in fractured coastal aquifers where fractures require specific numerical techniques to be accurately handled. The objective of this work is to explore the ability of COMSOL Multiphysics *[COMSOL, 2012]* for simulating SWI in fractured coastal aquifers. COMSOL is a comprehensive simulation software environment for a wide range of applications. It is a user-friendly tool that facilitates all the modeling steps (preprocessing, meshing, solving and postprocessing). The primary advantage of COMSOL is that it easily allows coupling several physical processes together to include all the necessary factors for a complete model. COMSOL includes a module called "Subsurface Flow" devoted to the simulation of fluids flow in saturated and variably saturated porous media. This module allows for considering fractures via the discrete fracture network (DFN) model *[Sebben et al., 2015]* which assumes that fracture size is small compared to a representative elementary volume. With DFN, fractures are represented in (n-1) dimensional element embedded in an n-dimensional computational domain *[Ramasomanana et al., 2018]*. The use of COMSOL in applications related to hydrogeology is increasingly frequent. To the best of our knowledge, this software has been never applied to SWI in fractured aquifers. Hence, the capacity of COMSOL to deal with the challenges related to SWI in fractured coastal aquifers is worthy of investigation. In this context, we set up a COSMOL model for the popular Henry problem *[Henry, 1964; Sebben et al., 2015; Fahs et al., 2016]*. The accuracy of COMSOL is highlighted by comparison against the semianalytical solutions for simple homogeneous domains *[Fahs et al., 2016]*. For fractured domains, the performance of COSMOL is evaluated by comparison against an in-house finite element model based on the discrete fracture model (DFM) *[Younes et al., 2009; Ramasomanana et al., 2018]* and against the results of the Hydro-GeoSphere code published in *Sebben et al. [2015]*.

PROBLEM DESCRIPTION, GOVERINING EQUATIONS AND SOFTWARE

To evaluate the performance of COMSOL for modeling SWI we used the Henry problem *[Henry, 1964]* which is an abstraction of SWI in a vertical cross-section of a confined coastal aquifer perpendicular to the shoreline. In this aquifer, an inland freshwater flow is in natural equilibrium with the seawater intruded from the seaside due to its higher density (Figure 1). Henry problem is widely used as a surrogate for the understanding of SWI processes in coastal aquifers. Because of the existence of semianalytical solutions, Henry problem has been accepted as one of the primary benchmark for the assessment of SWI numerical codes. A new semianalytical solution has been developed in *Fahs et al. [2016]* for velocity dependent dispersion. This solution is used here to test the performance of COMSOL in the case of simple porous media. The standard Henry problem considers homogenous porous media. Recently *Sebben et al. [2015]* presented a new configuration by considering fractured coastal aquifer. Several fractures configurations have been suggested in *Sebben et al. [2015]*. Two among these configurations are used in this work. The first one deals with one horizontal fracture as in Figure 2.a while the second one consists of a network of fractures as in Figure 2.b.

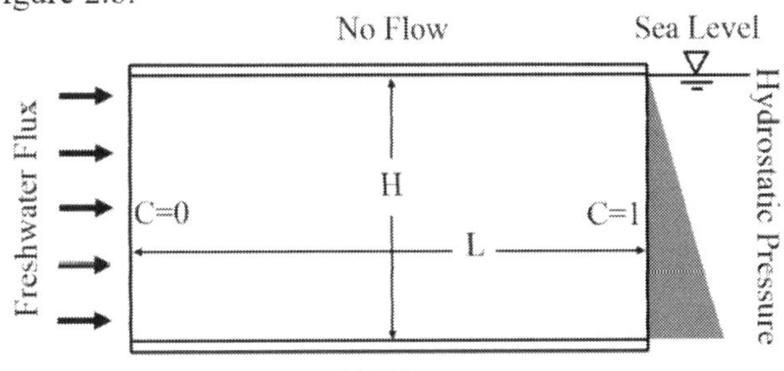

Figure 1. Henry problem domain and boundary conditions.

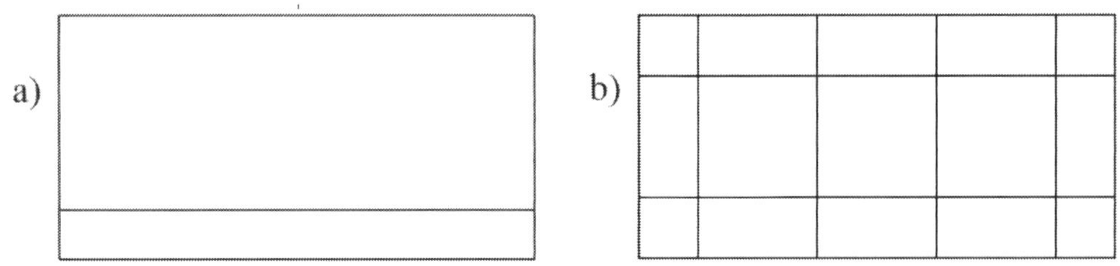

Figure 2. Two cases of fracture networks, a) A horizontal fracture, and b) Horizontal-Vertical fractures network.

In COMSOL, the Darcy's Law interface has been used for study water flow in the coastal aquifer. According to Darcy's Law, three factors affecting velocity field are the pressure gradient, the fluid viscosity, and the porous medium structure. Therefore, the net flux across a face of the porous surface is

$$u = -K\left(\nabla h + \frac{\rho - \rho_0}{\rho_0} e_z\right), \tag{1}$$

where u is the Darcy velocity *[m/s]*, h is hydraulic head *[m]*, K is hydraulic conductivity tensor *[m/s]*, e_z is a unit vector in the direction that gravity acts, ρ is fluid density *[kg/m^3]* that depends on concentration (c),

$$\rho = \rho_0\left(1 + \beta \frac{c}{c_s}\right), \tag{2}$$

where c_s is seawater concentration and $\beta = (\rho_s - \rho_0)/\rho_0$.
With COMSOL the fractures can be included using the DFN model. With this model the fractures are considered as lines (1D). The flow within the fractures is modeled using the Darcy's law with very high permeability.
In coupled interface with Darcy's law, the Transport of Diluted Species Interface provides a predefined modelling environment in which it is possible to examine the transport of chemical species by diffusion and convection, wherein diffusion follows Fick's law. The interface assumes that all existing species have been diluted. Due to the dilution, the properties of the mixture, such as density and viscosity, can be assumed to be consistent with solvents. The interface implements mass balance equation:

$$u.\nabla c = \nabla.(D\nabla c), \tag{3}$$

where c is the concentration of the species [mol/m^3], D denotes the hydrodynamic diffusion-dispersion tensor (m^2/s), u is the velocity vector (m/s), which is obtained using Darcy law as a coupled interface in this case. D is equal to $(D_D + \varepsilon D_m I)$, where D_D and D_m are, respectively, the mechanical dispersion (tensor) and molecular diffusion (scalar). I is an identity matrix and ε is the porosity. D_D is defined as below:

$$D_D^{xx} = \alpha_L \frac{u_x^2}{|u|} + \alpha_T \frac{u_y^2}{|u|}, D_D^{yy} = \alpha_T \frac{u_x^2}{|u|} + \alpha_L \frac{u_y^2}{|u|}, D_D^{xy} = D_D^{yx} = (\alpha_L - \alpha_T)\frac{u_x u_y}{|u|} \tag{4}$$

RESULTS

Simple porous media: Comparison against the semi analytical solutions

We first checked the performance of COMSOL for SWI in simple porous media. To do so we compared COMSOL against the semianalytical solutions presented in *Fahs et al. [2016]*. We checked the three cases presented in that paper. The parameters considered in the three cases are given in table 1. The main isochlors (25%, 50% and 75%) obtained with COMSOL as well as the semianalytical ones are given in Figure 3. This figure shows excellent agreement between COMSOL and the semianalytical solution and highlights the accuracy of COMSOL. The third test case is computationally challenged as the transport processes are dominated by advection. The finer grid level generated by the COMSOL meshing tool is used to obtain a solution matching with the semianalytical solution. For coarser mesh COMSOL can generate unphysical oscillations that lead to convergence problems.

Table 1. Physical parameters used in COMSOL.

Parameter	Value				
	Simple aquifer			Fractured aquifer	
Length (L, m)	3			2	
Height (H, m)	1			1	
Gravity (g, m/s^2)	9.81			9.81	
Freshwater density (ρ_f, kg/m^3)	1000			1000	
Seawater density (ρ_s, kg/m^3)	1025			1025	
Freshwater concentration (c_0, mol/m^3)	0			0	
Seawater concentration (c_s, mol/m^3)	1			1	
Porosity (ε, –)	0.35			0.2	
Viscosity (μ,kg/m/s)	10^{-3}			10^{-3}	
Permeability (κ,m^2)	1.0204×10^{-9}			-	
Freshwater inflow velocity (v, m/s)	6.6×10^{-5}			6.6×10^{-6}	
	Test case 1	Test case 2	Test case 3	Case A	Case B
Molecular diffusion coefficient (D_m, m^2/s)	18.86×10^{-6}	9.43×10^{-8}	9.43×10^{-8}	10^{-9}	10^{-9}
Matrix Longitudinal dispersivity (α_L, m)	0	0.1	0.001	0.1	0.1
Matrix Transverse dispersivity (α_T, m)	0	0.01	0.0001	0.01	0.01
Horizontal hydraulic conductivity (K_x, m/s)	-	-	-	2.5×10^{-4}	2.5×10^{-4}
Vertical hydraulic conductivity (K_y, m/s)	-	-	-	2.5×10^{-4}	2.5×10^{-4}
Fracture Longitudinal dispersivity (α_{LF}, m)	-	-	-	0.1	0.1
Fracture conductivity (K_f, m/s)	-	-	-	0.772	0.122
Fracture aperture (2b, m)	-	-	-	9.71×10^{-4}	3.85×10^{-4}

Fractured porous media

The parameters used for the fractured configurations are given in table 1. Steady state simulations have been performed. When regular triangular meshes are used COMSOL encounters convergence difficulties. To avoid this problem we use the option boundary later mesh. When creating a boundary layer mesh, COMSOL creates an initial mesh in which the boundary layer elements are inserted. Example of boundary layer mesh is given in Figure 4. COMSOL results have been compared against an advanced in-house numerical model based on the discrete fracture model in which the fractures are considered as 2D domain *[Younes et al., 2009; Ramasomanana et al., 2018]*. The results (not presented for sake of brevity) show excellent agreement between COMSOL and the in-house code which confirms the ability of DFN model to simulate SWI in fractured coastal aquifers. The COMSOL resulting isochlors for both configurations are given in Figure 5. This figure shows similar results as the ones obtained by *Sebben et al. [2015]* using the Hydro-GeoSphere code (see Figure 4 in their paper). It should be noted that for the comparison with Sebben et al. [2015], variable concentration boundary conditions are imposed at the seaside. Despite that this boundary condition is not common; it has been implemented directly in COMSOL via the interactive interface.

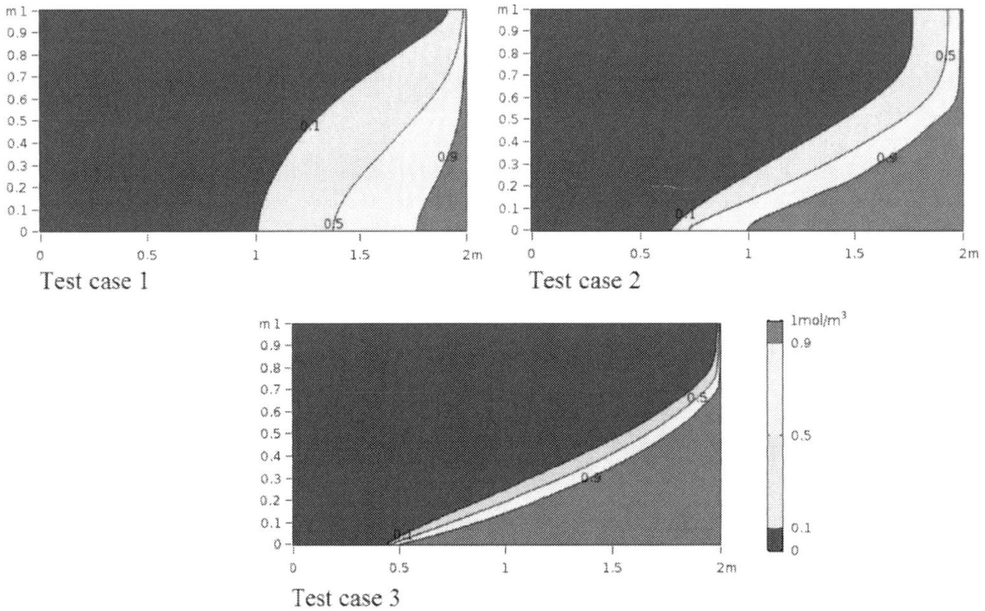

Figure 3. Comparison of COMSOL against the semi analytical solutions by Fahs et al. [2016]. Main isochlors (75%, 50% and 25%) with COMSOL (Flood) and semianalytical (lines).

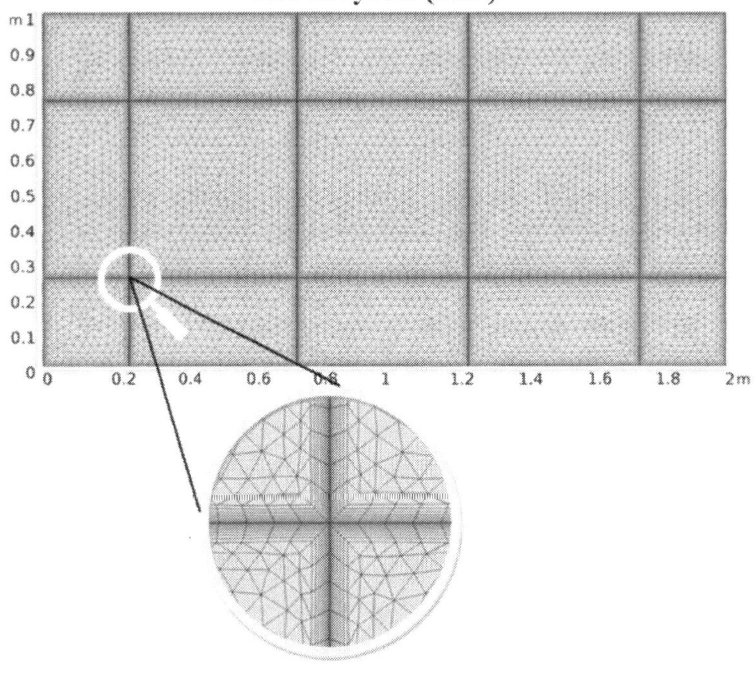

Figure 4. The boundary layer mesh used in COMSOL simulations.

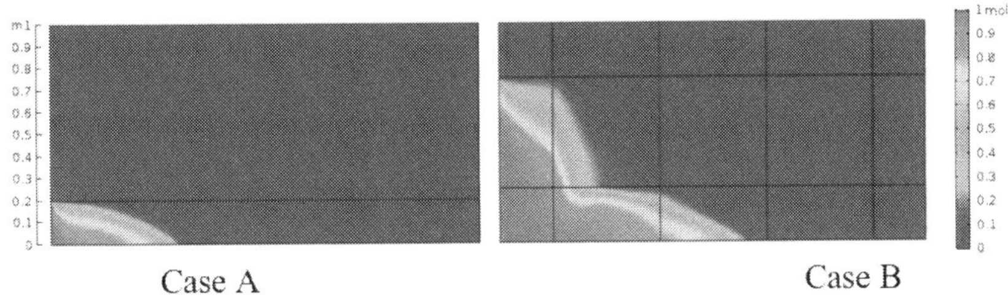

Figure 5. Concentration distribution obtained with COMSOL in the case of fractured porous media.

CONCLUSIONS

In this paper, Henry problem is used to explore the ability of COMSOL Multiphysics for simulating SWI in fractured coastal aquifers. COMSOL results are compared against semianalytical solutions, an in-house numerical code and existing results based on Hydro-Geosphere code. Results show that the DFN model allows accurate simulations of the SWI in coastal aquifer. The benchmark results show that the COMSOL software has the capability to simulate subsurface flow and it is a useful tool for the development of the conceptual models of SWI in fractured aquifers. The user interface in this software facilitates modelling process: defining geometry, mesh generation, PDE equations editing and coupling, solving, and post-processing. COMSOL is also flexible in including variable concentration boundary conditions at the seaside.

REFERENCES

Werner, A. D., M. Bakker, V. E. A. Post, A. Vandenbohede, C. Lu, B. Ataie-Ashtiani, C. T. Simmons, D. A. Barry, 2013. Seawater intrusion processes, investigation and management: Recent advances and future challenges. *Advances in Water Resources 51, 3–26,* doi:10.1016/j.advwatres.2012.03.004.

COMSOL Multiphysics, 2012. User's Guide, Version 4.3. Comsol Inc.

Sebben, M. L., A. D. Werner, and T. Graf. 2015. Seawater intrusion in fractured coastal aquifers: A preliminary numerical investigation using fractured Henry problem. *Advances in water resources, 85, 93-108.*

Ramasomanana, F., M. Fahs, H. M. Baalousha, N. Barth and S. Ahzi, 2018. An Efficient ELLAM Implementation for Modeling Solute Transport in Fractured Porous Media, Water Air Soil Pollut., 229:46, https://doi.org/10.1007/s11270-018-3690-8.

Fahs, M., B. Ataie-Ashtiani, A. Younes, C.T. Simmons, and P. Ackerer. 2016. The Henry problem: New semianalytical solution for velocity-dependent dispersion. *Water Resources Research, 52,* 7382–7407, doi:10.1002/2016WR019288.

Henry, H. 1964. Effects of dispersion on salt encroachment in coastal aquifer. *U.S. Geol. Surv. Water Supply Pap., 1613,* C70–C84.

Younes, A., M. Fahs, and S. Ahmed. 2009. Solving density driven flow problems with efficient spatial discretizations and higher-order time integration methods. *Advances in Water Resources, 32,* 340–352, doi:10.1016/j.advwatres.2008.11.003.

Contact Information: M. Fahs, Laboratoire d'hydrologie et de Géochimie de Strasbourg, 1 Rue Blessig, 67084, Strasbourg. Phone : 00 33 3 68 85 04 48. Email : fahs@unistra.fr

E3S Web of Conferences 54, 00021 (2018) https://doi.org/10.1051/e3sconf/20185400021
SWIM 2018

Monitoring the groundwater quality/quantity from your desktop – application to salt water intrusion monitoring EMI: Environmental data Management Interface

D. Neyens[1], M. Baïsset[1], H. Lovighi[1]
[1]imaGeau, Clapiers, France

ABSTRACT

More and more data of groundwater quantity and quality is acquired and exists. The means to collect it (GPRS, Radio, IoT...) have been developed during the last years. Storing this data is not a problem. But, having a mean to have all the different data gathered in the same platform with the possibility to display and analyse them in a simple and rapid way doesn't exist enough. A web application has been developed to answer this problem. This application enables to store all type of groundwater data (water table elevation, groundwater quality, borehole information, water extraction data…) in order to display it. The organisation of this application allows displaying all the data by site. Several tools to analyse the data are included (calculation of hydrogeological parameters, statistics, interpolation) allowing a first data interpretation, in particular for salt water intrusion issues. This web application was born from collaboration between IT experts and environmental engineers.

INTRODUCTION

Connected instrumentation development, ioT objects for sensors and monitoring devices for groundwater, has met spectacular progress in the recent years. Consequently, more and more groundwater quality/quantity data is easily available.

The new challenge is not to collect data but finding a way to manage this huge data amount and having a rapid view of them. It is also to find a tool giving the possibility of analysis and allowing global indices generation for aquifers monitoring.

A purpose design online software platform (using an Internet browser), the Environmental Monitoring Interface (***EMI***), has been developed to collate and display information about the aquifer. It includes full description of water dynamics and has specific tools for salt-water intrusion data management. Current and historic data are available 24/7.

ARCHITECTURE

EMI application is based on two identities:

- A database where all the data coming from sensors installed in different sites is stored. Different processes to import data are available (several format files, API…)
- A web application allowing viewing and processing data

 https://doi.org/10.1051/e3sconf/20185400021

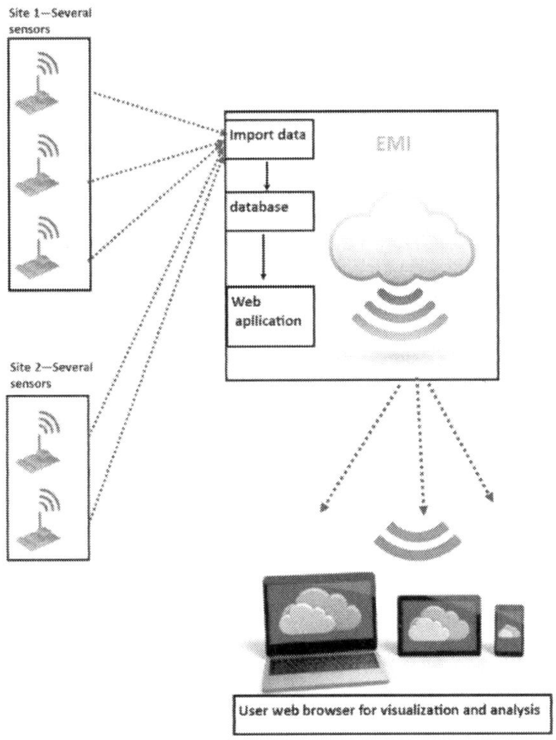

Figure 1: EMI architecture.

EMI can accept and integrate data from every monitoring tools already installed on the well field (whatever the tool and the manufacturing), such as flow meters, water level sensors, water quality sensors, pluviometers…

FEATURES

To enable accessing to the data online, an interface has been developed allowing to:

- viewing data on graphs (value parameter versus time)
- processing the data
    - calculate hydrogeological parameters
    - have statistics on data
    - interpolate the data on 3D graphs
- export, if necessary, a large samples of preformatted data in order to use it in other programs
- setting alerts and thresholds on some parameters (water level…)

EMI organization is based on tree architecture. For each defined site, several monitoring stations can be associated; and for each station, several sensors with several measured parameters can be defined.

The site overview can be used to locate the monitoring tools connected to EMI.

 https://doi.org/10.1051/e3sconf/20185400021

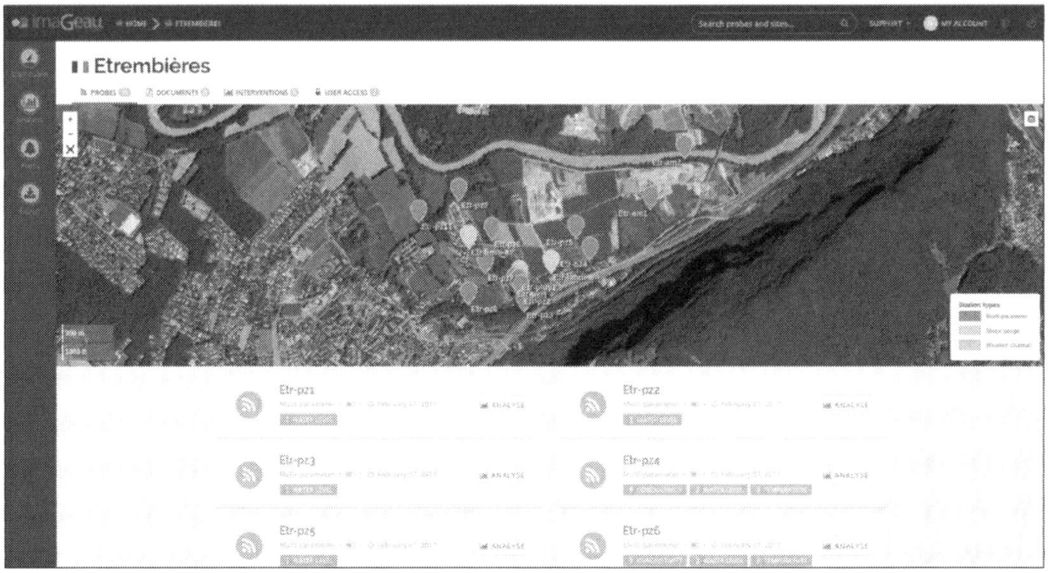

Figure 2: Site overview with different monitoring stations (dark blue: multi-parameters sensors (Cw, water level T°); light green: rain gauge; blue: water level sensor).

This organization allows having all the data associated to the same site in a same place in order to visualise and analyse it with different process. For example, on a site with several boreholes equipped with water level elevation sensors, all the water level elevation curves can be plotted on the same graph. If a rain gauge is installed on the site, the water level can be plotted in front of this rain gauge allowing having data cross-correlation.

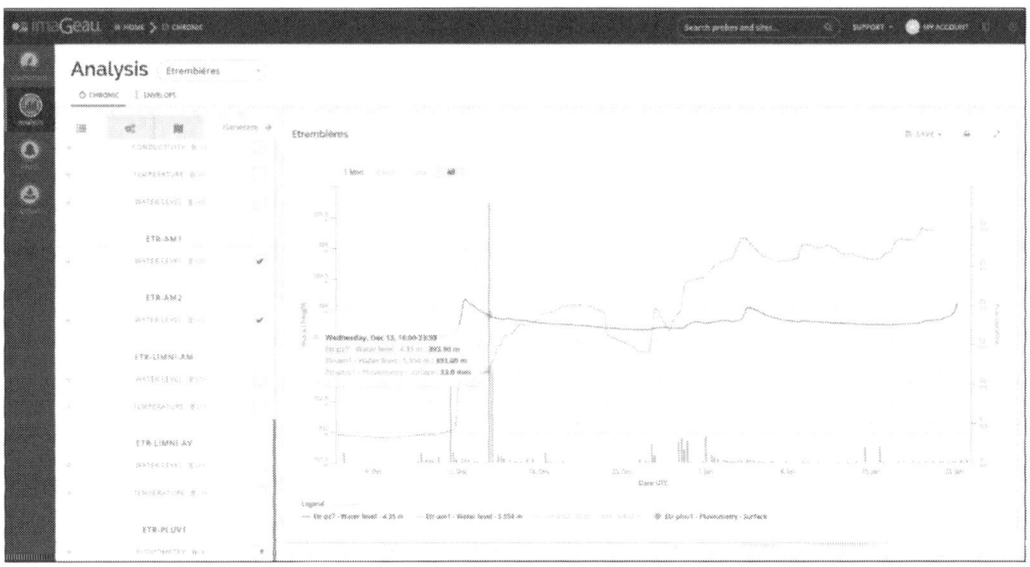

Figure 3: Site overview with different monitoring stations (dark blue: multi-parameters sensors (Cw, water level T°); light blue: rain gauge; green: water level sensor).

ANALYSIS TOOLS

Different viewing/analysis tools are available in order to better analyze the data.
Depending on the sensor type, different analyses are available.

For single point sensor (i.e. water level sensor, quality sensor…), data can be plotted with:

- **"Temporal Chronic"**: Used to view environmental data changes in graph form. Up to ten curves can be added to the graph and multiple parameters can be viewed through linked graphs.

- **"Envelop"**: used to plot actual data relative to historic data (i.e. mean to see the evolution of the water level relative to anterior years)

When multiple parameters are measured at the same measurement station, it is possible to display them simultaneously on superimposed graphs with the same timescales. The browser's cursor can be used to directly compare different parameters at the same "T time."

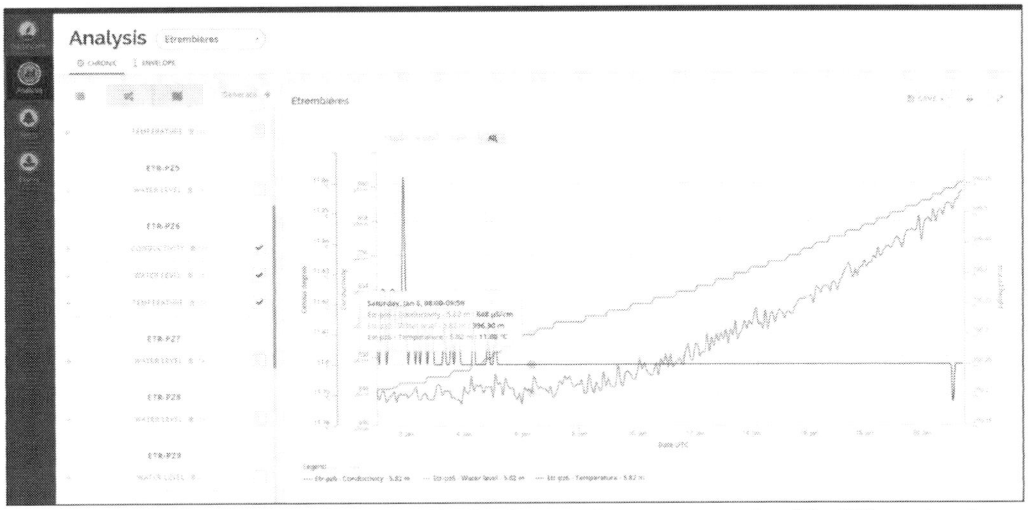

Figure 4: Mutli-parameters curves displaying on the same graph with different colours and scales depending of the parameter (black: T°, blue: conductivity, orange: water level).

For multi-points sensors (i.e. SMD), in addition, other analyses are available:

- **"Interpolation"**: Used to interpolate multi-point data in order to create temporary iso-value maps, example: display the map of water conductivity changes over the whole aquifer block over a one-year period.
- **"Log"**: Used to view multi-point data in the form of a profile at a given date or over a given time interval.
- **"Saltwater intrusion interface"**: In the case of a conductivity profile measured using imaGeau systems (SMD), the saltwater intrusion interface can be used to follow over time the depth of an interface defined by the user, e.g.: change in depth of the fresh water/salt water interface (limit 5000 µS/cm). If multiple conductivity sensors are measuring changes over the entire vertical of a piezometer, it is useful to track the evolution of the depth of the interface between two masses of water of differing salinities. In the case of saltwater intrusion tracking, the depth of the

https://doi.org/10.1051/e3sconf/20185400021

saltwater intrusion (freshwater/saltwater interface) is an essential parameter for understanding the evolution of the saltwater intrusion.

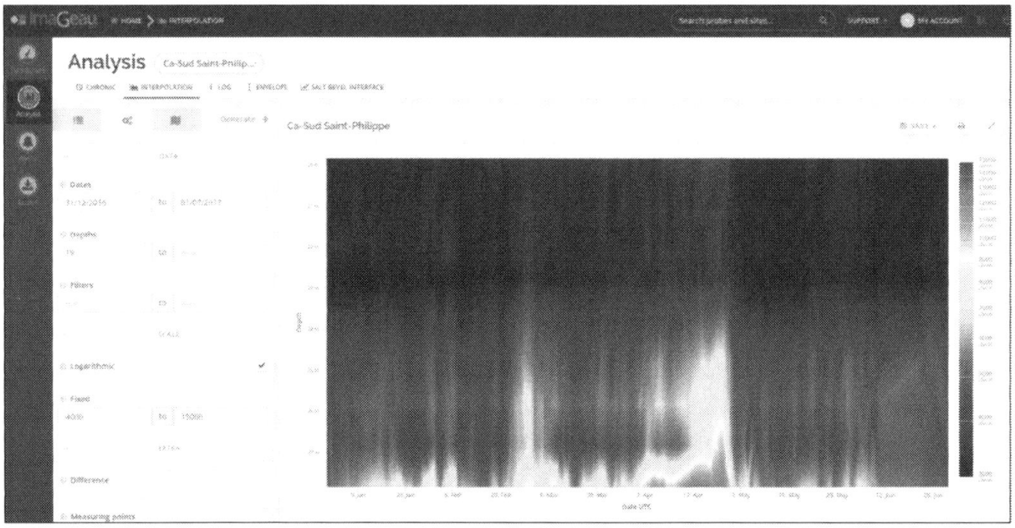

Figure 5: 3D interpolation generated from multi point data over vertical and time, example of SMD monitoring a salt water intrusion.

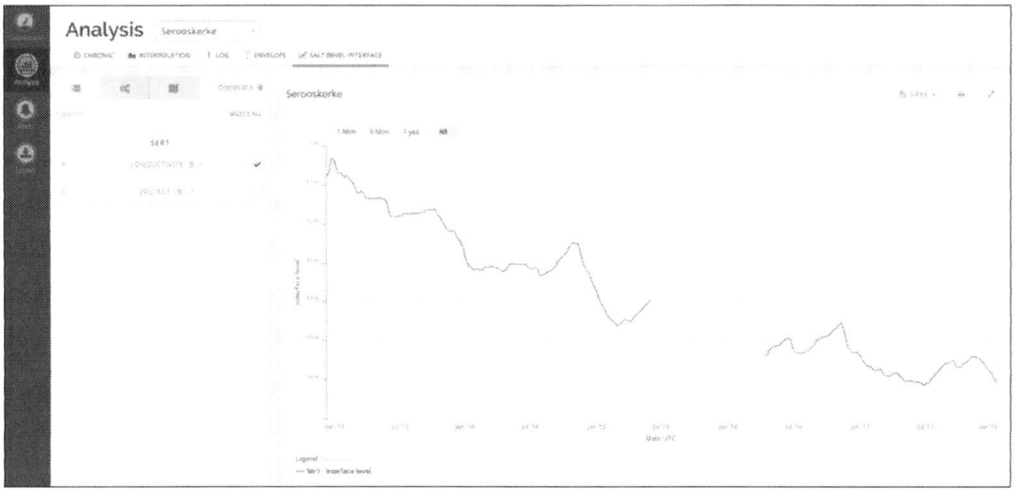

Figure 6: Freshwater/saltwater interface (8000 µS/cm value versus time) - calculated graph from a site in the Netherlands.

CONCLUSION

This web application was born out of collaboration between IT experts and environmental engineers so that its consistency and ergonomics would best meet the monitoring and operational needs. This service is used by water companies, water policy, governmental services and research institutes to monitor and manage coastal aquifers in real time, in particular in case of salt water intrusion monitoring. Several upgrades are in progress to take into account the groundwater data from a global point of view, working on water identity in regional scale. EMI uses "responsive" technology, which enables it to be used on any platform: PC, Mac, tablet, and Smartphone.

E3S Web of Conferences 54, 00022 (2018) https://doi.org/10.1051/e3sconf/20185400022
SWIM 2018

Can bomb-peak tritium persist in the transition zone? A case study from the German island of Langeoog

V.E.A. Post[1], G.J. Houben[1]
[1]BGR, Hannover, Germany

ABSTRACT

Tritium has been used as a tracer in hydrogeological studies to date young groundwater. It was released in massive amounts to the atmosphere by nuclear bomb testing in the 1950's and early 1960's. Its activity in the atmosphere peaked around 1963-1964 and has been steadily falling since then due to its half-life of 12.32 years. On the northern hemisphere, where the atmospheric tritium concentration reached much higher levels than on the southern hemisphere, elevated tritium activities in groundwater persist, and thus can still be used to identify groundwater that has recharged during the post-bomb era.

Tritium has also been encountered in brackish and saline groundwater in coastal aquifers. Stuyfzand *et al.* (2012) found ^{3}H values of 30.3 TU in recently intruded North Sea water directly below the beach near the city of The Hague in the Netherlands. These high values are linked to the discharge from nuclear reprocessing plants in the UK and France (Nies *et al.*, 2010). Lower, but still above natural background, values ($3.3 < {}^{3}H < 23.9$ TU) were measured in samples from a wedge of brackish groundwater at the same study site. This wedge was therefore believed to have intruded during the period of maximum groundwater extraction around 1956. Bryan *et al.* (2016) studied a freshwater lens system on Rottnest Island off the coast of Perth in Western Australia. All but one saline groundwater samples had tritium concentrations below the detection limit. The one exception had a ^{3}H activity of 0.67 TU, which was believed to indicate relatively recent intrusion of seawater.

Sivan *et al.* (2005) studied the distribution of tritium in the mixing zone of a coastal aquifer south of Tel Aviv, Israel. Similar to Bryan *et al.* (2016), they interpreted the presence of measurable tritium in their saline groundwater samples as an indication for seawater intrusion during the 60 years prior to their study. However, they also considered the possibility that the tritium in the transition zone was not derived from the intruded seawater but from the overlying freshwater. That is, old, tritium-free seawater may obtain tritium by mixing with young, tritium-containing freshwater, especially if the freshwater was impacted by the bomb peak.

On the island of Spiekeroog, located in the northern part of Germany, Röper *et al.* (2012) detected virtually no ^{3}H activity in saline groundwater samples from depths of 60 m below sea level or more, and they inferred an age of over 70 years. Chloride concentrations and stable water isotope values indicated that the samples contained up to 35% freshwater, but apparently this mixing did not contribute appreciable amounts of tritium. By contrast, previously-unpublished data from the neighbouring island of Langeoog show that groundwater in the transition zone beneath the freshwater lens has elevated ^{3}H activities. Interestingly, the freshwater tritium activities have been found to decrease with depth to values < 2 TU at the base of the lens (Houben *et al.*, 2014), while in the transition zone tritium values increased to up to 5.7 TU again (Figure 1c).

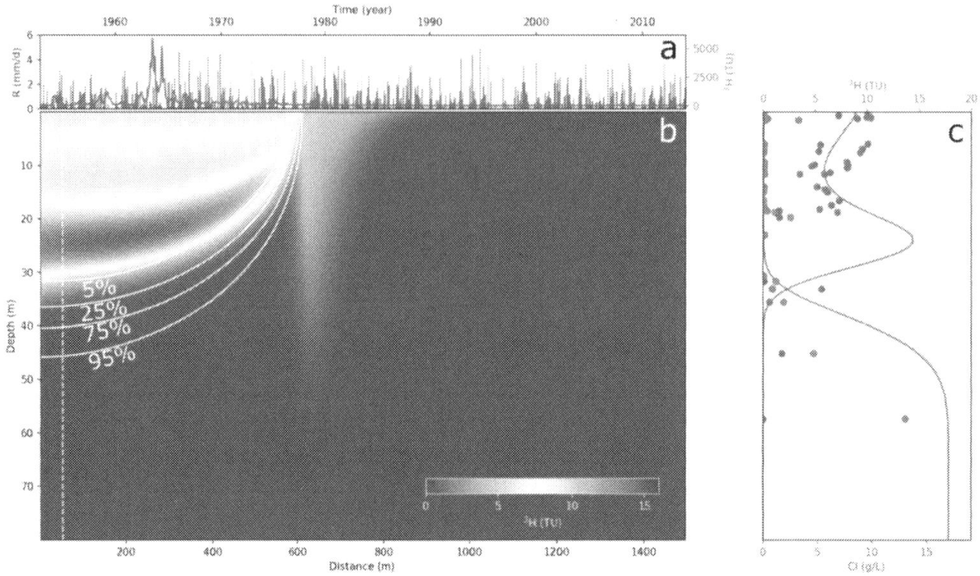

Figure 1. Graph showing the monthly recharge rate (blue) and ^{3}H activities (green) versus time. (b) Modeled contour plots of the ^{3}H activity (colours) and percentage of seawater (white contour lines). (c) Graph showing the ^{3}H activity (green) and chloride concentrations (red) as a function of depth along the white stippled line in (b). Lines indicate model results, data points represent samples from various locations across the lens from *Houben et al. (2014)*.

Numerical simulations were conducted to explore the possibility that the tritium in the transition zone is a remnant from the atmospheric bomb peak. Groundwater recharge was calculated based on rainfall and evaporation data and was varied using monthly time steps as described in Post & Houben (2017). A two-dimensional section across the freshwater lens was considered. The transient simulation spanned the 61 year period from 1 January 1953 (the earliest year for which the atmospheric tritium input could be estimated) to 31 December 2013. The meteoric water recharging the freshwater lens was assigned a ^{3}H activity based on measured activities of nearby rainfall stations of the global network of isotopes in precipitation (GNIP) network (IAEA/WMO, 2016). To generate a continuous input time series, the partial records of the German stations in Stuttgart, Braunschweig and Cuxhaven as well as the Dutch station in Groningen were collated. For the period prior to 1961 the data from Ottawa, Canada were substituted. The overall trends measured at the different stations are similar, but there can be appreciable differences in the ^{3}H activity values for any given month. Despite the uncertainty thus introduced, this approach was considered to be the most appropriate to generate the longest-possible input time series required for the model simulations.

The initial conditions for the model were determined by simulating the same 61 year period with all model parameters unchanged except for the ^{3}H activity of the meteoric recharge, which rather than being variable was assigned a constant value of 5 TU. The latter simulation also served to provide the initial concentrations for groundwater salinity.

Preliminary simulations show that it is indeed possible for atmospheric tritium to migrate to the bottom of the circa 35 m thick freshwater lens and mix with seawater within the duration of the simulation period (Figure 1b). The tritium bomb peak itself is still located within the freshwater part of the lens, but some tritium has mixed with the brackish water in the transition zone. The model under-estimates the thickness of the lens to some degree (Figure 1c), which is presumably due to the adopted recharge rates being too low. Nevertheless, the general trend of the ^{3}H activity with depth across the lens as well as the elevated ^{3}H activity at the upper limit of the transition zone displayed by the data points are also visible in the model simulations (Figure 1c).

Great uncertainty exists though about the effect of abstraction, which has not yet been considered, because of the two-dimensional nature of the model, yet it is believed to be an important factor on Langeoog. It is therefore not possible to draw site-specific conclusions for the island, but the model simulations are general proof-of-concept that tritium in the transition zone can be used to study mixing between fresh- and saltwater in coastal aquifers. This means that (as long as the bomb-peak can be detected) tritium may be used to infer timescales of mixing, but potentially also to better constrain the dispersivity in numerical models of coastal aquifers.

The model simulations also showed the development of a zone of elevated ^{3}H activities below the seafloor, which protrudes much deeper than the base of the freshwater lens. This is attributed to the intrusion of seawater, which is driven by the loss of solutes from the system, as dispersive mixing causes salt to be sequestered into the seaward directed groundwater flow that discharges near the coastline. The shape of this tritium plume shows that the intrusion can have a strong vertical component and thus that the determination of seawater intrusion velocities based on tritium activities measured along a shore-perpendicular transect needs to be done with great care. Interpretations that do not take this vertical motion into account could yield erroneous results. To our knowledge, field evidence of this deep zone has not been documented so far.

REFERENCES

Bryan E., K.T. Meredith, A. Baker, V.E.A. Post, M.S. Andersen, 2016. Island groundwater resources, impacts of abstraction and a drying climate: Rottnest Island, Western Australia. Journal of Hydrology 542: 704-718.

Houben G.J., P. Koeniger and J. Sültenfuß, 2014. Freshwater lenses as archive of climate, groundwater recharge, and hydrochemical evolution: Insights from depth-specific water isotope analysis and age determination on the island of Langeoog, Germany. Water Resources Research 50: 8227-8239.

IAEA/WMO, 2018. Global Network of Isotopes in Precipitation. The GNIP Database. Accessible at: http://www.iaea.org/water. Accessed on 16 December 2016.

Nies, H., I. Goroncy, J. Herrmann, R. Michel, A. Daraoui, M. Gorny, D. Jakob, R. Rachse, L. Tosch, S.P. Nielsen, M. Dawdall, A.L. Rudjord, T. Gäfvert, H.-A. Synal, M. Stocker and V. Alfimov, 2010. Kartierung von Tc-99, I-129 und I-127 im Oberflächenwasser der Nordsee [*Mapping of Tc-99, I-129 and I-127 in the surface water of the North Sea*], Bundesamt für Strahlenschutz, Salzgitter, 204 pp.

Post V.E.A. and G.J. Houben, 2017. Density-driven vertical transport of saltwater through the freshwater lens on the island of Baltrum (Germany) following the 1962 storm flood. Journal of Hydrology 551: 689-702.

Röper T., K.F. Kröger, H. Meyer, J. Sültenfuss, J. Greskowiak and G. Massmann, 2012. Groundwater ages, recharge conditions and hydrochemical evolution of a barrier island freshwater lens (Spiekeroog, Northern Germany). Journal of Hydrology 454-455: 173-186.

Sivan O., Y. Yechieli, B. Herut and B. Lazar, 2005 Geochemical evolution and timescale of seawater intrusion into the coastal aquifer of Israel. Geochimica et Cosmochimica Acta 69: 579-592.

Contact Information: Vincent Post, Federal Institute for Geosciences and Natural Resources (BGR), Stilleweg 2, 30655 Hannover, Germany, Phone: +49-511-6432393, Email: vincent.post@bgr.de

Assessing groundwater vulnerability to sea water intrusion in the coastline of the inner Puck Bay using GALDIT method

Dawid Potrykus[1], Anna Gumuła-Kawęcka[1], Beata Jaworska-Szulc[1], Małgorzata Pruszkowska-Caceres[1], Adam Szymkiewicz[1] and L. Dzierzbicka-Głowacka[2]
[1]Faculty of Civil and Environmental Engineering, Gdańsk University of Technology, Gdańsk, Poland
[2]Physical Oceanography Department, Institute of Oceanology PAS, Powstańców Warszawy 55, 81-712 Sopot, Poland

ABSTRACT

In this research, GALDIT method was used to assess seawater intrusion in the coastal aquifer of the inner Puck Bay (Southern Baltic Sea). The impact of potential sea-level rise on groundwater vulnerability for years 2081-2100 was also considered. The study area was categorized into three classes of vulnerability: low, moderate and high. The most vulnerable area is the Hel Peninsula with northern part of the Kashubian Coastland. Increased class of aquifer vulnerability is also adopted to glacial valleys. The results of this research revealed that about 18.9% of the analyzed area is highly vulnerable to seawater intrusion, 25.3% is moderately vulnerable and 55.8% is potentially at low risk. The simulated scenario of predicted sea level rise shows enlargement of high vulnerability areas.

INTRODUCTION

The deterioration of groundwater resources due to seawater intrusion is a significant problem for the water supply in coastal areas. Seawater intrusion is mostly a result of sea-level rise and groundwater overexploitation that causes coastal aquifers severely vulnerable to salinization. This phenomenon has a degrading influence on the availability of fresh water, ecosystems condition, coastal infrastructure and tourism. Potential implications of sea-level rise for coastline in Poland has been subject of several studies (Pruszak and Zawadzka 2008; Moskalewicz 2016; Paprotny and Terefenko 2017). In order to control seawater intrusion in coastal aquifers, a number of management strategies and prevention methods have been developed (Chachadi and Lobo-Ferreira 2001; Kallioras et al. 2012, Trabelsi et al. 2016). One of the principal approaches to predict the influence of seawater intrusion to groundwater is analysis of vulnerability maps. The most common worldwide method of assessing groundwater vulnerability to seawater intrusion is GALDIT. It is a numerical ranking system, based on six hydrogeological, hydrological and morphological parameters with appropriately assigned weights and ranges (Chachadi and Lobo-Ferreira 2001, 2007; Lobo-Ferreira et al. 2007). The method (or its modifications) has been applied in many coastal regions (Chachadi and Lobo-Ferreira 2007; Louma et al. 2017; Kazakis et al. 2018). In this paper, GALDIT method was used to assess seawater intrusion in the coastal aquifer of the inner Puck Bay (Southern Baltic Sea). The impact of potential sea-level rise to groundwater vulnerability for years 2081-2100 was also considered.

AREA OF STUDY

The research area is situated along the coastline of the inner Puck Bay, in the eastern part of the Kashubian Coastland and western part (13 km) of the Hel Peninsula, covering an area of 85,2 km^2 (Fig. 1). The vicinity of Baltic Sea creates a specific marine climate in this region, which characteristic features are moderate winters and mild summers. The annual temperature is around 7.4°C and the average amount of precipitation does not exceed 700 mm. The income of local inhabitants comes mainly from agriculture, fishery and tourism, related to the seaside location and unique environmental values of this region.

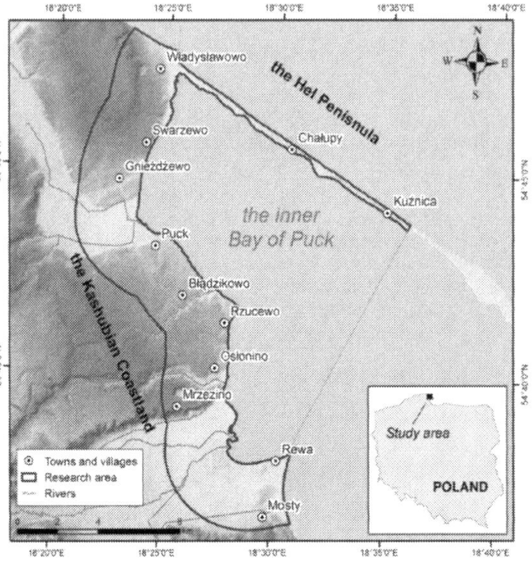

Figure 1. Research area.

The Kashubian Coastland mesoregion has a varying topography, typical for young glacial landscape. Morainic uplands are separated by deeply cut ice marginal valleys. Moraines consist of till deposits, which are divided by series of fluvioglacial sands and gravels. Glacial valleys are covered by extensive peat wetlands and consist of fluvioglacial sediments. The Hel Peninsula was formed during the Holocene under the influence of aeolian and marine processes. This structure is a sandbar, which separates the Puck Bay from the open Baltic Sea. With reference to the coastal system classification (©EUCC), the landscape of Hel Peninsula is a wave-dominated sedimentary plain with dune coasts in microtidal zone (Furmańczyk 2004). Hydrogeological conditions depend mostly on terrain relief and geological setting. In the Kashubian Coastland two aquifers have been identified (Frączek 1998). They consist of fluvioglacial sediments series (sands and gravels) which separate till deposits. Both aquifers are hydraulically connected. In the Hel Peninsula fresh groundwater occurs as a lens in sandy aquifer, which directly overlies pleistocene silty sand sediments. This groundwater cannot be used for water supply, however it is exploited locally and temporally.

APPLIED METHODOLOGY AND MATERIALS

Assessment of groundwater vulnerability in the studied area has been carried out using GALDIT method. Proposed by Chachadi and Lobo Ferreira (2001), GALDIT method is a numerical ranking system, based on the most important map-able factors that control the seawater intrusion: (*G*) groundwater occurrence, (*A*) aquifer hydraulic conductivity, (*L*) height of groundwater level above sea level, (*D*) distance from the shore, (*I*) impact of existing status of seawater intrusion and (*T*) thickness of the aquifer. Relative importance of seawater intrusion influence is determined by weight established for each factor and assigned numeric rating, based on its susceptibility to seawater intrusion. The rating and weight values used in calculations are presented in Table 1. Applying the related weight and

 https://doi.org/10.1051/e3sconf/20185400023

importance rating thematic maps were generated for each GALDIT parameter (Fig. 2). The final map of groundwater vulnerability to seawater intrusion (Fig. 3) was obtained by overlying six weighted maps with the following equation (Chachadi and Lobo-Ferreira 2007):

$$GALDIT_{index} = \frac{\sum_{i=1}^{6}(W_i \cdot R_i)}{\sum_{i=1}^{6} W_i} \quad (1)$$

where W_i is the weight of the ith indicator and R_i is the importance rating of the ith indicator. The final GALDIT index ranges from 2.5 to 10 and is divided into three vulnerability classes: high (>7.5), moderate (5-7.5) and low (<5). Higher values of importance ratings of the variable, corresponds to more vulnerable aquifers to seawater intrusion. Detailed description of GALDIT methodology was presented by Chachadi and Lobo-Ferreira (2001, 2007).

Table 1. Summary of GALDIT parameter weight, rates and ranges.

Indicator	*G*	*A*	*L*	*D*	*I*	*T*	Impor -tance Ratin
Weight	1	3	4	4	1	2	
Unit	[-]	[m/d]	[m]	[m]	[epm]	[m]	
Indicator variables	Confined Aquifer	>40	<1.0	<500	>2.0	>10	10
	Unconfined Aquifer	10-40	1.0-1.5	500-750	1.5-2.0	7.5-10	7.5
	Leaky confined Aquifer	5-10	1.5-2.0	750-1000	1.0-1.5	5-7.5	5
	Bounded Aquifer	<5	>2.0	>1000	<1.0	<5	2.5

Various multidisciplinary data are required for groundwater vulnerability assessment. The investigation was carried out using data from the Central Hydrogeological Data Bank – The HYDRO Bank of Polish Geological Institute – National Research Institute: 235 boreholes, digital elevation model (DEM) and hydrogeological maps. Spatial distributions of each GALDIT parameter in 3 km buffer of the inner Puck Bay coastline were integrated with GIS environment and mapped using geostatistical interpolation techniques and GIS analyst tools. Final vulnerability index were estimated by using *raster calculator* tool.

RESULTS AND DISCUSSION

Thematic maps of GALDIT parameters obtained for the research area are presented in Fig. 2. Groundwater occurs (*G*) as confined as well as unconfined aquifers. The hydrogeological system is confined mainly in the upland area and in that part highest vulnerability rank – 10 was assigned. Aquifers in the remaining area are unconfined thus in this part rating of 7.5 was applied (Fig. 2-G). The aquifers hydraulic conductivity (*A*) varies from 6 to 120 m/d. Therefore, the study area was classified into three classes of vulnerability with ranges 5, 7.5 and 10 (Fig. 2-A). The major part of studied area was rated with 7.5 and locally within upland area with 5 or 10. In reference to the height of groundwater level above mean sea level (*L*), the considered coastal area was categorized into four classes of vulnerability and rated between 2.5-10 (Fig. 2-L). The lowest values of rating are observed in upland area and increases in discharge area direction (Bay of Puck, open Baltic Sea and glacial valleys). The distance from the shore (*D*) was estimated according to three distances to the coastline ranges: 500, 750 and 1000 m. The maximum value (10) is assigned for areas situated close to the coast (distance lower than 500 m), whereas it decreases as distance inland increases (Fig. 2-D). According to poor availability of groundwater quality data it was difficult to reliably estimate impact of existing status of seawater intrusion (*I*) in entirely area. To evaluate this parameter the ratio of Cl^- / HCO_3^- (Lobo-Ferreira et al. 2007) were considered instead of recommended Revelle index ($Cl / [HCO_3^- + CO_3^{2-}]$ (Chachadi and Lobo-Ferreira 2007). The impact of parameter *I* was divided into four classes and rated from 2.5 to 10 (Fig. 2-I). According to the available laboratory analysis the maximum ratio of Cl^- / HCO_3^- was observed at the Hel Peninsula and northern part of the Kashubian Coastland. In the major part of the Kashubian Coastland the minimal value was adopted. In terms of the aquifer thickness (*T*), study area can be divided into two parts. The thickness of aquifer in the Hel Peninsula is 6 m and it was rated with 5 value, while aquifer thickness in the remaining area is higher than 10 m and thus it was rated with the highest score (Fig. 2-T).

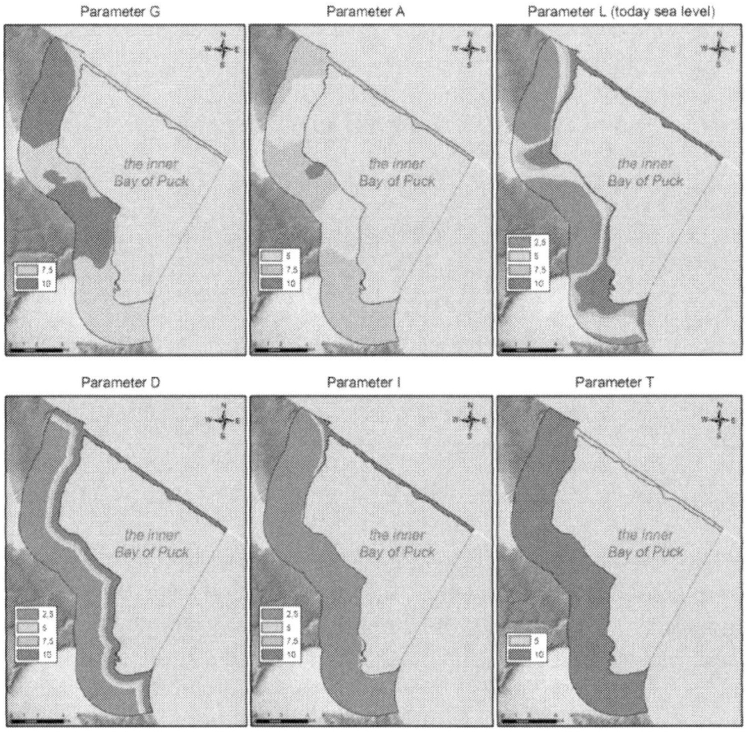

Figure 2. Thematic maps of GALDIT method parameters.

As the Polish coastline is thought to be the most susceptible to sea level rise in Europe (Paprotny and Terefenko 2017) the impact of predicted sea-level rise to groundwater vulnerability was checked. Actual trends of sea level changes may cause increase of coastal aquifers vulnerability to seawater intrusion in the future. Predictions of sea level rise under simulated climate change scenarios for a shallow low-lying coastal aquifer in southern Finland shows that some areas along coastline will be below the seawater level. Together with a coastal flooding, it may expand the saltwater intrusion extension, increasing the pollution of the aquifer with seawater (Luoma et al. 2017). According to the KLIMAT project (Wibig and Jakusik [eds.] 2012), the mean sea level of the inner Puck Bay in the period 2081-2100 is predicted to exceed the current level by 0.28 m (the most pessimistic scenario). The prediction was estimated based on the greenhouse gases emissions (A2), the impact of circulation factor and expected changes in the global sea level according to the IPCC. The scenario of sea water level rise was included in GALDIT method by decreasing the height of groundwater level above sea level by 0.28 m – *L* parameter. The range of the index score was categorized into three classes: <5, 5-7.5 and >7.5, respectively, as low, moderate and high. The spatial distribution of GALDIT index in the studied area for the current and raised sea level is illustrated Fig. 3.

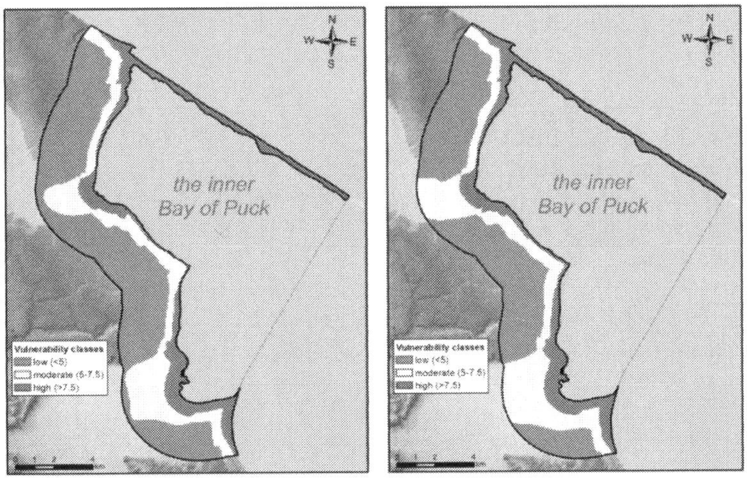

Figure 3. Comparison between GALDIT index for the current and future sea levels (left – current sea level; right–predicted sea level in 2081-2100 years with rise +0.28 m).

The percentage distribution of low, moderate and high vulnerability zones for the current sea level are 55.8, 25.3 and 18.9 %, respectively. The simulated scenario of predicted sea level rise shows changes in vulnerable zones percentage distribution, respectively: low 48.7%, moderate 30.5% and high 20.8% (Tab. 2).

Table 2. Areal comparison of vulnerability classes.

Vulnerability classes	Current sea level		Predicted sea level in 2081-2100 years (+0.28 m)	
	[km^2]	[%]	[km^2]	[%]
Low	47.5	55.8	41.5	48.7
Moderate	21.6	25.3	26.0	30.5
High	16.1	18.9	17.7	20.8

CONCLUSIONS

Assessment of the inner Puck Bay coastal aquifer vulnerability to seawater intrusion has been conducted using GALDIT method. The most vulnerable areas are the Hel Peninsula and northern part of the Kashubian Coastland. Increased class of aquifer vulnerability was also obtained for glacial valleys. Differences between actual and future aquifer vulnerability are noticeable. The predicted sea level rise of 0.28 m in the scenario for period 2081-2100 will increase vulnerability to seawater intrusion on 7% of the research area. It will be particularly significant in glacial valley areas, where the groundwater table is only slightly above the sea level. The final map of vulnerability can be a useful tool for management of the coastal zone of the inner Puck Bay.

ACKNOWLEDGMENTS

This work has been supported by National Centre for Research and Development, Poland, in the framework of the project BIOSTRATEG3/343927/3/NCBR/2017 "Modelling of the impact of the agricultural holdings and land-use structure on the quality of inland and coastal waters of the Baltic Sea set up on the example of the Municipality of Puck region – Integrated info-prediction Web Service WaterPUCK" – BIOSTRATEG Programme.

https://doi.org/10.1051/e3sconf/20185400023

REFERENCES

Chachadi A.G. and Lobo-Ferreira J.P. 2001. Sea Water Intrusion Vulnerability Mapping of Aquifers Using GALDIT Method. Proceedings of the Workshop on Modelling in Hydrogeology, Anna University, Chennai, 143-156.

Chachadi, A.G. and Lobo-Ferreira, J.P., 2007. Assessing aquifer vulnerability to seawater intrusion using GALDIT method: part 2, GALDIT indicators description. Water in Celtic Countries Quantity, Quality and Climatic Variability, IAHS Publ. 310, pp. 172–180.

Frączek E. 1998. Hydrogeological Map of Poland 1:50 000 – Puck sheet explanations (in Polish). Polish Institute of Geology.

Furmańczyk K. 2004. Hel Peninsula (Poland). Eurosion Case Study, 28 pp. Available (Feb 2018): http://copranet.projects.eucc-d.de/files/000143_EUROSION_Hel_peninsula.pdf

Kallioras A., Pliakas F-K., Schuth Ch. And Rausch R. 2012. Methods to Countermeasure the Intrusion of Seawater into Coastal Aquifer Systems. In book: Wastewater Reuse and Management (eds. Sharma S.K and Sanghi R.) 479-490.

Kazakis N., Spiliotis M., Voudouris K., Pliakas F-K. and Papadopoulos B. 2018. A fuzzy multicriteria categorization of the GALDIT method to assess seawater intrusion vulnerability of coastal aquifers. Science of the Total Environment 621: 524-534.

Lobo-Ferreira, J.P., Chachadi A.G., Diamantino C. and Henriques M.J. 2007. Assessing aquifer vulnerability to seawater intrusion using GALDIT method: part 1, application to the Portuguese Monte Gordo aqiufer. Water in Celtic Countries Quantity, Quality and Climatic Variability, IAHS Publ. 310, pp. 161–171.

Luoma S., Okkonen J. and Korkka-Niemi, K. 2017. Comparison of the AVI, modified SINTACS and GALDIT vulnerability methods under future climate-change scenarios for a shallow low-lying coastal aquifer in southern Finland. Hydrogeology Journal v. 25:1 203–222.

Moskalewicz D. 2016. Hel Peninsula – development and future of sandy barrier under sea level rise. In book: Quaternary geology of north-central Poland: from the Baltic coast to the LGM limit (eds. Sokołowski R.J. and Moskalewicz D.) 5-16.

Paprotny D. and Terefenko P. 2017. New estimates of potential impacts of sea level rise and coastal floods in Poland. Natural Hazards v. 85:2 1249-1277.

Pruszak Z. and Zawadzka E. 2008. Potential Implications of Sea-Level Rise for Poland. Journal of Coastal Research v. 24:2 410-422.

Trabelsi N., Triki I., Hentati I. and Zairi M. 2016. Aquifer vulnerability and seawater intrusion risk using GALDIT, GQI_{SWI} and GIS: case of a coastal aquifer in Tunisia. Environmental Earth Sciences 75:669

Wibig J. and Jakusik E. (eds.) 2012. Warunki klimatyczne i oceanograficzne w Polsce i na Bałtyku południowym (in Polish). Institute of Meteorology and Water Management – National Research Institute. Available (Feb 2018): http://klimat.imgw.pl/wp-content/uploads/2013/01/tom1.pdf?edmc=

Contact Information: Dawid Potrykus, Gdańsk University of Technology, Faculty of Civil and Environmental Engineering, 80-233 Gdańsk, ul. Narutowicza 11/12, Poland, Phone: +48 58 3472311
Email: dawpotry@pg.edu.pl

https://doi.org/10.1051/e3sconf/20185400024

Seawater intrusion due to pumping mitigated by natural freshwater flux: a case study in Władysławowo, northern Poland

Małgorzata Pruszkowska-Caceres[1], Adam Szymkiewicz[1], Beata Jaworska-Szulc[1], Maria Przewłócka[1], Anna Gomuła-Kawęcka[1], Dawid Potrykus[1]
[1]Gdańsk University of Technology, Faculty of Civil and Environmental Engineering

ABSTRACT

The paper presents a case study of seawater intrusion into a coastal aquifer, caused by a groundwater intake located close to the seashore in Władysławowo, northern Poland. Evolution of the basic hydrogeochemical parameters for the 50-year period from 1964 to 2014 indicates progressing encroachment of saline seawater into the aquifer. However, the spatial pattern of salinity was influenced by the variability of hydraulic gradient in fresh water discharging from aquifer to the sea. As a consequence, significant changes in salinity occurred in the directions both perpendicular and parallel to the coastline.

INTRODUCTION

The main goal of this study was investigation of temporal and spatial variability of the basic hydrogeochemical parameters in the vicinity of a coastal groundwater intake in Władysławowo, for the period from 1964 to 2014. Particular attention is paid to the role of natural hydrogeological conditions as a factor enhancing or limiting saltwater intrusion caused by groundwater abstraction.

Władysławowo is situated in northern Poland, at the base of the Hel Peninsula, on the border of the upland moraine area of Kępa Swarzewska (Fig. 1). There are two Quaternary aquifers. The shallow one is composed of Pleistocene and Holocene sandy and gravel marine and glacial sediments. Currently it is not exploited. The deeper aquifer, composed of Pleistocene sands and gravel of thickness 16 to 70 m, occurs at depth -8 to -44 m a.s.l., however in the area of the moraine upland the depth increases to about -100 m a.s.l. It is covered by 25 to 35 m thick detached block of Miocene clay-silt sediments transported by glacier and deposited over Pleistocene layer. Locally it is covered by 5 to 80 m thick layer of Pleistocene boulder clay (Fig. 1). The hydraulic conductivity of the deeper aquifer ranges from 0.43 m/h to 1.55 m/h. The potentiometric surface occurs at 3.7 to 6.2 m b.g.l. (0.3 to 6.1 m a.s.l.). However, due to intensive exploitation the level has decreased to -1.8 meters a.s.l., causing saltwater intrusion. Previous research based on the concentration of tritium (TU 5.8 to 13.5) showed that there is approximately 40% of marine origin water in the deeper Quaternary aquifer of the Władysławowo area (Kozerski and Pruszkowska, 1996). In the same analysis the possibility of another origin of the salinity was ruled out. Groundwater recharge of the aquifers takes place on Kępa Swarzewska - as direct infiltration and lateral inflow, and on the Żarnowiecka moraine upland - as lateral far distance inflow (Pruszkowska 2005). Both aquifers discharge to the Baltic Sea, the shallower one has direct contact with the seaside, while the deeper aquifer has an outcrop in the sea bottom, where submarine discharge predominates.

Groundwater from the deeper aquifer was abstracted continuously from 1964 to 2014 for the needs of a fish processing plant. Currently the wells are used as an emergency water supply. The intake consists of 6 wells aligned along the coast (Fig 1). The western wells (415, 423) worked with discharge up to 42 m^3/h, while the discharge of the other wells did not exceed 20 m^3/h.

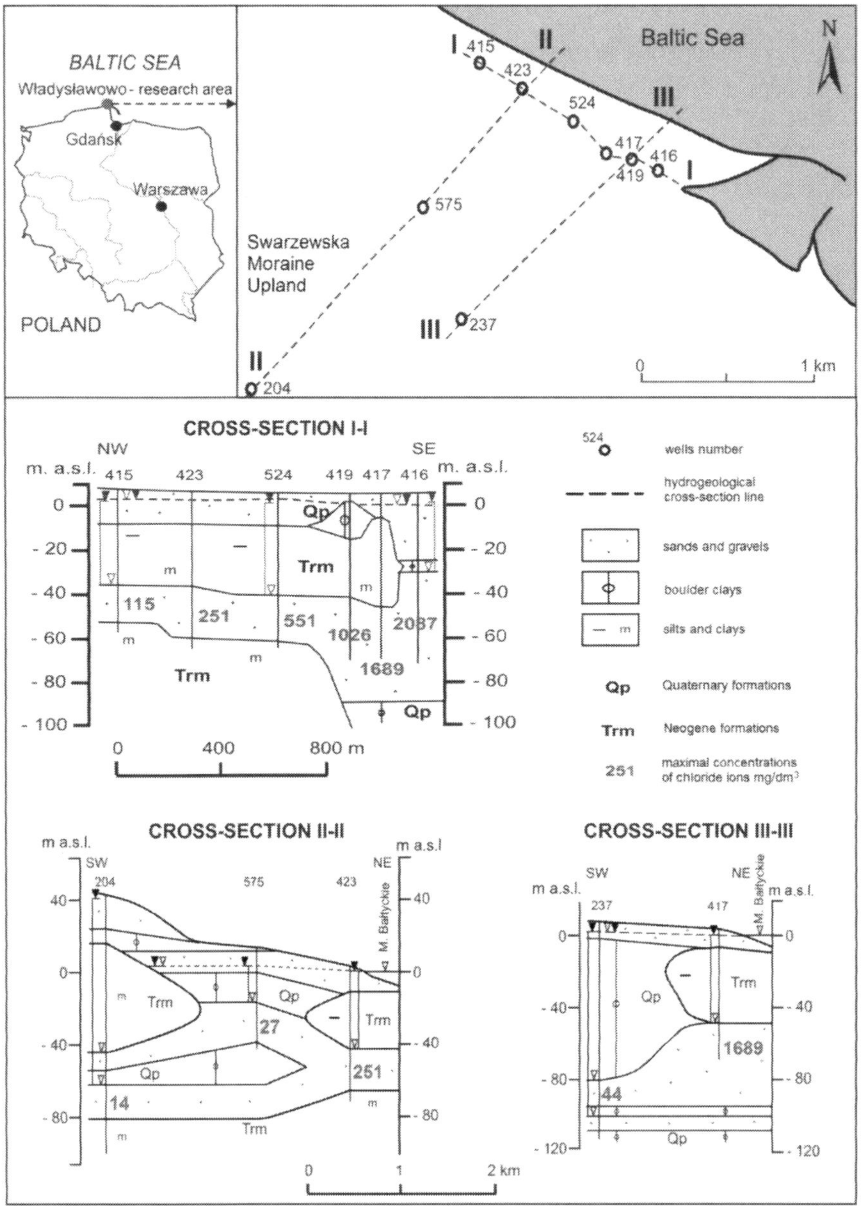

Figure 1. Hydrogeological cross-sections and location of the research area.

METHODS

We collected archival data on the chloride and sulphate concentrations measured in the 6 pumping wells of the intake between 1964 and 1994. The samples for chemical analyses were taken from the lowest part of each well in July each year. Between 1998 and 2014 we conducted sampling and analysis, also taking samples from the lowest part of each well. The chemical analyses included TDS and main dissolved ions (Na, K, Ca, Mg, Cl, SO4, HCO3), as well as Br, Sr, pH, Eh and temperature. Besides the abstraction wells, we included monitoring wells 204, 237, 575 (Fig. 1) and samples of water from the Baltic Sea. Additionally, in 2004 a more detailed series of measurements was carried out to capture the vertical salinity profiles In each of the wells samples were taken from a range of depths, starting from 2 m below the top of aquifer, down to the bottom of the well in 5 m intervals.

Based on the most recent measurements from 2014 the hydrogeochemical types of sampled groundwater were determined, according to the proposition of Stuyfzand (1989). This classification scheme distinguishes between fresh (F, < 300 Cl), fresh-brackish, brackish, brackish-saline (Bs), saline (S) and Hypersaline (H) waters. Further division is based on the dominant ion groups and dominant ions within each group. The Base Exchange Index (BEX) was calculated as BEX = Na+K+Mg - 1.071Cl, where the concentrations are given in mval/dm^3 (Stuyfzand 1989, 2008). Several other indices, frequently used in seawater intrusion studies were determined (Appelo & Postma 2004, Bear et al. 1999). They include the following ratios: Na/Cl, (Ca+Mg)/Cl, Ca/(HCO3+SO4), Cl/Br, Ca/Sr (all concentrations expressed in mval/dm^3).

RESULTS AND DISCUSSION

The temporal variability of Cl^- and SO_4^{-2} concentrations is presented in Figure 2 for each well of the intake. In 1964, before groundwater exploitation began, concentrations of each ion were similar in all wells. The chloride concentration varied in the range from 11 to 32 mg Cl/dm^3, sulphate from 12 to 40 mg SO_4/dm^3., which can be considered typical for Pleistocene aquifers of the region (Pruszkowska & Malina 2008). Due to continuous water uptake an increase in the ion concentrations was observed during the whole considered period. This increase was unevenly distributed along the line of wells (Fig. 2). The largest concentrations were measured in well 416 (in 2014: 2087 mg Cl/dm^3, 421 mg SO4/dm^3), while in well 415 at the opposite end of the line the increase was much smaller (in 2014: 115 mg Cl/dm^3, 32 mg SO4/dm^3).

This spatial variability can be explained by several factors. Well 416 is practically surrounded by sea, being situated not only close to the main shore line but also in the vicinity of a local harbor bay. Moreover, the hydrogeological cross sections in Fig. 1 show that the hydraulic gradient in the direction perpendicular to the sea coast is much larger in the north-west end of the intake (wells 415 and 423, cross-section II-II) than in the south east end (wells 417 and 416, cross-section III-III). Thus one can expect larger discharge flux of fresh water in the aquifer close to well 415, compared to the discharge flux in the vicinity of well 416. Additionally, the considered aquifer is significantly thicker in the eastern part, which possibly facilitates encroachment of saline water in its lower part. Yet another factor which may affect the non-uniform evolution of groundwater salinity is the variable confinement of the deeper aquifer. As shown in Fig. 1, near well 416 the confining stratum is relatively thin and possibly semi-permeable. Such a situation can potentially facilitate the contact with the saline water from the sea and the shallow aquifer. Note that the pattern of salinity was completely opposite to the distribution of the well discharge, with western well pumping rates twice as large as the eastern ones.

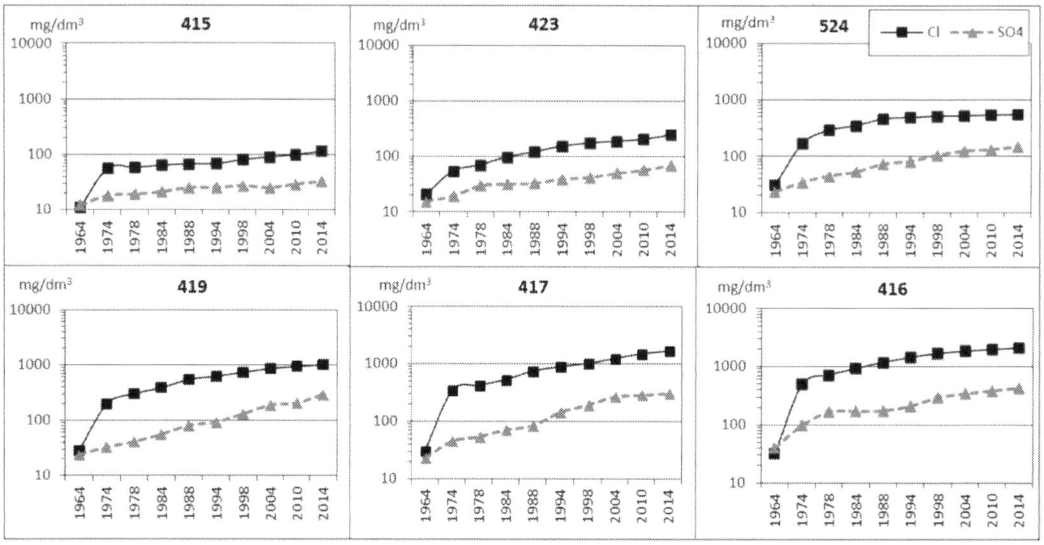

Figure 2. Evolution of the chloride and sulphate concentrations measured in the pumping wells.

Cross sections in the direction perpendicular to the sea coast (Fig. 3) show that the spatial extent of saltwater intrusion is limited to a relatively narrow zone along the coast. The concentrations of Cl in observation wells 204, 237, 575 in 2014 were at the same level as measured in the intake wells in 1964, before the abstraction started.

Vertical variability of Cl distribution is clearly seen in the eastern wells 416, 417, see Fig. 4. Concentration increases approximately linearly with depth and the relative differences in chloride concentration between the top of the aquifer and well bottom are about 20%. This is consistent with the well-known pattern of seawater intrusion, where the more saline water occupies the lower part of the aquifer. In contrast, the results for western wells, where the salinity is lower, do not show any appreciable differences of vertical chloride distribution.

Hydrogeochemical groundwater types were determined following Stuyfzand (1989) and listed in Table 1. They are listed in Table 1. With respect to the major ion composition, in 5 of 6 pumping wells water was of NaCl type, which indicates considerable sea influence. The only exception was the westernmost well 415 where water was of CaHCO3 type, characteristic for shallow terrestrial groundwater. The salinity, represented by the main type of water changed gradually from brackish-saline in well 416 (similar to the Baltic Sea) to fresh in well 415. The three monitoring wells located away from the sea had fresh CaHCO3 water. All groundwater samples represented 8 alkalinity type (extreme alkalinity). Furthermore, we calculated the Base Exchange Index, as an indicator of cation exchange related to salinization or freshening of the aquifer (Walraevens & van Camp 2004). The calculation was based on the procedure outlined by Stuyfzand (2008). In almost all cases BEX was 0, indicating equilibrium conditions with respect to cation composition. The only exception occurred in well 204, where BEX was positive. Thus it seems that the composition of water is a result of simple mixing due to dispersion and diffusion, without much ion exchange.

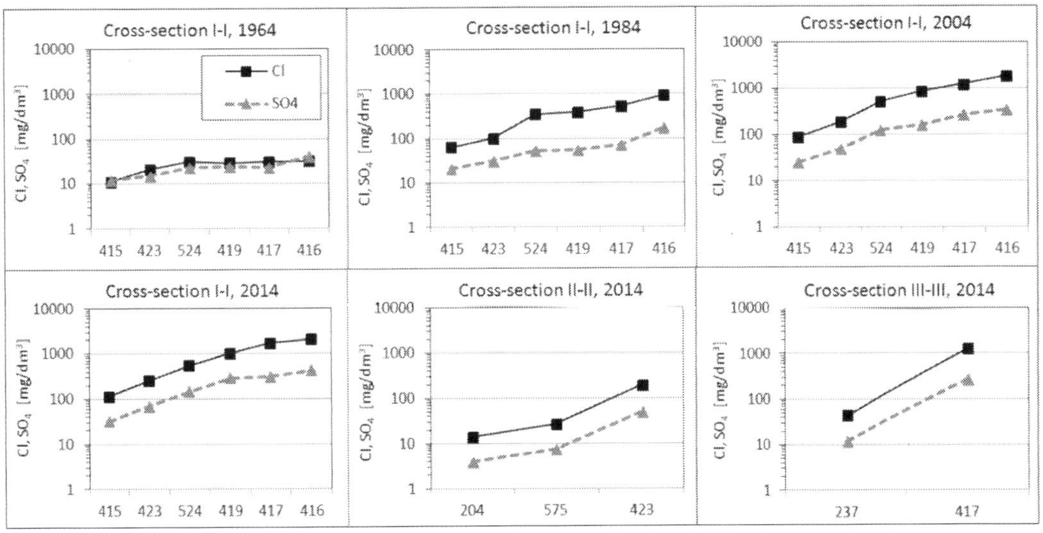

Figure 3. Concentration distribution along cross-sections.

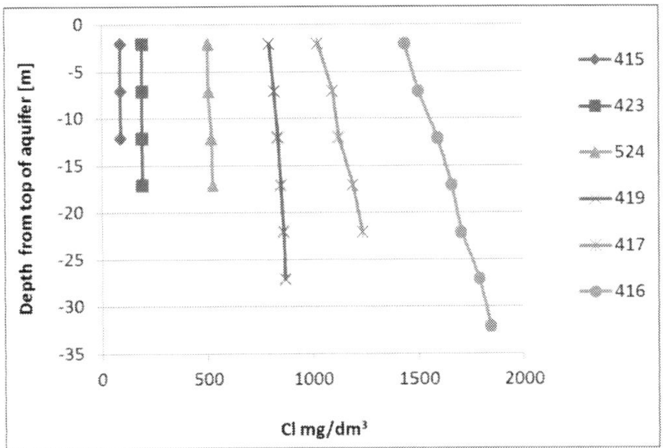

Figure 4. Vertical profiles of chloride concentration measured in pumping wells in 2004.

Table 1. Hydrogeochemical parameters of water samples (2014).

Well number	Type	Na/Cl	(Ca+Mg)/Cl	Ca/(SO_4+HCO_3)	Ca/Mg	Cl/Br
Baltic Sea	b7-NaCl(0)	0.86	0.27	0.37	0.33	297
416	b8-NaCl(0)	0.86	0.30	0.44	0.49	310
417	b8-NaCl(0)	0.86	0.36	0.47	0.57	318
419	b8-NaCl(0)	0.86	0.47	0.53	0.82	320
524	B8-NaCl(0)	0.87	0.47	0.60	1.31	326
423	f8-NaCl(0)	0.88	0.75	0.66	2.29	332
415	F8-CaHCO3(0)	0.90	1.35	0.69	3.58	337
237	F8-CaHCO3(0)	1.05	4.68	0.71	5.18	338
575	F8-CaHCO3(0)	1.06	7.00	0.72	5.80	338
204	g8-CaHCO3(+)	1.26	11.88	0.72	6.56	350

Other indices reported in Table 1 also show a clear gradation of chemical composition along the line of wells. The indices of water in wells 416 and are quite similar to seawater, while the characteristics of water in well 415 are closer to the fresh groundwater from inland wells.

The ratios Na/Cl, (Mg+Ca)/Cl, Ca/(SO_4+HCO_3), Ca/Mg and Cl/Br increase gradually along the line of wells from east to west. The inland monitoring wells have distinctly different water with Na/Cl > 1 and (Ca+Mg)/Cl >> 1 and Ca/Mg > 5. The data in Table 1 is consistent with salinity profiles in Fig. 3, showing a gradual, significant changes of composition between along the line of pumping wells and clear differences between the pumping wells and the inland monitoring wells.

CONCLUSIONS

Significant temporal and spatial variability of groundwater chemical composition was observed in the considered aquifer in Władysławowo. The salinity of groundwater, which is associated with the marine water intrusion, has a limited spatial extent, and does not penetrate to a large distance into the aquifer in the landward direction. The zone of saltwater intrusion is located near the groundwater intake, in a flat coastal area, about 600-1000 m from the sea. Remarkable variability of chemical composition was observed along the line of pumping wells, parallel to the coast. The largest encroachments of seawater occurred in the eastern well despite their low pumping rates. On the other hand, the westernmost well showed only a slight influence of seawater. These differences can be explained by local variability of geological structure and natural hydraulic gradient in the aquifer.

REFERENCES

Appelo, C. A. J., & Postma, D. (2004). Geochemistry, groundwater and pollution. CRC press.

Bear, J., Cheng, A. H. D., Sorek, S., Ouazar, D., & Herrera, I. (Eds.). (1999). Seawater intrusion in coastal aquifers: concepts, methods and practices (Vol. 14). Springer Science & Business Media.

Kozerski, B., Pruszkowska, M., 1996. The Origin of groundwater salinity on the Polish Baltic Sea coast. (in Polish). Inżynieria Morska i Geotechnika, 1, 46-48.

Macioszczyk, A., 1987. Hydrogeochemia (in Polish). Wyd. Geol., Warszawa.

Pruszkowska, M., 2005. Hydrogeological conditions and quality of groundwater on the tertiary sediments on Kępa Swarzewska (in Polish with English summary). Inżynieria Morska i Geotechnika, 5, 373-376.

Pruszkowska, M., Malina, G., 2008. Hydrogeochemistry and vulnerability of groundwater in the moraine upland aquifers of the Gdańsk region (Northern Poland). Geological Quarterly, 52 (3), 291-300.

Stuyfzand, P.J. (1989) A new hydrochemical classification of water types. in: Regional Characterization of Water Quality, IAHS Publ. 182.

Stuyfzand, P.J. (2008) Base exchange indices as indicators of salinization or freshening of (Coastal) Aquifers. Proceedings of the 20th Salt Water Intrusion Meeting, Naples, FL, USA.

Walraevens K., van Camp M. (2004) Advances in understanding natural groundwater quality controls in coastal aquifers. Proceedings of 18th Salt Water Intrusion Meeting, Cartagena, Spain.

Contact Information: M. Pruszkowska-Caceres, Gdańsk University of Technology, Faculty of Civil and Environmental Engineering, 80-233 Gdańsk, Narutowicza 11/12, Email: mpru@pg.edu.pl

Freshening of salinized groundwater in Gdańsk Quaternary aquifer

Maria Przewłócka[1], Beata Jaworska-Szulc[1], Bohdan Kozerski[2], Małgorzata Pruszkowska-Caceres[1], Adam Szymkiewicz[1], Andrzej Kwaterkiewicz[3]
[1] Gdańsk University of Technology, Faculty of Civil and Environmental Engineering, Gdańsk, Poland
[2] emeritus, formerly at the Gdańsk University of Technology, Gdańsk, Poland
[3] emeritus, formerly at Polgeol company, Gdańsk, Poland

ABSTRACT

The main Quaternary aquifer in the Gdańsk region connects directly with salt waters of the Martwa Wisła River and the Gulf of Gdańsk. This aquifer has been used for many years by large municipal and industrial intakes located on the marine lowlands, with maximum output 6236 m^3/h in 1985. Owing to an intensive exploitation, groundwater in the area of the Old Gdańsk has been salinized by intrusion of salt water from the Martwa Wisła River. Concentration of chlorides in some wells reached 2000-3000 $mgCl/dm^3$, leading to degradation of groundwater resources in this area. As a consequence the "Grodza Kamienna" groundwater intake had to be put out of operation. Salt water intrusion from the Gulf of Gdańsk was provoked in 1970s and 1980s by overexploitation of groundwater resources by three municipal intakes situated on the marine terrace. Concentration of chlorides in some piezometers of the "Czarny Dwór" groundwater intake amounted to 700 – 1000 $mgCl/dm^3$, however in the pumping wells of the intake it never exceeded the drinking water standards. In 1990s groundwater pumping rates in both areas were lowered. As a result, the chloride concentration decreased and the resources started to recover. Current observations prove that groundwater of the main Quaternary aquifer is freshening and salt water intrusion has stopped. However, elevated concentrations of chlorides are observed in the shallow aquifer, due to anthropopressure.

INTRODUCTION

Costal lowlands around the city of Gdańsk – (the marine terrace and the north-western part of the Vistula River Delta) together with the Gulf of Gdańsk constitute the discharge zone for the regional aquifer system called the Gdańsk Aquifer System (GAS). The recharge area of GAS in on the moraine plateau of Kashubian Lake District. The extent of the system is determined by the confined Upper Cretaceous aquifer which forms the major regional water bearing stratum (Sadurski 1989). Groundwater in the Cretaceous formation occurs mainly in fine-grained glauconitic sands, at the depth about 150 meters below sea level. On the coastal plains and in the western part of the Vistula River Delta the original piezometric level of the Cretaceous aquifer reached 15-18 m a.s.l. The overlying Paleogene- Neogene aquifer consists of Oligocene glauconitic sands and Miocene sands separated by layers of silts and clays with brown coal. The natural piezometric level in this aquifer was about 5-12 m a.s.l. on the costal lowlands.

However, it is the Quarternary aquifer which has the greatest importance for the water supply in Gdańsk. It spreads on the marine terrace, in the Vistula Delta and also offshore in the Gulf of Gdańsk. The major water bearing series of this aquifer consist of Pleistocene and Holocene fluvioglacial sands and gravels. In the north-eastern part of the Delta fine grained sands of marine origin (Eemian interglacial) have replaced fluvioglacial sediments. The thickness of the Quaternary sediments reaches up to 60 m on the coastal terrace and up to 100 m in the Vistula Delta. In almost the whole area of the Vistula Delta and partly on the marine terrace the main aquifer is covered with mud sediments of the thickness varying from a few to 30 m. The natural potentiometric surface sloped from the moraine hills to the sea and was about 2 m a.s.l. along the coastline of the marine terrace (Fig. 1), which indicates that the discharge took place a few kilometers from the shoreline in the Gulf of Gdańsk.

Groundwater of the Quaternary aquifer has been exploited for more than a hundred years. The problem with salt water intrusion appeared in the 1960s, due to significant increase of water abstraction by municipal and industrial wells located in close vicinity of the Martwa Wisła (one of the branches of Vistula) and port channels. Concentration of chlorides in this area reached 2000-3000 mgCl/dm^3 with total output from Quaternary aquifer on the marine lowlands amounting to 6236 m^3/h in 1985. Five main municipal intakes located here (Fig. 1) – "Lipce", "Grodza Kamienna", "Zaspa", "Czarny Dwór" and "Bitwy pod Płowcami" abstracted 4700 m^3/h.

The influence of intensive exploitation on groundwater salinity became a subject of thorough research and observations after 1970, including development of a monitoring system. The results were presented in numerous reports and publications (Kozerski & Kwaterkiewicz 1984, 1990, 1997; Kozerski et al. 1992). By the end of 20th century groundwater abstraction has been significantly lowered, which resulted in reducing or even stopping of salt water intrusion.

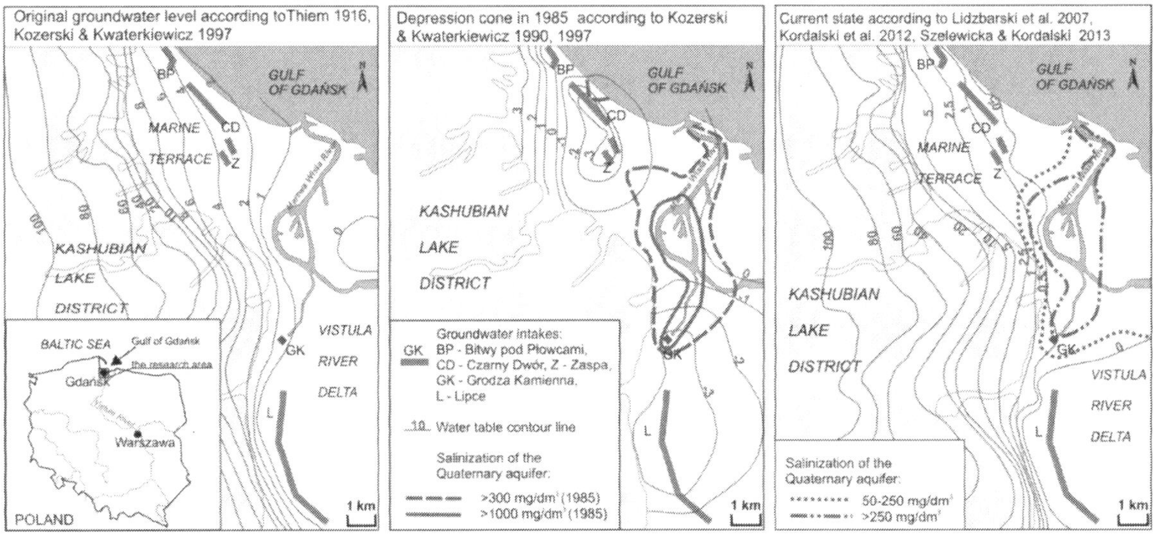

Figure 1. Hydraulic head distribution for the Quaternary aquifer with the reach of groundwater salinization zone at different periods.

EVOLUTION OF GROUNDWATER SALINITY

The history of groundwater exploitation is similar in both areas – on the marine terrace and in the western part of the Vistula River Delta. Up to 1960 the total output of two main groundwater intakes – "Grodza Kamienna" and "Zaspa" was quite low and didn't exceed 1000 m^3/h. In the next period, together with building new municipal groundwater intakes – "Lipce", "Czarny Dwór", "Bitwy pod Płowcami", the abstraction from the Quaternary aquifer rose considerably, reaching 4700 m^3/h in 1985 (6236 m^3/h including industrial intakes).

Because of alarming changes in groundwater quality the exploitation was reduced after 1985. In 1993 pumping at the "Grodza Kamienna" intake had to be stopped due to salinization. After a period of decreasing exploitation lasting till the end of the 20th century, the abstraction stabilized at a more or less constant level 1200 to 1400 m^3/h for the last 15 years. In 2017 it was 1247 m^3/h with 476 m^3/h in the western part of Vistula Delta and 771 m^3/h on the marine terrace (Fig. 2).

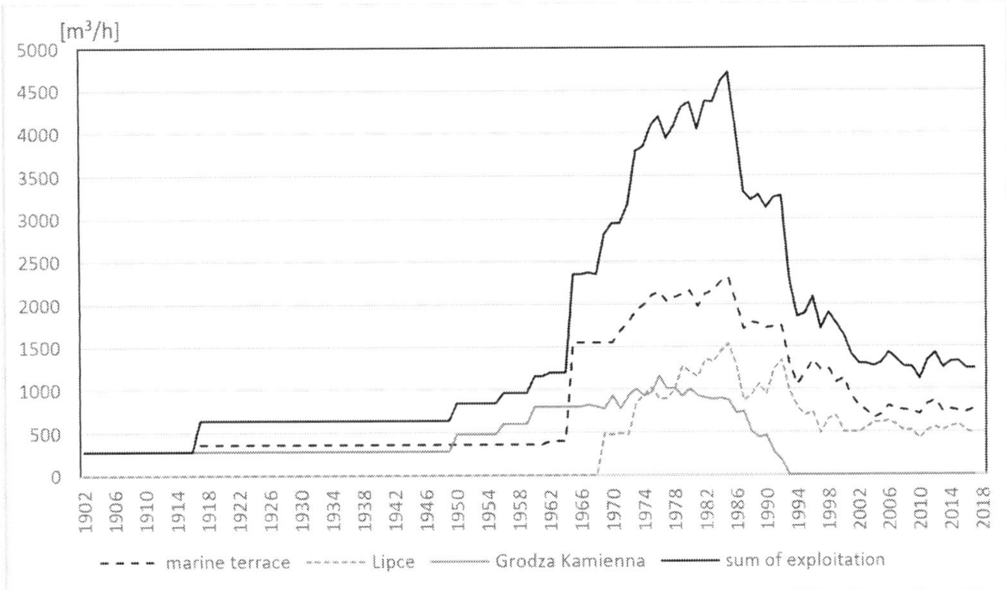

Figure 2. The output of the Gdańsk municipal groundwater intakes from the Quaternary aquifer.

Groundwater salinity changes were tightly connected to the pumping rates, thus three stages of chlorides concentration can be distinguished. **The first period** of moderate abstraction up to 1960s can be described as a stage of natural quality. According to Prehn (1931), at the beginning of 20th century, the concentration of chlorides in groundwater of the marine lowlands varied within the range 13.7 – 19.2 mgCl/dm^3. These values are consistent with natural or rather semi-natural hydrogeochemical background determined by Przewłócka (2003) for the period 1954 – 70 as 5- 40 mg Cl/dm^3 for the marine terrace and 3 – 25 mgCl/dm^3 for the western part of the delta.

The second stage is characterized by increasing salinization up to maximum concentrations exceeding 2000 mg Cl/dm^3 in 1980. Deep and extended depression cones developed around the main groundwater intakes disturbed natural circulation and provoked salt water intrusion (Fig. 1) – from the bay of Gdańsk and from Martwa Wisła and port channels. The highest concentrations of chlorides occurred in wells of the Grodza Kamienna intake and also in industrial wells situated in this area (concentrations amounting to 2350 mg Cl/dm^3). It is important to emphasize that safe yield of Grodza Kamienna intake estimated as 800 m^3/h was exceeded for a dozen of years contributing to degradation of exploited groundwater

(Fig. 3). Salinized industrial wells were successively switched off and the saline waters migrated southwards reching wells of the Grodza Kamienna intake. The exploitation of Grodza Kamienna was reduced by the end of 1980s and in 1993 the wells had to be switched off due to high concentrations of chlorides. It is an exceptional case of groundwater degradation of such an abundant municipal intake.

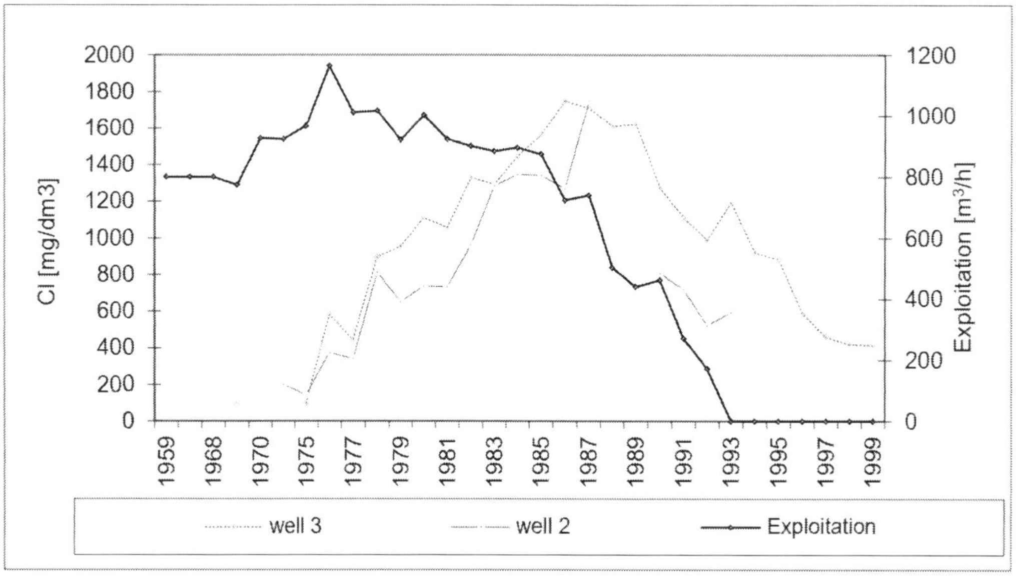

Figure 3. Grodza Kamienna intake; exploitation versus concentration of chlorides in chosen wells.

On the marine terrace salt waters from the Gulf of Gdańsk intruded the exploited aquifer over a limited area indicated in Fig. 1. Zone of saline waters (concentration over 300 mgCl/dm^3) approached the Czarny Dwór municipal intakes to the distance 300 m from the line of wells. In piezometers situated 100 m from the shore line salinization reached maximum value 1000 mgCl/dm^3. Water extracted by wells however, demonstrated law chlorides content even over next years, thanks to intensive lateral recharge from the upland. Concentrations rarely exceeded 100 mgCl/dm^3 in individual wells. In most wells concentrations amounted 40 - 60 mgCl/dm^3. The state of groundwater quality on the marine terrace at that time was the subject of many publications (Kozerski & Kwaterkiewicz, 1984, 1990, 1997; Kozerski et al., 1992) and it is presented in Fig.1.

In the third stage concentrations of chlorides dropped (both on marine terrace and in western Delta) thanks to decreased output. The most distinctive desalinization occurred in wells of Grodza Kamienna (Fig.3) where exploitation of groundwater from Quaternary aquifer was stopped for 25 years and groundwater level rose to its natural state. Groundwater monitoring carried out in 8 piezometers (upper and bottom part of the aquifer observed in 4 points) shows concentrations 40 – 70 mgCl/dm^3. Only in one observation well (bottom part of the aquifer) it oscillates around 200 mgCl/dm^3 (Fig. 4).

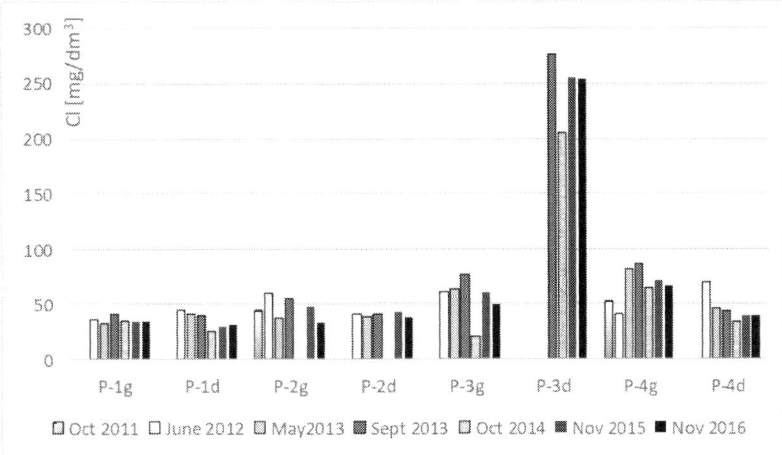

Figure 4. Concentrations of chlorides in observation wells of the Grodza Kamienna intake, recent monitoring.

In the area of the Czarny Dwór intake concentration of chlorides in observation wells situated between the sea shore and the line of wells considerably decreased. The lowest amount is observed at the closest distance (80 m) from the shore line (20 – 30 mgCl/dm^3); in other piezometers and in most of the wells concentrations are in the range 50 – 80 mgCl/dm^3 except for one well which currently has 160 mgCl/dm^3 due to anthropogenic contamination (not seawater). Figure 5 presents the decreasing amount of chlorides in piezometers and moderate concentration in a representative well.

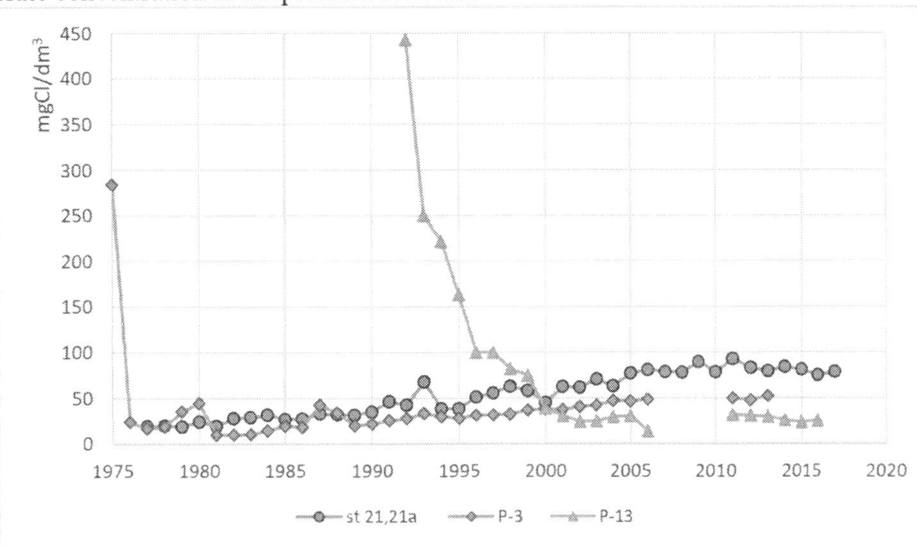

Figure 5. Concentrations of chlorides in a well located 750 m from the shore line (st 21, 21a) through piezometer P-3 (580 m from the sea) to the piezometer P-13 situated 80 m from the shore line.

DISCUSSION AND CONCLUSIONS

Starting from the 1960's groundwater of the Quaternary aquifer was salinized by intrusions from the Gulf of Gdańsk, from Martwa Wisła River and also from port channels due to intensive pumping from wells. As a result groundwater resources in the area of the Old Town of Gdańsk were degraded. However, the reduction of well output led to freshening and recovery of the resources. From 1990s decreasing concentrations of chlorides is observed, which can be considered a proof of high and fast renewability of the resources in Gdańsk. At the same time chloride concentrations appear in high and variable amounts especially in the upper part of the aquifer, up to 500 – 800 $mgCl/dm^3$ in newly drilled observation wells on the opposite side of the line of wells of the Czarny Dwór intake than the sea shore and also in southern part of Gdańsk in western part of the Vistula River delta. It is an effect of developing town with its pollution sources on the surface, especially roads of high traffic and their winter maintenance. Over the last ten years groundwater monitoring system has been considerably developed. New observation wells were drilled in respect to detect possible pollution sources and to protect and monitor groundwater quality.

REFERENCES

Kordalski Z. et al. 2012. Assessment of the dynamics and quality of groundwater in the area of Gdańsk and Sopot, geoMonitoring – Raport. PGI NRI.

Kozerski B., Kwaterkiewicz A., 1984. Strefowość zasolenia wód podziemnych a ich dynamika na obszarze delty Wisły. Archiwum Hydrotechniki, 31 (3), s. 231-255.

Kozerski B., Kwaterkiewicz A., 1990. The origin and state of Quaternary groundwater salinization in Gdańsk region. Proceedings of the 11th Salt Water Intrusion Meeting, Gdańsk 1990.

Kozerski B., Kwaterkiewicz A., Sadurski A., 1992. Zagrożenia wód podziemnych strefy brzegowej morza w rejonie Gdańska. W Służbie Polskiej Geologii. Wydawnictwa AGH, Kraków, s. 117-131.

Kozerski B., Kwaterkiewicz A., 1997. O zmianach zasolenia wód podziemnych czwartorzędu Gdańska. Współczesne Problemy Hydrogeologii tom VIII, 345-347. WIND-J. Wojewoda, Wrocław.

Lidzbarski M., Kachnic J., Kachnic M., Kozerski B., Kreczko M., Pomianowska H., Prussak E. , Pruszkowska M., 2007. Region dolnej Wisły. Subregion Żuław Wiślanych. W: Hydrogeologia regionalna Polski. T. 1. Wody Słodkie (Red. B. Paczyński, A. Sadurski); 239-246.

Prehn B., 1931. Chemische und geologische Untersuchung von wassern des Freistaates Danzig. Dissertation Genehmigt von der Technischen Hochschule der Freien Stadt Danzig zur Erlangung der Wurde eines Doktor-Ingenieurs.

Przewłócka M. 2003. Zmiany zasolenia na ujęciu „Czarny Dwór" w świetle eksploatacji czwartorzędowego poziomu wodonośnego. Współczesne Problemy Hydrogeologii tom XI, cz. 2. 273-280.

Sadurski A., 1989. Górnokredowy system wód podziemnych Pomorza Wschodniego. Zesz. Nauk. AGH, Kraków.

Szelewicka A,, Kordalski Z., 2013. Hydrodynamics changes in the drainage area of Gdańsk hydrogeological system on the basis of recent research. Biuletyn PIG 456: 595-600, 2013R.

Thiem G., 1916. Expertise on water-supply system development in Gdansk, with two sketches and one plan. Gdansk University of Technology, Gdansk, 15 pp.

Contact Information: M. Przewłócka, Gdańsk University of Technology, Faculty of Civil and Environmental Engineering, 80-233 Gdańsk, Narutowicza 11/12, Email: mprzew@pg.edu.pl

Characterization of a regional coastal zone aquifer using an interdisciplinary approach – an example from Weser-Elbe region, Lower Saxony, Germany

Mohammad Azizur Rahman[1], Eva González[2], Helga Wiederhold[1], Nico Deus[2], Jörg Elbracht[2] and Bernhard Siemon[3]
[1] Leibniz Institute for Applied Geophysics (LIAG), Hannover, Germany
[2] Geological Survey of Lower Saxony (LBEG), Hannover, Germany
[3] Federal Institute for Geosciences and Natural Resources (BGR), Hannover, Germany

ABSTRACT

In this study, interdisciplinary approaches are considered to characterize the coastal zone aquifer of the Elbe-Weser region in the North of Lower Saxony, Germany. Geological, hydrogeological, geochemical and geophysical information have been considered to analyze the current status of the aquifers. All the information collectively states that the salinity distribution in the subsurface is heterogeneous both horizontally and vertically. Early age flooding also contributed to this heterogeneity. No general classification of groundwater quality (according to some piper diagrams) could be identified. Helicopter-borne electro-magnetic data clearly show the presence of freshwater reserves below the sea near the west coast. Groundwater recharge largely happens in the moraine ridges (west side of the area) where both the surface elevation and the groundwater level are high. Consequently, submarine groundwater discharge occurs from the same place. All these information will facilitate to develop the planned density driven groundwater flow and transport model for the study area.

INTRODUCTION

Alike some other countries in the world, the impacts due to climate change in Germany have been investigated by several researchers (e.g., Brasseur et al. 2017; Frondel et al. 2017; Rannow et al. 2010) and institutions (e.g., Climate Service Center Germany (GERICS); Potsdam institute for climate change impact research (PIK)). According to the Umweltbundesamt (2015), the yearly average temperature in Germany may increase by 0.5°C and more for the period of 2021-2050. The precipitation in future might decrease during June to August and increase during December to February. The northern half of Germany, surrounded by several rivers (Elbe, Weser, Ems etc.) is considered to be the most potentially affected area (Umweltbundesamt 2015). The coastal zone of the northern part of Lower Saxony is vulnerable to the potential adverse impacts (such as sea level rise) (Sterr 2008) due to climate change. The drinking and industrial water supply and agriculture largely depend on the fresh groundwater reserve at this region. Understanding the impact of climate change on the groundwater reserve is very much essential at this place. Efficient management of available freshwater reserves in a coastal aquifer needs proper characterization of the groundwater catchment. For better understanding and detailed characterization of coastal groundwater aquifers, several scientific and technical issues should be investigated intensively. An efficient integration between traditional and innovative methods/techniques can facilitate the aquifer characterization enormously. This study applies an interdisciplinary approach to characterize the coastal zone aquifer in Northern Germany.

Background and objective

This study is a part of an ongoing EU financed project - TOPSOIL (2015-2020). TOPSOIL project (www.topsoil.eu) focuses to develop diverse methods and techniques to monitor, model and improve the management of the subsurface. Sixteen pilot projects from five EU countries such as Belgium, Denmark, the UK, Germany and the Netherlands were selected to develop the approaches jointly. These will help to predict climate change impacts, including urban flooding, saltwater intrusion, groundwater quality and available quantity, as well as the impact on changing near surface soil conditions. The main research focus of this pilot area is to assess the salt water and fresh water dynamics in the aquifer that is important for estimating future fresh groundwater reserve. Hence, development of a density driven groundwater flow and transport model have been planned. This paper reports about the characterization of the groundwater aquifers that will facilitate the development of groundwater model. For the aquifer characterization, an interdisciplinary approach has been formulated. This paper also briefly demonstrates how the interdisciplinary approach supports efficiently the catchment characterization.

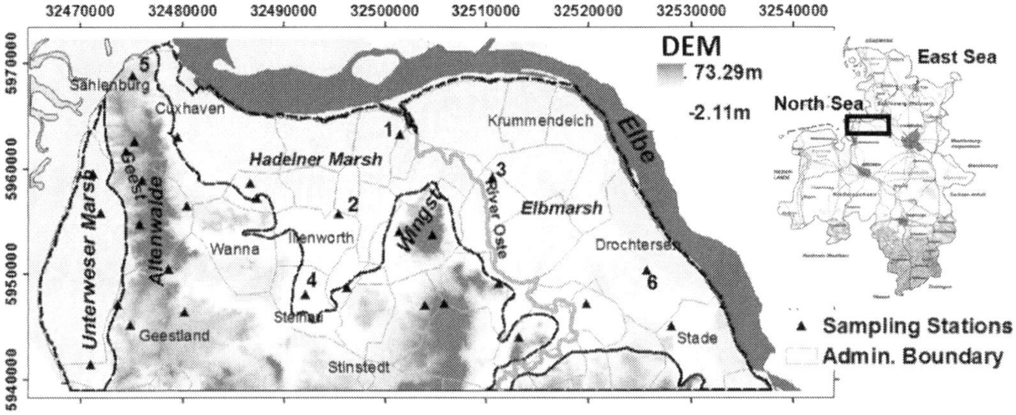

Figure 1. Study area showing topography, major rivers (Elbe and Oste) and groundwater (GW) quality sampling stations for Figure 4.

Study area

The study area (ca. 1700 km^2) is surrounded by the North Sea in the North and the West (ca. 81 km coastline). Elbe River constitutes the east boundary with 35 km long reach. Another major river within the area is the Oste. The topography is relatively flat with some hills (Geest) in the West and the South (ranges between -2.11 m ASL (above sea level) to 73.3 m ASL). The study area is between the moraine ridges of Altenwalde (W) and the River Elbe (E) including the Hadelner Marsh that is widely used for agriculture. It is characterized through low groundwater recharge (ranges between 51 mm/year and 150 mm/year) with saltwater intrusion in the marsh area and high groundwater recharge (ranges between 101 mm/year and 400 mm/year) with a high water table (>10 m ASL) (NIBIS® Kartenserver, 2014a) in the moraine ridges. Most of the recharge occurs during November to January.

The geology of Elbe-Weser region is characterized by Quaternary glacial and interglacial periods. During the Elsterian and the Saalian glaciation, the glacial maximum reached the Central German Uplands, thus the Elbe-Weser region was completely covered by ice several times (Ehlers et al. 2011). During this time massive glaciofluvial sediment bodies were deposited, mainly consisting of sand and gravel. Besides, during the Elsterian glaciation tunnel valleys up to 500 m depth were formed by subglacial erosion and filled predominantly by glaciofluvial sand. The valleys are covered by so called 'Lauenburger Clay' (Kuster and Meyer 1979). Due to sea level rise in the following Eem interglacial, the sea extended into the region bounded by the moraine ridges of Land Hadeln (Höfle et al. 1985). The maximum of the Weichselian glaciation did not reach the Elbe-Weser region (Streif and Köster 1978), so that the area is only influenced by proglacial processes. In the Holocene, the brackish-marine sediments of the marshlands were deposited (Streif and Köster 1978).

Pleistocene, and partly Pliocene sand and gravel form the upper aquifer of the Elbe-Weser region with local variations. The uppermost groundwater storage in the Elbmarsh consists of Weichselian fluviatile sand and glaciofluvial sand of the Drenthe glacial (Saalian). In the Unterweser Marsh and the Geest area, Pleistocene sand forms the upper aquifer. The tunnel valley aquifer consists of Elsterian glaciofluvial sediments. The hydraulic contact between the tunnel valley aquifer and the surrounding Pliocene sand is discontinuous. The Lauenburger Clay is the most important separator in the region between the different aquifers (Elbracht et al. 2016). Hydraulic conductivity values at the region ranges between 10^{-4} and 10^{-3} m/s (Reutter 2011). Cohesive Holocene sediments in the marshlands work as a protective layer for the groundwater, whereas the protection potential of groundwater in the moraine ridges is very low. Groundwater in marshlands has a high salinity in contrast to the moraine ridges where groundwater is fresh (Elbracht et al. 2016).

METHODS

In this study, an interdisciplinary approach has been adopted to characterize the coastal groundwater catchment. This interdisciplinary approach combines several geological, geochemical and geophysical analysis of secondary and primary data. Though data from several geophysical methods are available at the study area, in this study we have limited our analysis to only helicopter-borne electromagnetic (HEM) data (Siemon et al. 2014) to map the horizontal and vertical variability of salinity. Literature and lithological information have been considered to analyze the aquifer structures of the study area. The role of geochemical information is to determine the extent of groundwater pollution and to analyze the salinization process. All these information can be used for density driven groundwater flow and transport model calibration and validation. Finally, this model will be used for the development of an adaptation strategy for climate change, agricultural water management, planning for groundwater withdrawal strategy etc.

HEM data have been obtained from the Federal Institute for Geosciences and Natural Resources, (BGR), Germany. Geological, and hydrogeological information have been taken from the archive of the Geological Survey of Lower Saxony (LBEG), Germany. LBEG's database of mGROWA (Herrmann et al. 2013) provided useful information about the spatial and temporal variation of recharge. Lower Saxony department of water, coastal and nature conservation (NLWKN), Germany, provided information on groundwater level and groundwater quality.

RESULTS

Geophysical investigation of the study area

HEM data provides information about the resistivity of the subsurface that can be interpreted for salinity distribution. Low resistivity values corresponds to saline to brackish water or clay and high resistivity values corresponds to fresh water or sand (Siemon et al. 2018). Figure 2 shows spatial distribution of resistivity at 35 to 40 m depth from the surface. Saline water already occurs in large areas of the north-eastern part of the groundwater catchment. From Figure 2 it is clear that horizontal distribution of salinity is not uniform and it does not correlate with the coastline. Resistivity maps of other depths confirm the inhomogeneity in vertical salinity distribution as well. The occurrence of low resistivities from the North Sea and the Elbe River towards the inland correlates with the early age flooding pattern (Figure 2).

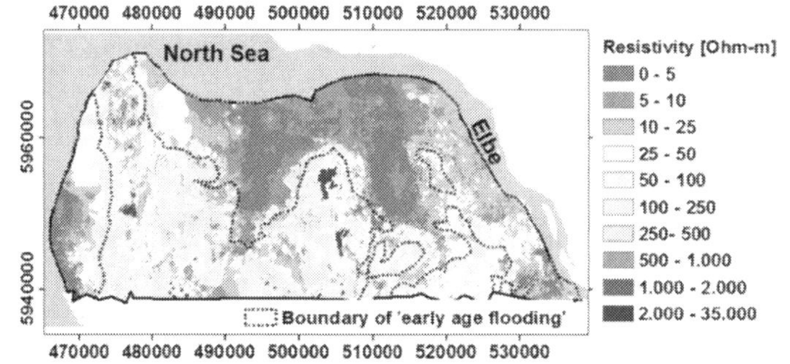

Figure 2. Resistivity distribution at 35 m to 40 m depth. Dotted line showed the boundary of 'early age flooding' (digitised after *NIBIS® Kartenserver 2014b).*

Geochemical analysis

Figure 3 shows a piper trilinear diagram of major anions (Ca^{+2}, Mg^{+2}, Na^{+}, K^{+}) and cations (HCO_3^-, SO_4^{-2}, Cl^- NO_3^-) in groundwater at three different depths from the surface (year 2014). The diagram indicates that the groundwater quality in the study area is vertically heterogeneous. Piper diagrams from other year's data also show the similar pattern.

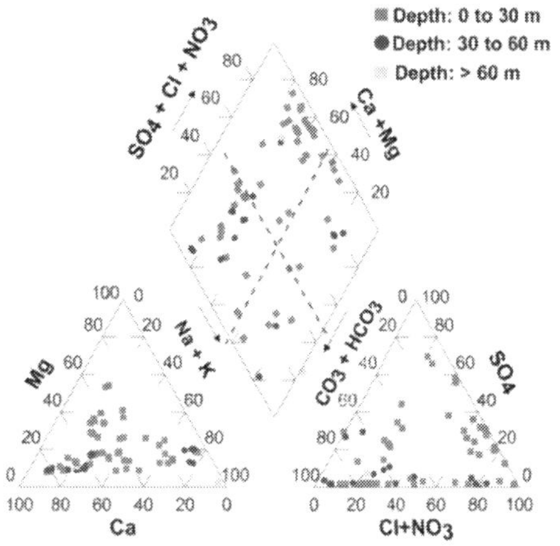

Figure 3. Piper diagram representing the geochemical evolution of groundwater.

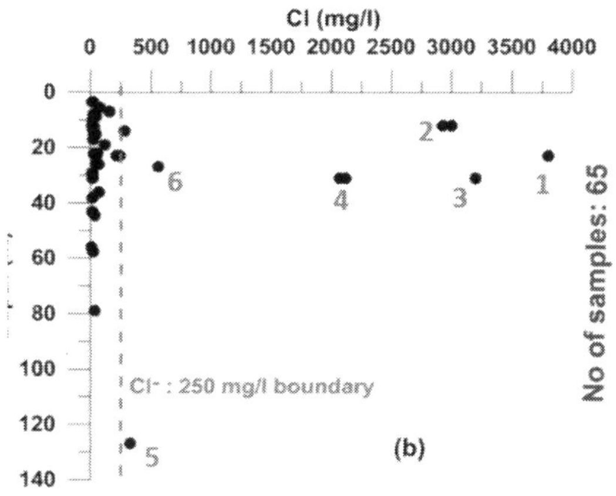

Figure 4. Chloride concentration across the depth of the aquifers (data source: NLWKN).

Not only the vertical distribution but also the horizontal distribution of GW quality is heterogeneous in the study area. Chloride concentration from 67 monitoring wells have been analyzed to observe the chloride distribution and corresponding groundwater pollution due to salinity in the area (Figure 4). In general, the groundwater quality is suitable for drinking purpose. Only at some places, the chloride values exceeded the WHO limit (250 mg/l). It is to be noted that the north-eastern part of the aquifer does not have any monitoring station, hence the saline groundwater has not been represented in the figure. In the Figure 4, some monitoring stations are marked in red color and numbered, who show relatively high chloride content in the groundwater. Stations 1 (at Neuhäuserfelde), 3 (at Moorstrich) and 6 (at Drochtersen) are close to the coast, River Oste and River Elbe, respectively, and the chloride contents are above WHO limit. Stations 2 (at Ilienworth North) and 4 (at Steinau) are relatively far from the coastline and the major rivers, but they show high chloride content within a shallow depth (ranges between 10 m and 36 m below surface). Though station 5 (at Sahlenburg) is close to the coast, it shows relatively low (330 mg/l) chloride content even at ca. 125 m depth. Historical data indicates a slight increase in salinity at this station: from 290 mg/l in 1997 to 330 mg/l in 2014.

Submarine groundwater discharge and freshwater reserves below the sea

Although SDG at the North Sea is reported by some authors (e.g. Moosdorf and Oehler 2017), very few scientific information is available. In this study, geophysical and groundwater level data were used to study the SDG and freshwater reserves close to the North Sea.

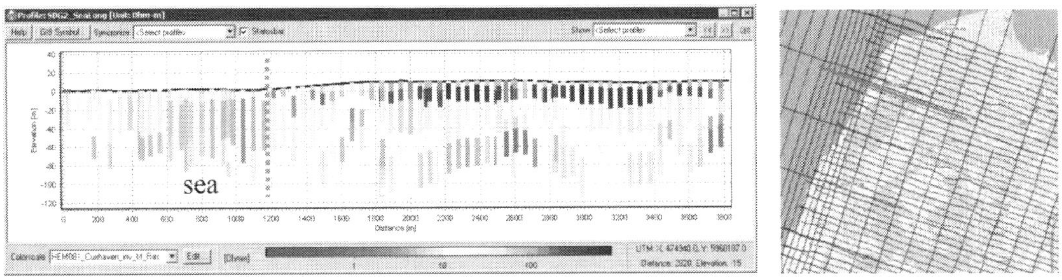

Figure 5. Fresh groundwater reserve near the coastline of the study area. Left: 2D cross sectional profile, right: position of the profile.

Figure 5 gives a clear indication of freshwater reserves near the western coastline of the study area. Spatial maps of resistivity distribution and several vertical cross-sections (not shown here) show that the horizontal and vertical distribution of freshwater reserve is not uniform. Groundwater level at the Geest area (near the west coast) is relatively high (<10 m ASL) and the groundwater gradients allows a substantial amount of water to flow towards the sea contributing to the global SDG.

DISCUSSIONS

The geological and hydrological condition of the area is complex. Representation of geological structures and hydrogeological properties in the groundwater model needs careful consideration and some simplifications (e.g., drainage canals). The salinity distribution is not uniform both horizontally and vertically. As groundwater quality is not uniform throughout the aquifers, transport model simulation will be complex. Early age flooding also contributed to the heterogeneous salinity distribution within a shallow depth. The high chloride content at some stations represents the infiltration of saline water due to early age flooding. SDG will contribute a vital role in the planning process of groundwater artificial recharge at the Geest hills as injection of water into the aquifer might change groundwater flow pattern at the place. Therefore, site selection will be an important point in order to avoid loss of freshwater. The planned groundwater model will be helpful for the selection of managed aquifer recharge injection and recovery location at the region.

ACKNOWLEDGEMENTS

The TOPSOIL project is co-funded by the Interreg North Sea Region Programme, project J-No. 38-2-27-15.

REFERENCES

Brasseur, G.P., D. Jacob, and S. Schuck-Zoeller (editors). 2017. Klimawandel in Deutschland: Entwicklung, Folgen, Risiken und Perspektiven (Climate change in Germany: development, consequences, risks and perspectives). Springer Spektrum. ISBN No: 978-662-50396-6. 352p.

Ehlers, J., A. Grube, H.-J. Stephan, and S. Wansa. 2011. Pleistocene Glaciations of North Germany - New Results. Developments in Quaternary Science 15: 149 - 161, Amsterdam.

Elbracht, J., R. Meyer, and E. Reutter. 2016. Hydrogeologische Räume und Teilräume in Niedersachsen – GeoBerichte 3 (Hydrogeological areas and subareas of Lower Saxony- Geo report 3); 117 S.; Hannover, Germany.

Frondel, M., M. Simora, and S. Sommer. 2017. Risk perception of climate change: empirical evidence for Germany. Ecological economics 137: 173-183.

Herrmann, F., S. Chen, L. Heidt, J. Elbracht, N. Engel, R. Kunkel, U. Müller, H. Röhm, H. Vereeckem, and F. Wendland. 2013: Zeitlich und räumlich hochaufgelöste flächendifferenzierte Simulation des Landschaftswasserhaushalts in Niedersachsen mit dem Model mGROWA. - Hydrologie und Wasserbewirtschaftung. – 57 (5): 206 – 224; Koblenz. Germany.

Höfle, H.-C., J. Merkt, and H. Müller. 1985. Die Ausbreitung des Eem-Meeres in Nordwestdeutschland. - Eiszeitalter und Gegenwart (The Limits of the Eemian Sea in Northwest Germany) 35: 49-59, Hannover.

Kuster, H. and K.-D. Meyer. 1979. Glaziäre Rinnen im mittleren und nordöstlichen Niedersachsen. - Eiszeitalter und Gegenwart (Glacial Channels in Middle and Northeastern Lower Saxony) 29: 135-156, Hannover.

Moosdorf, N. and T. Oehler. 2017. Societal use of fresh submarine groundwater discharge: An overlooked water resource, Earth-Science Reviews, 171, 338-348.

NIBIS ® Kartenserver. 2014a. Hydrogeologische Übersichtskarte Karte 1:200.00 „Grundwasserneubildung, Methode mGROWA" (Generalised hydrogeological map of Lower Saxony, 1: 200000 "Groundwater recharge, mGROWA), LBEG, Hannover.

NIBIS ® Kartenserver. 2014b. Geologie: Frühgeschichtliche Hochwasserereignisse. Massstab 1:500000 (Geology: Early age flood events, scale 1:500000), LBEG, Hannover.

Rannow, S., W. Loibi, S. Grieving, D. Gruehn, and B.C. Meyer. 2010. Potential impact of climate change in Germany- identifying priorities for adaptation activities in spatial planning. Landscape and Urban Planning 98, No 3-4: 160-171.

Reutter, E. 2011. Hydrostratigraphische Gliederung Niedersachsens – Geofakten (Hydrostratigraphical divisions of Lower Saxony- Geo Facts) 21; 11 S.; Hannover.

Siemon, B., A. Steuer, N. Deus, J. Elbracht, and H. Wiederhold. 2018. Comparison of manually and automatically derived fresh-saline groundwater boundaries from helicopter-borne EM data at the Jade Bay, Northern Germany. 25th Salt Water Intrusion Meeting, 17.-22.6.2018, Gdańsk, Poland.

Siemon, B., H. Wiederhold, A. Steuer, M.P. Miensopust, W. Voß, M. Ibs-von Seht, and U. Meyer. 2014. Helicopter-borne electromagnetic surveys in Northern Germany. In proceedings of 23rd Salt Water Intrusion Meeting 2014, ed Wiederhold, et al., 375-378, 15.-22.6.2014, Husum, Germany.

Sterr, H. 2008. Assessment of Vulnerability and Adaptation to Sea-Level Rise for the Coastal Zone of Germany. Journal of Coastal Research 24, No 2: 380 – 393.

Streif, H. and R. Köster. 1978. Zur Geologie der deutschen Nordseeküste. - Die Küste (The Geology of the German North Sea Coast - the coast) 32: 30-49, Heide i. Holstein.

Umweltbundesamt (Federal office of Environment). 2015. Germany's vulnerability to climate change. Summary Report (No. 002226). Project No. (FKZ) 24309. 62p.

Contact Information: M. Azizur Rahman, Leibniz-Institut für Angewandte Geophysik, Stilleweg 2 , 30655 Hannover, Germany, Tel: +49-(0)511-6433234, Fax: +49-(0)511-6433665; Email: MohammadAzizur.Rahman@liag-hannover.de

New Data on Seawater Intrusion in Liepāja (Latvia) and Methodology for Establishing Background Levels and Threshold Values in Groundwater Body at Risk F5

I. Retike[1,2], J. Bikše[1]
[1]Faculty of Geography and Earth Sciences, University of Latvia, Riga, Latvia
[2]Hydrogeology division, Latvian Environment, Geology and Meteorology Centre, Riga, Latvia

ABSTRACT

Intensive water consumption in former decades caused formation of large depression cone near city Liepāja and resulted with seawater intrusion into Upper Devonian Mūru-Žagaru confined freshwater aquifer. Area affected by seawater intrusion is delineated as separate groundwater body at risk (F5) and according to Groundwater Directive threshold values for groundwater bodies at risk must be established to assess the status of a body and identify possible trends. Correct estimation of background levels is significant for determination of threshold values. This study shows an updated so called "BRIDGE" methodology for determination of background levels. A two-step approach how to establish background levels in much stricter manner is presented. Also, data on major ion chemistry, biogenic and trace elements in groundwater and seawater from sampling campaign in 2017 are displayed. Dataset include unique seawater sample taken from Baltic Sea. Finally, the calculated seawater fraction results in groundwater samples shows up to 50 % presence of seawater which decrease with increasing distance from the coast and increasing screen interval.

INTRODUCTION

Groundwater body is a management unit established by Water Framework Directive (WFD 2000). Good chemical status of groundwater body shall be based on compliance to existing Community quality standards and on threshold values (TV) to be established by Member States at the most appropriate level (local, regional or national) (Marandi and Karro 2008).

TV must be established for all pollutants and indicators of pollution which characterized groundwater bodies as being at risk of failing to achieve good groundwater chemical status in accordance with WFD (2000). Regarding saline concentrations resulting from human activities, Member States may decide to establish TV values either for sulphate and chloride or for conductivity (GWD 2006).

The determination of TV should be based on several aspects including hydrogeological characteristics and information on background levels (BL) (GDW 2000). Normally BL are estimated following so called "BRIDGE methodology" (Müller et al. 2006). Methodology proposes approach for derivation of TV which includes application of both natural BL and environmental quality standards (Urresti-Estala et al. 2013).

The aim of this study was to develop and present updated methodological approach for determination of BL and TV for groundwater body at risk F5- seawater intrusion in Liepāja (Latvia). As well the results from sampling campaign (year 2017) are presented demonstrating data on major ion chemistry, biogenic and trace elements taken from monitoring wells and Baltic Sea. Finally, the results from survey were used to calculate the proportion of seawater fraction (%) in groundwater.

METHODS

Materials and data preprocessing

Information about monitoring and abstraction wells and springs was gathered from the largest Latvian hydrogeological database "Wells" (limited access) (Urbumi 2017). Data about major ion chemistry (Ca, Mg, Na, K, Cl, SO_4, HCO_3) and nitrates (NO_3) were extracted and limited to: (1) the area of groundwater body F1 part of which is the area affected by seawater intrusion (F5) and (2) to aquifers of interest - Upper Devonian Mūru-Žagaru (D_3*mr-žg*).

BRIDGE methodology (Müller et al. 2006) suggests a list of minimum criteria which should be used to avoid usage of anthropogenically influenced samples (Marandi and Karro 2008). Data preprocessing included: (1) removal of historical samples which reported sodium and potassium as a sum (NaK) (an additional criterion); (2) removal of samples with ionic balances error greater than ± 10% as suggested by Müller et al. (2006) and (3) where such information was available samples with nitrate content exceeding 4 mg/l were removed as potentially affected by human activities. Much stricter criterion than suggested 10 mg/l by Müller et al. 2006 was chosen based on most recent study about geochemical composition of groundwater in Latvia (Retike et al. 2016).

Determination of background levels

BL for chloride ion was calculated in two steps to minimize the error of visual identification of the inflection point. Firstly, freshwater samples were separated from seawater affected samples - the value of the inflection point on groundwater samples was detected by applying probability plots (Panno et al. 2006). According to BRIDGE methodology samples with NaCl > 1000 mg/l should be removed. Much stricter criteria were used, and value of the inflection point for chloride was set 18 mg/l. Results were compared with values obtained by Retike et al. (2016). Next, BL for chloride ion was determined as 90th percentile of all freshwater samples below the inflection point value according to BRIDGE methodology (Müller et al. 2006). This step was accomplished for two reasons: (1) the validation results from previous study suggested that 18 mg/l for chloride might be too high (Retike et al. 2016) and (2) visual observation of the inflection point is subjective and may hold some uncertainty. Similarly, BL were set for sulphate and sodium.

Calculation of threshold values

TV for chloride, sulphate and sodium were calculated according to BRIDGE methodology (Müller et al. 2006) which suggests deriving TV on the basis of the ratio between the estimated BL and relevant reference value (REF). In this case BL < REF, therefore Equation 1 was used.

$$TV=(REF+BL)/2 \tag{1}$$

Drinking water standard from Latvian legislation (CR No-671 2017) was chosen as REF, respectively 250 mg/l for chloride and sulphate, and 200 mg/l for sodium.

Sampling campaign

Groundwater samples were collected from 9 monitoring wells during sampling campaign accomplished from 08.06.2017 till 19.06.2017. Samples were taken by the staff of Latvian Environment, Geology and Meteorology Centre (LEGMC) responsible for national groundwater monitoring. Groundwater samples were taken according to LVS ISO 5667-11:2011 standard. Sample from Baltic Sea aquatory was taken in 11.05.2017 by the stuff of Latvian Institute of Aquatic Ecology. The seawater sampling area and depth (9 m) was chosen based on existing knowledge about possible area where seawater intrudes freshwater aquifers. All samples were analyzed in LEGMC laboratory which is accredited according to the standard LVS EN ISO/IEC 17025.

Concentrations of Ca, Na, K and Mg were determined by inductively coupled plasma optical emission spectrometry; SO_4, Cl and Br by ion chromatography; NO_3 and NH_4 by segmented flow analysis; As by atomic absorption spectroscopy; HCO_3 by titration; TOC by catalytic oxidation; P_{tot} by persulphate digestion, molybdenum blue colorimetry. Sampling methods are in accordance with the procedure laid down in Article 21 of WFD (2000) and meet the requirements of the European Commission "Guidance on Groundwater Monitoring" (European Commission 2007).

Calculations of seawater fraction

Seawater fractions f_{sea} in groundwater samples were calculated based on both chloride and bromide ions as conservative tracers (Appelo and Postma 2005) by Equation 1:

$$f_{sea} = \frac{m_{X(sample)} - m_{X(freshwater)}}{m_{X(seawater)} - m_{X(freshwater)}} \times 100\% \tag{1}$$

where m_X – concentration of either chloride or bromide concentration in either freshwater, seawater or groundwater sample. Chloride concentration for freshwater sample (3.4 mg/l) was calculated as average chloride value from samples taken from wells No.9322 and No.2254. These wells are inland background monitoring stations for *D₃mr-žg* aquifer (samples not affected by seawater intrusion). Sample taken from Baltic Sea represented overall seawater chemical composition.

RESULTS

Background levels and threshold values

Final BL for chloride was set as 13.2 mg/l, for sulphate 42.5 mg/l and for sodium 22.3 mg/l. Calculated TV for chloride, sulphate and sodium was respectively 131.6 mg/l, 146.3 mg/l and 111.2 mg/l. TV values are officially introduced in national level by an order of the Minister (Order-257 2016).

New data on seawater intrusion

Results from sapling campaign are presented in Table 1. Ionic balance for 6 samples (5 groundwater and 1 seawater) was in range from -0.3% till 6.9%. For three samples sodium was removed due to erroneous values derived from laboratory. All nitrates were under detection limit (0.091 mg/l).

Table 1. Results for major ion chemistry of groundwater samples taken from Upper Devonian Mūru-Žagaru aquifer and Baltic Sea.

Sample No	Coordinates (LKS-92)		Screen interval	Ca	Mg	Na	K	HCO_3	Cl	SO_4
	X	Y	m	mg/l						
862	316954	270095	40-50	186	77	45.7	17.3	151	87	534
2642	316749	267688	47-57	221	107	-	41	224	1270	186.3
2645	316398	268166	72-77	57.4	31.2	24.8	11.8	267	8.6	63
2647	316425	268173	45-58	240	129	-	47	213	1800	203
9322	379051	279726	38-52	74	24.3	5.1	5.2	340	3.08	10
2254	405437	302631	40-47	101	22.8	4.65	1.59	420	3.71	0.03
8849	318470	266461	54-67	63	32.8	21	12.3	370	9.1	16
8850	317574	267496	51-66	134	67.8	-	24.7	276	435	78
8851	318105	267105	51-65	68	35.4	30.8	14.2	310	62.1	31
Baltic Sea sample	310789	270506	9	94	286	2400	70	107	4100	537

Table 2. Results for *in situ* measurements, biogenic and trace elements of groundwater samples taken from Upper Devonian Mūru-Žagaru aquifer and Baltic Sea.

Sample No	pH	EC	T	Fe_{tot}	Br	As	P_{tot}	NH_4	TOC
		µS/cm (20°C)	°C	mg/l					mg C/l
862	7.65	685	10.4	1.45	0.31	1.19	0.006*	0.24	0.19
2642	7.6	3747	9.6	1.06	0.38	0.2*	0.011	0.6	1.27
2645	7.95	587	9.8	0.85	0.05*	3.4	0.098	0.3	0.71
2647	7.57	6971	10.4	1.12	6.1	0.2*	0.013	0.73	1.6
9322	7.67	234	7.9	1.66	0.05*	1.69	0.027	0.22	1.19
2254	7.52	360	7.4	2.13	0.05*	0.2*	0.037	0.13	7.7
8849	7.63	367	10.2	0.44	0.05*	0.2*	0.004*	0.27	1.38
8850	7.64	1593	10.1	0.83	1.33	0.2*	0.006	0.44	1.37
8851	7.75	495	10	0.86	0.19	0.2*	0.007	0.3	0.95
Baltic Sea sample	-	13300	5.86	0.02	12.2	0.92	0.024	0.042	4.4

*under detection limit

Seawater fraction in groundwater samples

Seawater fraction in groundwater samples from D_3*mr-žg* aquifer was calculated based on chloride and bromide ions as conservative tracers. Calculation by both tracers yield comparable results for less mineralized groundwater, however, more saline groundwater samples from wells No.2647 and No.2642 yield different results (Figure 1).

Seawater fraction reaches 50% in groundwater sample at the central part of seawater affected zone (distance from coastline about 1.3 km). The seawater fraction significantly decreases with increasing distance from coastline- at 3.4 km from the coastline the fraction is only 1% (well No.8851). Wells No.2647 and 2645 are both representing the same aquifer and located in the same area (one station), about 1.3 km from the coastline- but they have different screen intervals (Table 1). Well No.2647 with shallower screen interval shows seawater fraction up to 50%, but the well No.2645 with deeper screen interval reflects no seawater presence. The cause for such difference is more than 6 m thick clay layer separating D_3*mr-žg* upper part from lower part of the aquifer.

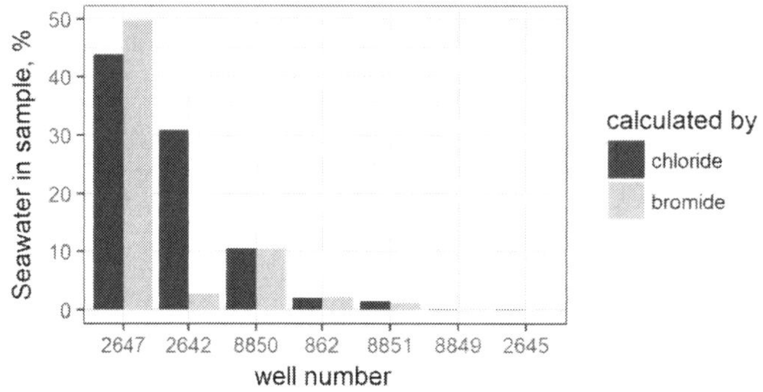

Figure 1. Seawater fractions in groundwater samples from D_3*mr-žg* aquifer.

DISCUSSION AND CONCLUSIONS

Up to 50% of groundwater in study area consists of seawater. Such high seawater fraction was promoted by continuous groundwater abstraction from the D_3*mr-žg* aquifer. Seawater fraction decreases with increasing distance from coastline as well as with increasing depth.

Background levels established by two-step approach are strict and accounts for worst case scenario. Such approach yields lower threshold values therefore it is more sensitive to water quality changes and it takes more time to reach good chemical status. However, this leads to more sustainable water management in coastal areas where groundwater resources are limited.

ACKNOWLEDGEMENT

The study was supported by performance-based funding of University of Latvia Nr.AAP2016/B041//ZD2016/AZ03 within the "Climate change and sustainable use of natural resources" programme.

REFERENCES

Appelo C.A.J, Postma D. 2005. Geochemistry, Groundwater and Pollution, second ed. A.A. Balkema Publishers.

CR No-671 2017. Ministru kabineta 2017. gada 14. novembra noteikumi Nr. 671 "Dzeramā ūdens obligātās nekaitīguma un kvalitātes prasības, monitoringa un kontroles kārtība" (Cabinet Regulation No 671 of 14 November 2017 "Mandatory Harmlessness and Quality Requirements for Drinking Water, and the Procedures for Monitoring and Control thereof" (In Latvian).

European Commission 2007. Guidance Document No. 15: Guidance on Groundwater Monitoring. European Communities, Luxembourg.

GWD 2006. Directive 2006/118/EC of the European parliament and of the Council of 12 December 2006 on the protection of groundwater against pollution and deterioration.

Marandi, A. and Karro, E. 2008. Natural background levels and threshold values of monitored parameters in the Cambrian-Vendian groundwater body, Estonia. Environmental Geology. 54: 1217–1225.

Müller, D., Blum, A., Hart, A., Hookey, J., Kunkel, R., Scheidleder, A., Tomlin, C., Wendland, F. 2006. Final proposal for a methodology to set up groundwater threshold values in Europe, Deliverable D18, BRIDGE project.

Order-257 2016. Order 257 on limit of values for pollutants and their groups in groundwater bodies at risk (03.10.2016). The Minister of Ministry of Environmental Protection and Regional Development of the Republic of Latvia.

Panno, S.V., Kelly, W.R., Martinsek, A.T., Hackley, K.C. 2006. Estimating background and threshold nitrate concentrations using probability graphs. Ground Water, 44: 697-709.

Retike, I., Kalvans, A., Popovs, K., Bikse, J., Babre, A., Delina, A. 2016. Geochemical classification of groundwater using Multivariate statistical analysis in Latvia. Hydrology Research, 47: 799-813.

Urbumi 2017. Derīgo izrakteņu atradņu reģistrs (Register of Mineral Resources). Latvian Environment, Geology and Meteorology Centre. Available from: https://www.meteo.lv (In Latvian).

Urresti-Estala, B., Carrasco-Cantos, F., Vadillo-Pérez, I., Jiménez-Gavilán, P. 2013. Determination of background levels on water quality of groundwater bodies: A methodological proposal applied to a Mediterranean River basin (Guadalhorce River, Málaga, southern Spain. Journal of Environmental Management, 117: 121-130.

WFD 2000. Directive 2000/60/EC of the European parliament and of the Council Establishing a Framework for Community Action in the Field of Water Policy.

Contact Information: I.Retike, University of Latvia, Faculty of Geography and Earth Sciences, Jelgavas Street 1, LV-1004, Riga, Latvia, Email: inga.retike@lu.lv

Deep geoelectrical investigation to bound a coastal thermal outflow area

Enzo Rizzo[1], L.E. Zuffianò[2], F. Santaloia[2] and **M. Polemio**[2]
[1] CNR-IMAA, National Research Council – Institute of Methodology for Environmental Analysis, C.da S. Loja, 85050 Tito Scalo, Italy
[2] CNR-IRPI, National Research Council – Research Institute for Hydrogeological Protection Via Amendola 122 I, 70126 Bari, Italy

ABSTRACT

The coastal carbonate Apulian aquifers, located in southern Italy, feed several coastal fresh springs and constitute the main local source of high quality water. The Santa Cesarea Terme cave system is almost unique case of hypogenic coastal spring caves, located along the Adriatic Sea coastline and hosting spring coastal outflow of mixed groundwater (from 22°C to 33°C) mainly of thermal groundwater due to infiltration offshore, in the sea bottom, and pure fresh groundwater due rainfall infiltration.

Thermal springs and the outflow system are strictly controlled by both the discontinuity network and the karst processes involving the foreland environment. Detailed geoelectrical prospecting were carried out to bound the upflow continental area of this system, considering the geoelectrical effects of deep water mixing with different salinity and temperature close the Adriatic coast.

INTRODUCTION

The sulphurous and thermal waters of Santa Cesarea Thermal springs are known from ancient times (e.g. Aristotle in IV century BC). They flow out along a well-defined, 500 m-long, coastal sector of the Salento peninsula, which is part of the Apulia carbonate platform, (i.e. the foreland of the Apennines-Dinarides orogeny; Ricchetti et al. 1988), consisting of Jurassic-Cretaceous limestone and dolostone, over 5 km-thick in the study area. This succession rests above Late Triassic evaporites (Burano Fm), and is unconformably overlain by Cenozoic calcareous successions with stratigraphic lacunae.

The thermal water of Santa Cesarea Terme area origins offshore with the infiltration of fresh seawater, flowing down to several hundreds of meters below the sea level, almost directly into the deeply fractures Apulian carbonate platform (Santaloia et al 2016). The seawater flows downward along steeply inclined faulted zones and interacts with different rocks that are geothermally heated at least at 80-85 °C. Thereafter, this sulphurous water is driven upwards crossing the coastal transition zone of the seawater intrusion by density difference phenomena. Then, it outflows through the coastal springs of the Santa Cesarea Terme system, mixing itself with the continental pure fresh groundwater and the seawater. The whole flow system of thermal water should be considered a convective cell process.

Detailed geoelectrical prospecting were carried out to bound the upflow continental area of this cell, considering the geoelectrical effects of deep water mixing with different salinity and temperature close the Adriatic coast.

METHODS

In the investigated area, n. 6 high-resolution Electrical Resistivity Tomographies (ERTs) were performed, as shown in Figure 1.

The ERT method consists in the determination of the apparent resistivity parameter (Ohm*m), through combined measurements of electric current intensity, injected in the subsoil by means of a pair of electrodes fixed in the ground, and drop of potential measured on two different electrodes. The acquired data are memorized and subsequently elaborated by ad hoc software, in order to obtain the electrical resistivity distribution. In the last decade, the use of ERT was very intensive for the investigation of complex geological areas (geothermal areas, volcanoes, etc.). In particular, the geoelectrical technique of surface tomography along a profile consists in the realization of an extreme detail image of the areal behavior of the electrical resistivity along the plane of the vertical section passing through the chosen profile. The high resolution obtained by this technique makes it possible to discriminate much more effectively the resistivity contrasts existing in the subsoil, thus providing more reliable information on the subsoil: geological discontinuity (Caputo et al. 2003; Rizzo et al. 2004), groundwater, landslide systems (Lapenna et al. 2003; Perrone et al. 2004), buried archaeological finds (Rizzo et al. 2004).
The instrumentation used was the Syscal R2 with a 48-channel multi-electrode system and an electrode spacing of 20 m. Each profile was long 940 m and the investigation depth was about 180 m. The used geoelectrical methodology was the Wenner-Schlumberger (WS), which allows to obtain a high S/N ratio and good information on vertical structures (i.e. faults) and horizontal layers (i.e. geological stratigraphy). The apparent resistivity data were processed by the RES2DINV software (Loke 2001) to obtain in real time tomographic subsurface images of the electrical resistivity pattern. The 2D inversion routine applies a Gauss-Newton least squares method (Loke and Barker 1996), based on the finite-difference model of the subsurface, automatically adjusted in an iterative process. The Root Mean Squared (RMS) error provides a measurement of the quality of the inversion process.

GEOLOGICAL AND HYDROGEOLOGICAL SETTING

The Santa Cesarea Terme area is located along the south-eastern coast of the Salento peninsula, a carbonate platform defined by a wide, WNW–ESE trending, antiform structure, dissected by a series of extensional and strike-slip faults (Tozzi 1993).
As shown in the geological map of Figure 1, the Upper Cretaceous Altamura Limestone represents the calcareous bedrock at Santa Cesarea Terme, consisting of well-bedded, peritidal limestone and dolomitic limestone. Locally the Late Eocene-Early Oligocene coral reef limestone, belonging to the Castro Limestone, overlays the Cretaceous formation through a marked angular unconformity (Figure 1; Bosellini et al. 1999; Bossio et al. 2005). Both Altamura Limestone and Castro Limestone are unconformable overlain by the bentonic foraminifer-bearing Oligocene calcarenite (Porto Badisco Calcarenites). The youngest formation cropping out at Santa Cesarea Terme is the Salento Calcarenites (Pleistocene), consisting of massive to poorly bedded, weakly cemented calcarenites, related to a slope environment, as testified by the presence of slumpings and submarine slides. The offshore equivalent of the Salento Calcarenites are well imaged in the seismic lines and form a series of prograding units settled during a forced regression (Aiello and Budillon 2004). In the studied area, groundwater flow within a deep aquifer that occurs within the intensely fissured and karstified Altamura Limestone, Castro Limestone and Porto Badisco Calcarenites (Romanazzi et al. 2015, De Giorgio et al. 2018). These units have the same hydrogeological characteristics and play an important common role in the groundwater circulation. The borehole n. 4 (VIGOR well, Abate et al. 2015) has crossed the rocky succession of the deep aquifer. As shown by the rock coring, this succession is composed primarily of white fine and medium grained micritic limestones (L in Figure 1c), locally interbedded with dolomitic limestone and dolomite (L/DL in Figure 1c i.e. 135-145 m, 285-290 m and 295-300 m below

the ground level). Biogenic structures, as biostromal rudists (rb in Figure 1), have been found between 200 m and 215 m below the ground level. Many karst forms (microcavities, karst fractures, calcitic concretions, etc.) have been intercepted even at considerable depths. They seem to decrease starting from 230 m from the ground level (about -115 m a.s.l.). Moreover, it is important to note that the evolution of hypogenic karst forms should have been strongly influenced by the fluctuations of the karstic base level. This level has been linked to the sea-level oscillations occuring during the Quaternary. Morever, in the coastal study area, the development of the karst cavity should have been is also influenced by the fluctuation of the zone of dispersion where the seawater mix with the fresh groundwater (De Waele and Piccinni, 2008).

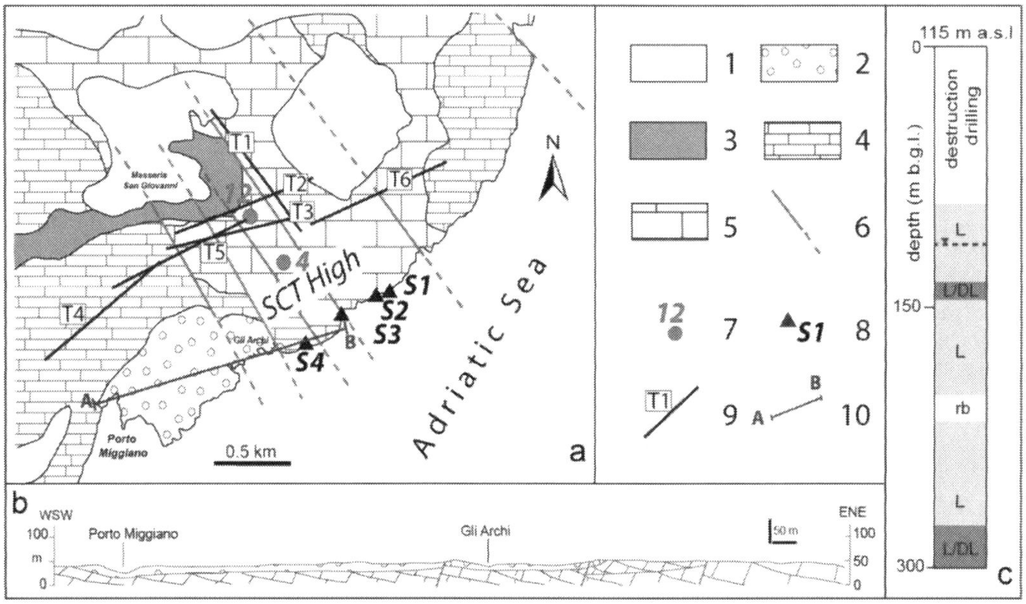

Figure 1. a) Simplified geological map. Legend: 1) red clays, 2) Salento Calcarenites, 3) Porto Badisco Calcarenites, 4) Castro Limestone, 5) Altamura Limestone, 6) fault (a-transtensional, b-normal, dashed when inferred), 7) borehole 12, 8) thermal spring, 9) total Electrical Resistivity Tomography (ERT) data profile, 10) trace of the geological section; b) geological section; c) stratigraphic column of borehole 4. Legend: L=Limestone, L/DL=Limestone and dolomitic limestone, rb= Microcrystalline limestone with traces of macrofossils.

RESULTS

The Figure 2 shows the total ERT profile obtained from the elaboration of the apparent electrical resistivity data coming from the profiles T3, T4, T5 and T6 in figure 1. Along the profile two empty data are well showed (white triangle), due to a lack of an overlap of the performed measurement. In general, the ERT image shows high electrical resistivity values (>500 Ohm*m). In details, the ERT image highlights two electrical resistivity distribution zones: the western one, from the starting point to around 1000 m, and the eastern one, from 1000 m to the end of the profile. The western zone shows a heterogeneity resistivity distribution with lateral and vertical variation. In details, between 700 m to 1000 m some vertical high electrical resistivity structures were recognized, while the first part of the profile shows horizontal electrical resistivity layers with low values (<500 Ohm*m).

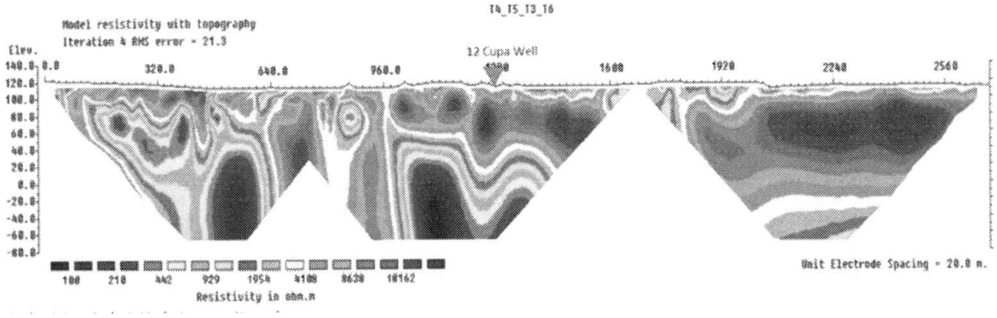

Figure 2. The total GRT profile inverted from the T4, T5, T3, and T6 data profile. The position of the GRTs profile is showed on Figure 1.

The eastern electrical resistivity zone displays a horizontal layer (>2000 Ohm*m) that increases in thickness going to the East (from 20 m to -80 m asl); below this layer, between 1000 m to around 1500 m a relative low resistivity layer (< 400 Ohm*m) is well defined.

DISCUSSION AND CONCLUSIONS

The complex electrical resistivity distribution could be interpreted from the geological point of view combining the geological knowledge of the area with the hydrogeological conceptualization. The geological formations (calcareous rocks) present in the area have very similar electrical characteristics (high resistivity values). However, the presence of some elements of erosion or fracturing and porosity variation due to the fault system assumes an important effect on the electrical resistivity heterogeneity. Therefore, some relatively lower resistivity values of the rocks are well defined, at least for the superficial parts on the western investigated zone, that should be associated to filled karst structures. Furthermore, in depth, their variability may depend on the presence of water circulations with salt characteristics and/or higher temperatures, as described by Archie's law. The thermal well (12, Cupa Well) shows thermal water at 0 m a.s.l., therefore the presence of a saline or heat fluid safely affects the resistivity distribution (i.e. water with high salt content or a high temperature produces a lowering of the resistivity values). Finally, the strong lateral variation (vertical structures) should be associated to the main fault system which delimited the deep geothermal zone well, as hypothesized on the basis of some well data.

REFERENCES

Abate, S., Aldighieri, B., Ardizzone, F., Barnaba, F., Basso, A., Botteghi, S., Caielli, G., Calvi, E., Caputi, A., Caputo, M. C., Cardellicchio, N., De Carlo, L., Casarano, D., Desiderio, G., De Franco, R., De Leo, M., Donato, A., Dragone, V., Festa, V., Giocoli, A., Inversi, B., Limoni, P.P, Liotta D., Lollino, P., Lombardo, G., Manzella, A., Masciale, R., Minissale, M., Montanari, D., Montegrossi, G., Mussi, M., Pagliarulo, R., Palladino, G., Parise, M., Perrone, A., Petrullo, A., Piemonte, C., Piscitelli S., Polemiol, M., Rizzo, E., Romanazzi, A., Romano ,G., Santaloia, F., Scrocca, D., Wasowski, J., and Zuffianò, L.E. 2015. VIGOR: Sviluppo geotermico nella regione Puglia – Studi di Fattibilità a Bari e Santa Cesarea Terme. Progetto VIGOR – Valutazione del Potenziale Geotermico delle Regioni della Convergenza, POI Energie Rinnovabili e Risparmio Energetico 2007-2013, CNR-IGG, ISBN: 9788879580168. www.vigor-geotermia.it

Aiello, G. and Budillon, F. 2004. Lowstand prograding wedges as 4th order glacio-eustatic cycles in the Pleistocene continental shelf of Apulia. SEPM Special Publication n. 81, Cyclostratigraphy: Approaches and Case Histories, ed. D'Argenio B et al., 215-230.

Bosellini, A., Bosellini, F.R., Colalongo, M. L., Parente, M., Russo, A. and Vescogni, A. 1999. Stratigraphic architecture of the Salento coast from Capo d'Otranto to Santa Maria di Leuca (Apulia, southern Italy). Riv. Ital. Paleont. Strat., 105 (3): 397-416.

Bossio, A., Mazzei, R., Monteforti, B. and Salvatorini, G. 2005. Stratigrafia del Neogene e Quaternario del Salento sud-orientale (con rilevamento geologico alla scala 1:25.000). Geologica Romana, 38: 31 – 60.

Caputo, R., Piscitelli, S., Oliveto, A., Rizzo, E., Lapenna, V. 2003. High-resolution resistivity tomographies in active tectonic studies. Examples from the Tyrnavos Basin, Greece. Journal of Geodynamics 36: 19–35.

De Giorgio, G., Chieco, M., Zuffianò, L.E., Limoni, P.P., Sottani, A., Pedron, R., Vettorello, L., Stellato, L., Di Rienzo, B., Polemio, M. 2018. The Compatibility of Geothermal Power Plants with Groundwater Dependent Ecosystems: The Case of the Cesine Wetland (Southern Italy). Sustainability, 10, 303.

Lapenna, V., Lorenzo, P., Perrone, A., Piscitelli, S., Rizzo, E., Sdao, F. 2003. High-resolution geoelectrical tomographies in the study of the Giarrossa landslide (Potenza, Basilicata). Bull. Eng. Geol. Environ. 62: 259–268.

Loke, M.H. 2001. Tutorial: 2-D and 3-D electrical imaging surveys. I: Course notes for USGS workshop 2-D and 3-D inversion and modelling of surface and borehole resistivity data. Storrs, CT, pp. 13-16.

Loke, M.H. and Barker, R.D., 1996. Rapid least-squares inversion of apparent resistivity pseudosections by a quasi-Newton method. Geophysical Prospecting, 44; 131-152.
Perrone, A., Iannuzzi, A., Lapenna, V., Lorenzo, P., Piscitelli, S., Rizzo, E., Sdao, F. 2004. High-resolution electrical imaging of the Varco d'Izzo earth flow (southern Italy). J. Appl. Geophys. 56: 17 –29.

Ricchetti, G., Ciaranfi, N., Luperto Sinni, E., Mongelli, F. and Pieri, P. 1988. Geodinamica ed evoluzione sedimentaria e tettonica dell'Avampaese Apulo. Memorie della Società Geologica Italiana, 41: 57-82.

Romanazzi, A., Gentile, F., Polemio, M. 2015. Modelling and management of a Mediterranean karstic coastal aquifer under the effects of seawater intrusion and climate change. Environ Earth Sci 74; 115-128.

Santaloia, F., Zuffianò, L. E., Palladino, G., Limoni, P. P., Liotta, D., Minissale, A., Brogi, A. and Polemio, M. 2016. Coastal thermal springs in a foreland setting: The Santa Cesarea Terme system (Italy). Geothermics 64: 344–361.

Tozzi, M. 1993. Assetto tettonico dell'avampaese apulo meridionale (Murge meridionali-Salento) sulla base dei dati strutturali. Geologica Romana, 29: 95 – 111.

Contact Information: Maurizio Polemio, Consiglio Nazionale delle Ricerche, Research Institute for Hydrogeological Protection (CNR-IRPI), Via Amendola 122 I, 70126 Bari, Italy, Phone: +39-080-5929584, Email: m.polemio@ba.irpi.cnr.it

Spreading of brine in the Puck Bay in view of in-situ measurements

Małgorzata Robakiewicz
Department of Coastal Engineering and Dynamics, Institute of Hydro-Engineering, Polish Academy of Sciences, Gdańsk, Poland.

ABSTRACT

Since autumn of 2010 in the north-eastern part of Poland underground gas stores are under construction by diluting salt deposits. A by-product of the technology applied is brine, which is discharged into the coastal waters of the Puck Bay (south Baltic Sea). In the pre-investment study a theoretical analysis of the mixing conditions in the near-field and far-field of the proposed installation was conducted. An extensive monitoring programme carried out since 2010 shows a good mixing of brine with the surrounding waters. Excess salinity due to the continuous discharge of brine estimated using data measured in situ is generally lower than permitted, i.e. not exceeding 0.5 psu in the near-field of installation.

INTRODUCTION

Increasing demands for gas storage capacity encouraged Polish Gas and Oil Company (PGNiG) to make use of salt deposits located in the north-eastern part of Poland, in the area bordering on the Puck Bay (the inner part of the Gulf of Gdańsk, South Baltic Sea), to create underground gas stores. A complex of 10 chambers (total volume of $250\text{x}10^6$ m^3) was designed for construction in Kosakowo, near Gdynia. Owing to local geological conditions, the chambers are created at a depth of 800-1600 m. The construction site (GSS, Figure 1) is located about 4 km away from the Baltic Sea coast. The drilling of boreholes and diluting of salt rock was proposed as a method of creating the chambers. Purified water from the treatment plant located in the vicinity (Dębogórze WWTP) is used for dilution purposes. For ecological reasons, the total volume of brine is limited to 300 m^3/h while its saturation cannot exceed 250 kg/m^3. The aim of this paper is to present results from the monitoring programme of brine spreading in the Puck Bay carried out in the years 2010-2017, in the context of theoretical analysis carried out in the pre-investment study.

THE STUDY SITE AND BRINE DISCHARGE SYSTEM

The Gulf of Gdańsk, situated in the south-eastern part of the Baltic Sea, is the area limited by the imaginary line between the Cape of Rozewie and the Cape of Taran (Figure 1). The Puck Bay is a shallow western sub-region of the Gulf of Gdańsk separated from the open sea by the Hel Peninsula. In the middle of the Hel Peninsula there is a shallow sandbank (called Rybitwia Mielizna) which divides the Puck Bay into two parts differing in circulation patterns: the eastern part, called the Outer Puck Bay (av. depth ~20.5 m), and the western part, called the Puck Lagoon (av. depth ~3 m). Currents in the region are generated mainly by wind and the accompanying water level variations in time and space. In addition, circulation in the Puck Lagoon is influenced by water exchange with the eastern part of the bay. Circulation patterns were described by Nowacki (1993) on the basis of occasional field observations of currents. It is quite well documented that water circulation patterns in the Puck Bay depend on the wind direction. The direction of surface currents is usually in accordance with the wind direction. The direction of bottom currents is frequently in accordance with surface currents, although in the south-eastern part of the Outer Puck Bay the opposite current direction can be observed. Scarce field measurements of the currents are insufficient to present a detailed picture of flow patterns that are generated by spatially and temporarily varying wind conditions in the region.

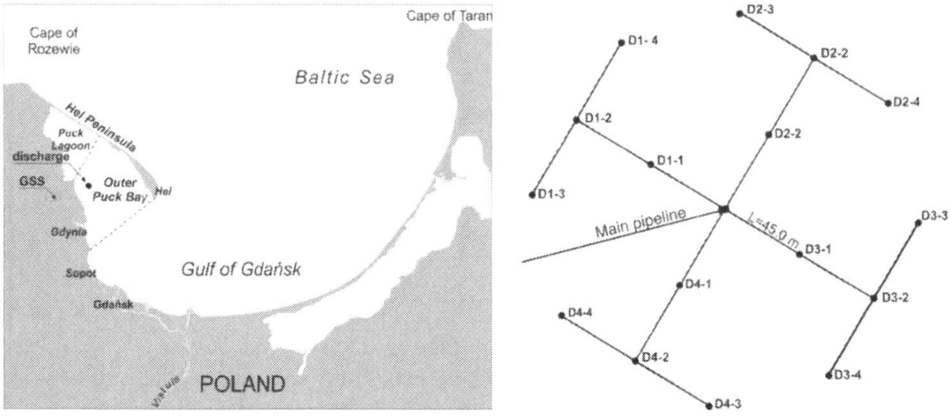

Figure 1. General view on the Gulf of Gdańsk, including the locations of the gas storage system (GSS) and the discharge system (left), and a general scheme of discharge system (right).

Water salinity in the Puck Bay is determined by interactions between marine and fresh waters. In the Outer Puck Bay, salinity is closely related to the inflow of more saline water from the Gdańsk Basin, predominantly through the bottom layer. In addition, the surface layer is modified by fresh water from the Vistula River. In the Puck Lagoon, salinity depends mainly on the intensity of water exchange with the Outer Puck Bay. The influence of small rivers is limited to their outlets.

The discharge system is located approx. 2300 m off-shore. It consists of 16 heads, each with 3 nozzles of 0.009 m diameter, distributed every 120° of circumference. The installation covers the area of 180x180 meters (Figure 1). Based on the pre-investment (Robakiewicz 2009) study brine is discharged 3 meters above the bottom at an angle of 45^0 to the ambient water. This system was designed to limit excess salinity to 0.5 psu in the near-field of installation.

THE MONITORING PROGRAMME

Execution of the monitoring programme was divided into two steps:

- **early-stage** (2010 – 2012), conducted to investigate the mixing of brine discharged into the coastal waters of the Puck Bay by a system of diffusers, and thus to verify theoretical assumptions made in the pre-investment study (Robakiewicz 2014, 2016).
- **basic** (2013 -), conducted to control the mixing of brine in the nearfield of discharge installation.

The analysis of the influence of brine discharge on the surrounding environment requires some general knowledge of hydrodynamic conditions in the region where the installation is located. To acquire such knowledge, in the early-stage of the monitoring continuous measurements of salinity, temperature, and water currents were carried out in two locations,

A and B (Figure 2), situated on the eastern and western sides of the installation. The measurements covered three periods, 13 October – 26 November 2010; 12 July –26 August 2011; and 22 May – 9 August 2012, which represented three stages of brine discharge (early, intermediate, target) characterized by different brine saturation and density. Current meters equipped with conductivity and temperature sensors were installed 1 meter above the bottom. Sample results for those are presented in Figure 3. Water currents in the region are generated by wind. The measured flow velocity at a depth of 1 meter above the bottom was low, hardly ever exceeding 0.1 m/s. It is characteristic that the predominant flow directions were to the north and south, which resulted from the local bathymetry and configuration of the coastline. In the autumn of 2010 (the early stage), when brine was discharged for only a few hours a day, brine saturation was relatively low. At that time, the values of salinity in locations A and B were very similar, differing at most by 0.1 psu. In the summer of 2011, the amount of brine increased to 250 m^3/h, and its saturation to 225 kg/m^3. During periods of weak dynamics, salinity differences between the two gauges did not exceed 0.25 psu. In the summer of 2012, discharged brine reached the maximum permissible parameters (total discharge – 300 m^3/h, saturation – 250 kg/m^3). Under such conditions, the maximum salinity differences between the two gauges occasionally reached 0.35 psu. The variation in salinity observed over time represents not only natural salinity changes in the region, but also anthropogenic influence. On the basis of the data collected, natural variability can be estimated as 6.6-6.8 psu in the autumn of 2010, 6.6-7.0 psu in the summer of 2011, and 7.3-7.8 psu in the summer of 2012. In-situ measurements conducted in the years 2010-2012 confirmed earlier observations that a two-directional flow dominates in this specific part of the Outer Puck Bay. They also showed that under weak wind conditions, the local water currents were low, although at no time were they completely absent.

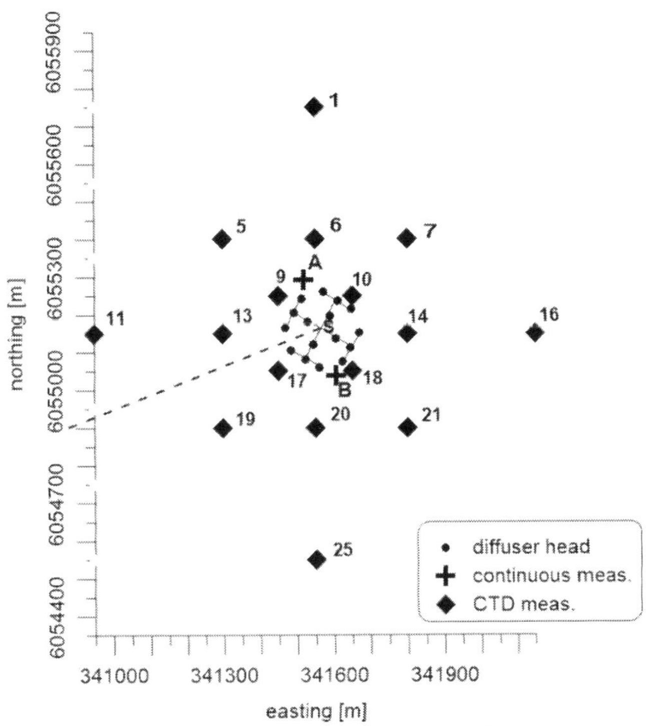

Figure 2. Location of points of the monitoring programme.

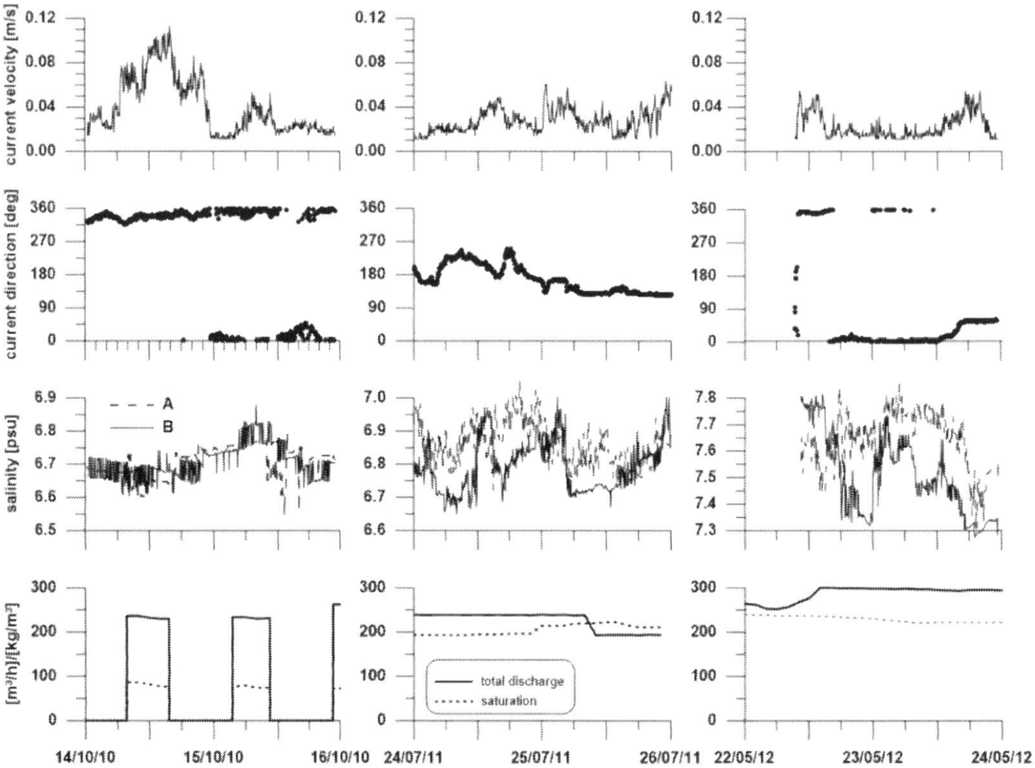

Figure 3. Continuous measurements in the early (left), intermediate (center) and target (right) stages of brine discharge – exemplary results.

The in-situ measurements of the spatial distribution of salinity in the nearfield of the installation started in October 2010. Up till now 42 series of CTD measurements were executed; in most cases they covered 17 verticals (Figure 2). To illustratre spreading of brine in the nearfield of installation two exemplary results, from the early (26.08.2011) and basic (23.10.2017) monitoring programme, are presented. Figure 4 shows spatial distributions of salinity in the bottom layer, and vertical profiles of salinity in chosen locations (S, 9, 10, 17, 18 – see Figure 2). It is very characteristic that in both cases salinity in the upper part of the water column is uniform (26.08.2011 – approx. 7.05 psu; 23.10.2017 – approx. 7.55 psu). Differences in salinity distribution in both cases are closely associated with wind conditions preceding the in-situ measurements, and natural salinity variablity in the Gulf of Gdańsk. The first in-situ measurement (26.08.2011) was executed after a three day period of a very gentle wind (av. velocity - 2.5 m/s; predominant direction – SE), while in the second case (23.10.2017) more severe conditions were present (av. velocity - 4.7 m/s, predominant direction – S). In a consequence, in the second case local currents were higher and assymetric salinity pattern in the vicinity of installation can be observed. Differences in mixing conditions resulted in substantial differences in vertical structure of salinity in the chosen dates (see Figure 4 – bottom).

To confirm good mixing conditions of a single jet of brine with the ambient water detailed CTD measurements in the vicinity of a single head (D1-1, Figure 1) were carried out on 26 August 2011. They covered 71 verticals in an area of 17x17 meters (Figure 5 left). The position of each vertical was registered by a GPS gauge. During in-situ measurements, brine saturated to 237.07 kg/m^3 was discharged through the D1 arm at a rate of 61.1-61.4 m^3/h, which means that the initial velocity at the nozzle was 22.20-22.45 m/s (Robakiewicz 2016). On the basis of the CTD data collected, it is possible to determine the spatial distribution of the highest salinity values in measured verticals (Figure 5 left). In this distribution, three lines of salinity increase can be distinguished, tracing the three jets ejected from the D1-1 head. The above-mentioned distribution was used to select verticals for the AA cross-section through a single jet (Figure 5). The comparison of the estimated, theoretical shape of a single jet with its actual shape measured on 26 August 2011, shows good agreement in terms of the terminal rise height. The horizontal distance from the source of the jet to its outer boundary at the bottom appears to be underestimated, which can be related to the fact that a gentle water current (~0.04 m/s) was observed during the in-situ measurements.

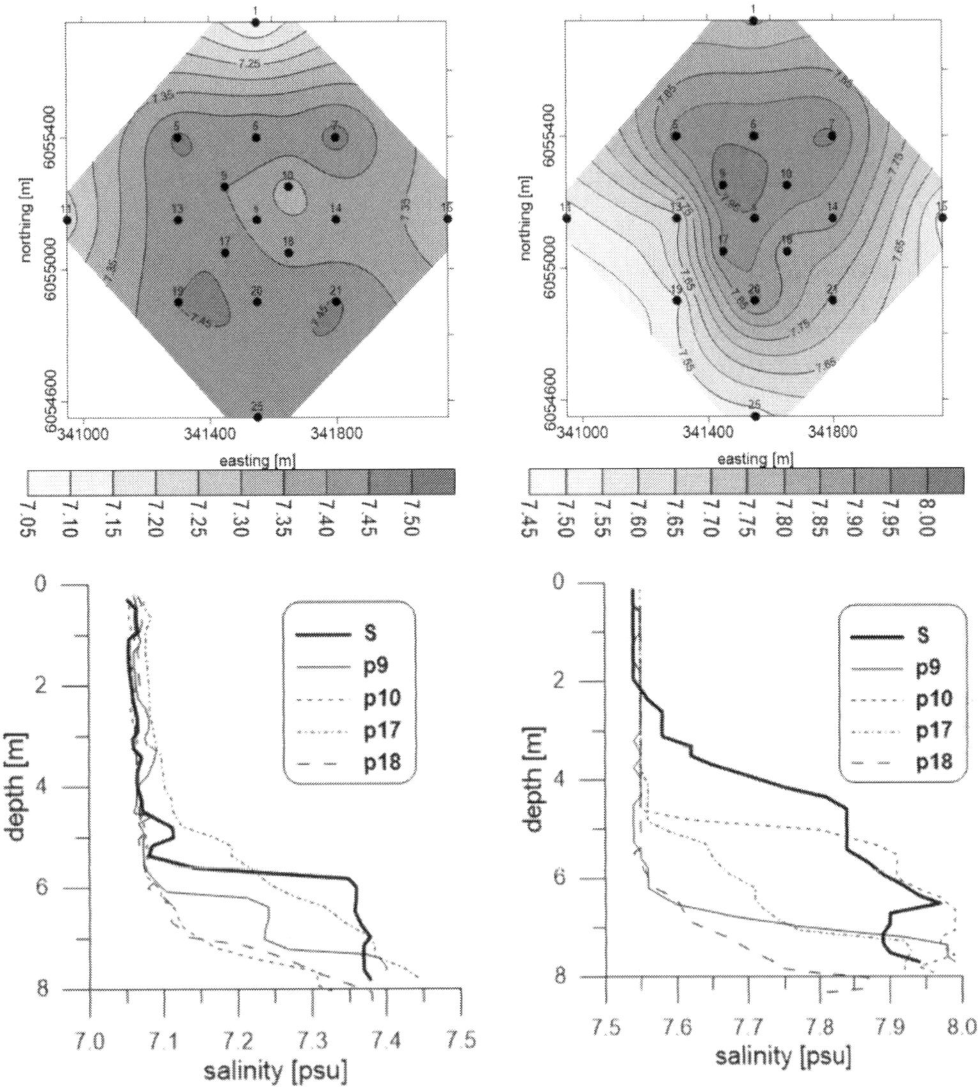

Figure 4. Salinity [psu] spatial distribution in the bottom layer (top) and profiles in selected verticals (bottom).

https://doi.org/10.1051/e3sconf/20185400029

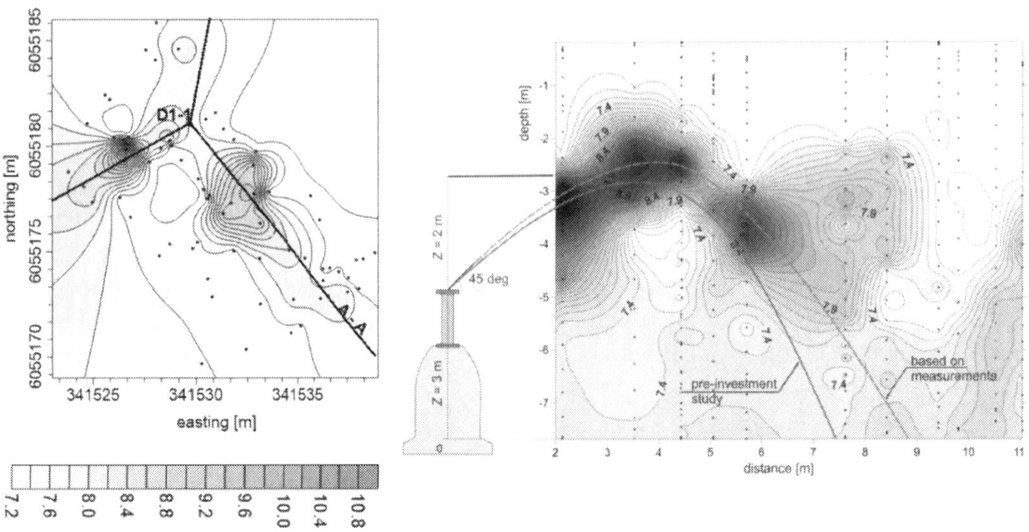

Figure 5. Spatial distribution of the highest salinity in verticals (left) and salinity in cross-section A-A in the vicinity of a single head carried out on 26.08.2011 (right).

DISCUSSION AND CONCLUSIONS

The extensive monitoring programme was carried out since the beginning of brine discharge into the Puck Bay. Continuous and short-term measurements confirmed significant natural variability of salinity in the Puck Bay region. Both, inflows of more saline water from Gdańsk Deep in the bottom layer, and traces of riverine water from Vistula have been registered. Measurements in the vicinity of a single head have shown that single jets introduced to the ambient water with high velocity mix intensively. In addition, the predicted trajectory of a single jet was confirmed by in-situ measurements.

Results from the monitoring programme confirm very good mixing of brine with the ambient water as predicted in the pre-investment study. They allow to conclude that the discharge system works in accordance with the obtained permission.

REFERENCES

Nowacki, J. 1993. Hydrophysics of the Puck Bay. In: Puck Bay (in Polish), ed. K. Korzeniewski, 79-111, 181-205. Institute of Oceanography of Gdańsk University.

Robakiewicz, M. 2009. Modelling of brine spreading in marine environment – Puck Bay example (in Polish), Przegląd Geologiczny. vol. 57: 778-779.

Robakiewicz, M. 2014. Salinity changes in the Bay of Puck due to brine discharge based on in-situ measurements. Oceanological and Hydrobiological Studies. vol. 43: 191-199.

Robakiewicz, M. 2016. Mixing of brine waste in the Puck Bay (south Baltic Sea) in light of in-situ measurements. Oceanological and Hydrobiological Studies, vol. 45: 42-54.

Contact Information: Małgorzata Robakiewicz, Institute of Hydro-Engineering PAS, Department of Coastal Engineering and Dynamics, Kościerska 7, 80-328 Gdańsk, Poland, Phone: +48585222945, Fax: +48585524211, Email: marob@ibwpan.gda.pl

Development and application of diagnostic tools for seawater intrusion analysis in highly heterogeneous coastal aquifers

M. Adil Sbaï[1] and N. Amraoui[1]
[1]Water Environment and Ecotechnologies Division, BRGM, Orléans, France

ABSTRACT

We developed novel, computationally efficient, methods as effective screening tools to analyze seawater intrusion processes in highly heterogeneous coastal aquifer systems. They enable delineation of pumping wells capture zones and swept zones associated to injection wells for remediation of seawater encroachment. Forward or backward travel times and residence time distributions are robustly simulated and visualized on the computational grid. These steady-state indicators, precomputed at fine grids, are used to generate optimal locally refined grids for efficient transient solute transport simulations.

INTRODUCTION

A large array of sweater intrusion (SI) processes may develop in hydrogeological environments. These effects include among others dispersive mixing, density and viscosity effects, unstable convective mixing, tidal effects, aquifer recharge variability, surface-subsurface interactions, and geological controls. The interplay between all these processes is a challenge for current generation of groundwater flow and transport models. One of the most challenging and still largely unresolved aspects of SI prediction is representation of the local geological features controlling SI. Only a handful set of well-documented case studies tackled the problem to highlight the influence of aquifer heterogeneity on SI processes.

Therefore, new methods have to be developed yet to address the shortcoming of traditional modeling approaches. Here we present a suite of diagnostic tools to fill existing gaps when analyzing SI processes in highly heterogeneous coastal aquifers. These methods are computationally cheap and equipped with novel visualization capabilities to enhance interpretation and understandings of flow and transport processes. They could be easily used at different stages of building, calibration, or updating seawater intrusion models. Two examples are given to illustrate their advantages when used in conjunction with a sharp interface or coupled flow and transport approaches.

APPROACH

The computational approach is an extension of a newly proposed grid-based alternative method to advective particle tracking (Sbai 2018). This is a powerful screening tool to accurately delineate and visualize capture zones around abstraction wells or outflow boundaries, the swept zones formed by injection wells or inflow boundaries, and the partitions associated with injection-pumping well doublets or inflowing-outflowing boundary pairs. The forward or backward travel times and residence time distributions are robustly simulated and visualized on the computational grid with little computational effort. This method is comparable to the particle-tracking approach (Pollock 1988). However, it holds more promise for complex groundwater applications because the interpretation and visualization of travel and residence times are easier.

The first extension proposed in this paper involves the interactions between the sea boundary and all sources/sinks of a sharp-interface groundwater flow model. A second addition is a novel way to generate optimal locally refined grids (LGR) for transient solute transport problems based on precomputed steady-state indicators derived from a fine scale steady-state flow and travel times solutions.

RESULTS

Two examples are presented to demonstrate the usefulness of the proposed extensions for seawater intrusion analysis in coastal aquifers.

Example 1

This first demonstration example considers four pumping wells in a highly heterogeneous hypothetical coastal aquifer system. As illustrated in Figure 1A the permeability distribution spans more than five orders of magnitude. An injection well is placed in the south of the domain near the coastline at the lower boundary. The effectiveness of this hydraulic barrier, over time, could be studied by computing and visualizing the expected spatial-temporal patterns of the flow field. A more effective way to understand the dynamics of hydraulic interactions between the injection well, pumping wells, and the salt-fresh water interface is to analyze the shape of swept zones corresponding to forward tracer partitions from the injector.

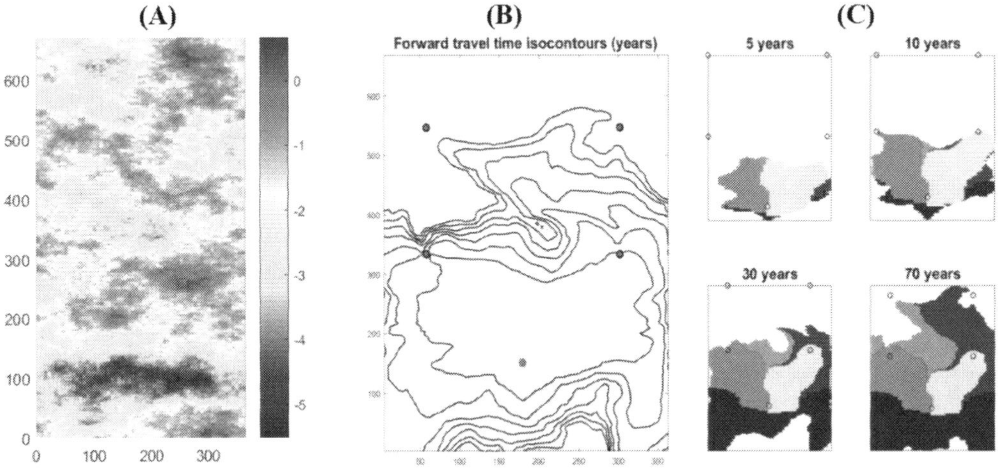

Figure 1. Permeability distribution in Log_{10} scale. (B) Forward time iso-values from 10 to 70 years with 10 years interval. Injection well is red and pumping wells are black dots. (C) Spatial-temporal patterns of swept zones formed by the injection well showing the strength and magnitude of hydraulic interactions between injection-pumping well pairs and its potential to remediate saltwater encroachment.

Figure 1B shows the contour lines of the forward arrival times simulated by a finite-difference grid-based method (Sbai 2018). Injected water reaches the pumping wells in the domain center and north positions after 10 and 70 years, respectively. Figure 1C shows the expansion of interaction zones between the injector and other sources/sinks such as the salt source from the sea and the pumping stations. In particular, we can conclude that with the prescribed flow rates remediation of seawater intrusion is starting to be effective after 30 years (notice the expansion of the dark blue partition in Figure 1C).

Because these results are obtained very efficiently (in less than a second in a laptop computer) the underlying method could be used for operational management of water resources in coastal aquifers. It could also be used in combination with a constrained optimization algorithm to design the optimal number, positions and flow rates of injection wells.

Example 2

In this second demonstration example we consider a binary distribution of the hydraulic conductivity field as shown in Figure 2A where permeable and impervious cells are at 1 and 10^{-3} m/d, respectively. Groundwater flow is controlled by a couple of injection and pumping wells placed at the lower left and upper right cells of the computational grid, respectively. Pumping and withdrawal rates are equal to 1 m^3/day. Effective porosity equals 25 %. Groundwater flow and finite-difference based forward and backward travel times are simulated on a fine grid with 1 m uniform grid spacing in each direction. Next, a fine scale spatial distribution of $I_\tau = -[\log(\tau_f) + \log(\tau_b)]$ is computed and shifted to the origin as illustrated in Figure 2B.

A general-purpose algorithm is developed to optimize the generation of LGR grids for transient solute transport simulations in heterogeneous porous media. The novelty of this method is its reliability on physical (flow and transport) dynamics for grid generation unlike generally available tools standing rather on geometrical constraints. This algorithm involves two distinct steps. First, the fine scale grid is coarsened by a regular agglomeration of the cells following an nxm pattern, where n and m are the number of successive rows and columns to agglomerate. This step involves eventually a flow-based upscaling procedure to get representative hydraulic conductivity values for the obtained coarse grid. In a second phase, coarse blocks whose cumulative I_τ index exceed a given threshold value τ_c are taken as local grid refinement targets. A user-selected grid refinement level is therefore applied to all these cells leading to a fine child grid having inactive cells which overlap with active cells of its parent coarse grid, and vice-versa. Flow and transport simulations involve an iterative loop between the parent and child grid(s) where jumps in flow rates and concentration fluxes at their interfaces are minimized until convergence. This is similar to a domain decomposition method at the algebraic level. An example of such generated grid is shown is Figure 2C for the presented problem where $n = m = 5$ and the refinement level inside the child grid is 5-by-5.

As shown in Figures 3A and 3B groundwater heads computed with the fine and LGR grids look very similar in sensitive parts to solute transport dynamics namely inside the fine child grid of the composite LGR grid. As expected pure solute advection simulations results computed on fine and LGR grids are undisguisable as illustrated in Figures 3C and 3D resulting in overall reduction of the computational time and memory storage. This conclusion is confirmed by comparing the breakthrough curves of the solute at the pumping well shown in Figure 3E where an excellent agreement is obtained.

We expect a more impressive speedup for three dimensional problems in heterogeneous aquifers exhibiting high flow channeling or localized solute transport spreading. This is also the case for seawater intrusion problems based on coupled viscosity- and density-dependent flow and transport models because repeatedly solving the flow problem on an LGR grid could dramatically reduce the computational time and accelerate the outer (e.g. nonlinear)

 https://doi.org/10.1051/e3sconf/20185400030

loop convergence. All these issues are being under investigation and will be reported in the near future.

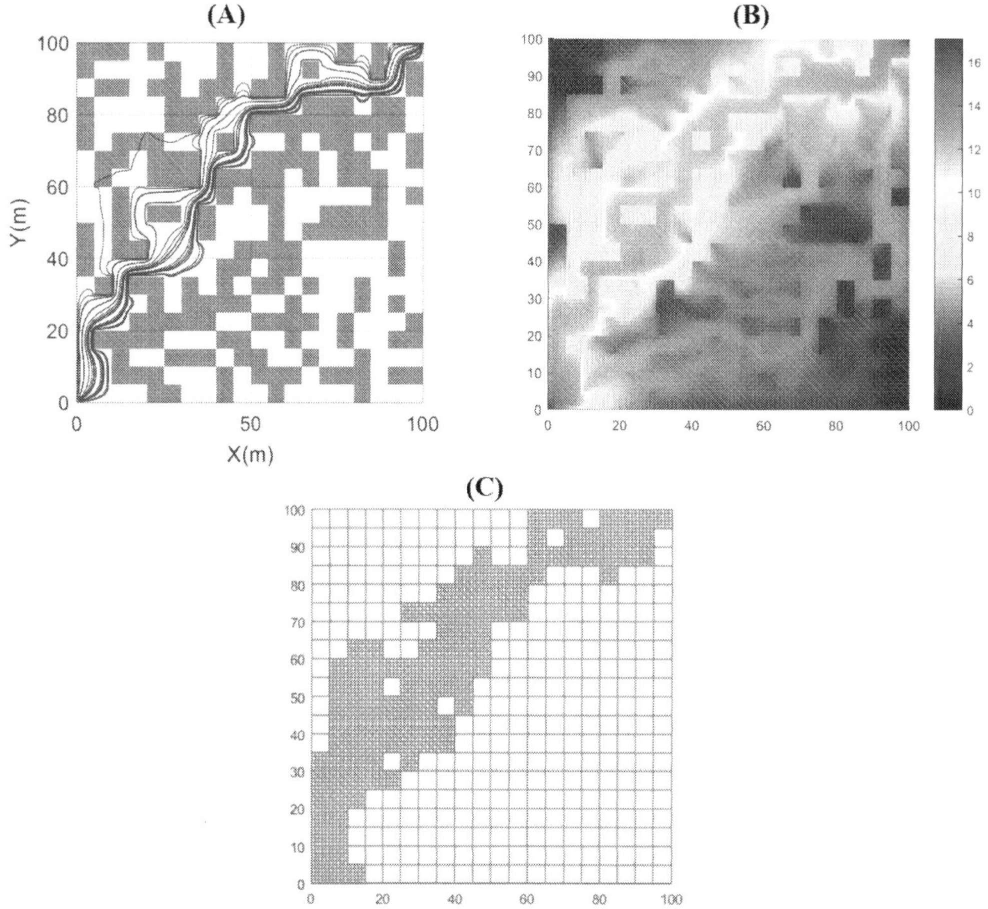

Figure 2. Hydraulic conductivity distribution showing impervious cells in gray. Flow pathlines of 20 particles released near the injection well are visualized. (B) Spatial distribution of I_τ indicator used for local grid refinement. (C) The obtained LGR grid to be used for the following transient solute transport simulation with τ_c= 30.

CONCLUSIONS

We introduce novel screening tools to analyze seawater intrusion processes in highly heterogeneous coastal aquifer systems. Illustrative examples were given to demonstrate:

- The capability to delineate the spatial-temporal interactions between the sea boundary and all sources/sinks of a sharp-interface groundwater flow model.
- The capability to generate optimal locally refined grids for transient solute transport problems based on precomputed steady-state indicators. Thus, we introduce a groundwater flow- and transport- based method for grid generation targeting finite difference models.

We are using these diagnostic tools to analyze and improve conceptual models of existing Mediterranean seawater intrusion pilot sites in South of France.

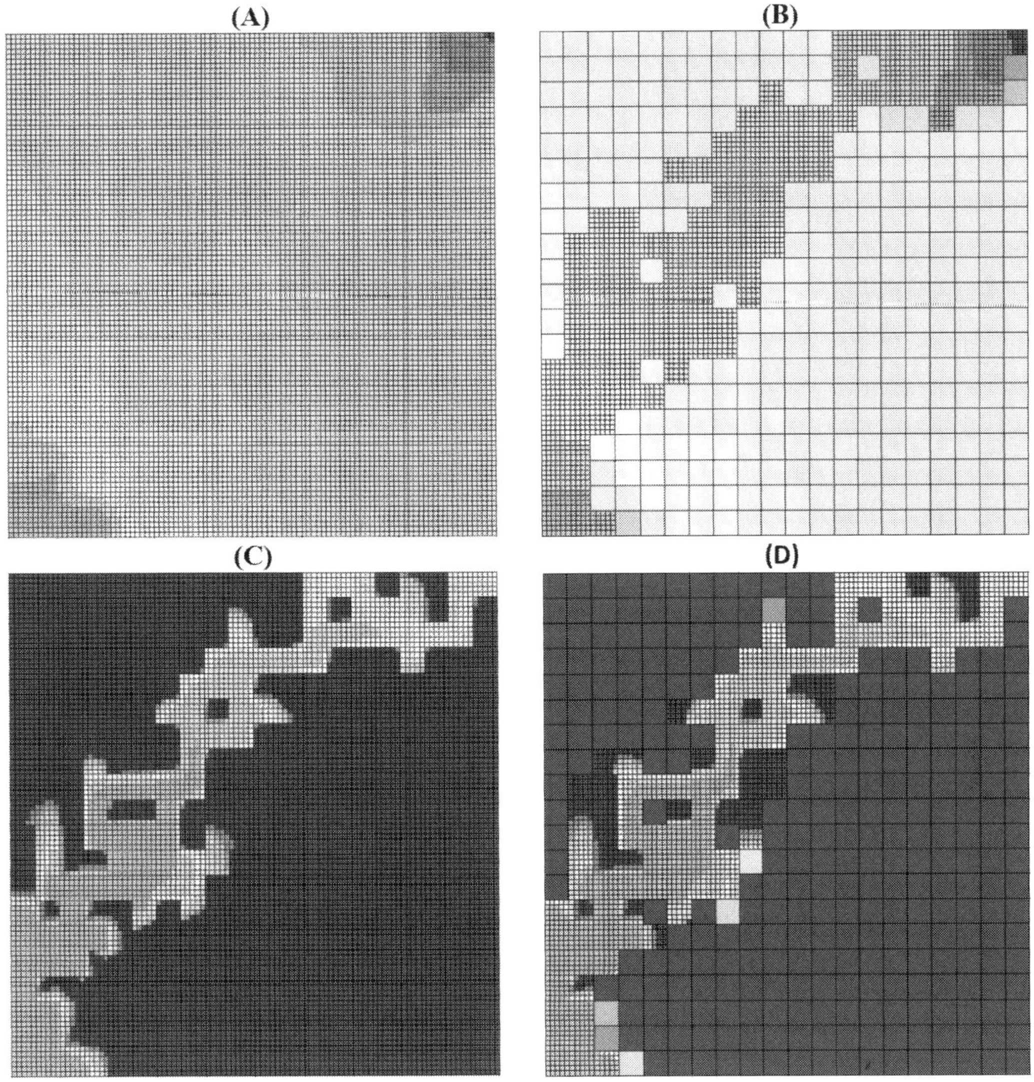

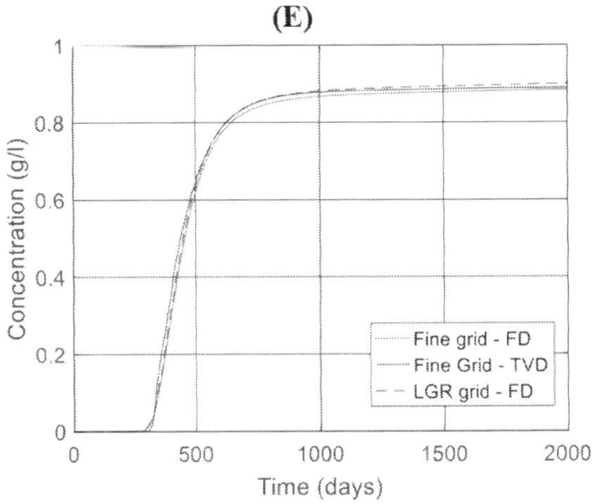

Figure 3. Spatial distributions of computed hydraulic heads on (A) fine and (B) LGR grids. Spatial distribution of simulated concentrations after 600 days of continuous injection/pumping using (C) fine and (D) LGR grids. A CPU time speedup close to two is obtained. (E) BTC's at the pumping well are shown for fine and LGR grids, respectively. Two fine grid solutions are produced with Finite-Difference (FD) and TVD schemes. An excellent agreement is obtained between the three simulations.

https://doi.org/10.1051/e3sconf/20185400030

REFERENCES

Pollock, D.W. 1988. Semianalytical computation of path lines for finite difference models, Groundwater 26(6), 743-50.

Sbai, M.A. 2018. A practical grid-based alternative method to advective particle tracking. Groundwater, accepted.

Contact Information: M. Adil Sbaï, BRGM (French Geological Survey), Water Environment and Ecotechnologies Division, Water Management Group, 3 Avenue Claude-Guillemin, BP 36009, 45060 Orléans Cedex 2, France, Phone: +33.2.38.64.35.27, Fax: +33.2.38.64.37.19, Email: a.sbai@brgm.fr

Modeling saltwater intrusion scenarios for a coastal aquifer at the German North Sea

A. Schneider[1], H. Zhao[1], J. Wolf[1], D. Logashenko[2], S. Reiter[2], M. Howahr[3], M. Eley[4], M. Gelleszun[4] and H. Wiederhold[5]
[1]Gesellschaft für Anlagen- und Reaktorsicherheit gGmbH (GRS), Braunschweig, Germany
[2]G-CSC, Goethe University, Frankfurt, Germany
[3]Oldenburg-Ostfriesian Water Board, Brake, Germany
[4]Leichtweiß Institute, Technical University, Braunschweig, Germany
[5]Leibniz Institute for Applied Geophysics, Hannover, Germany

ABSTRACT

A 3d regional density-driven flow model of a heterogeneous aquifer system at the German North Sea Coast is set up within the joint project NAWAK ("Development of sustainable adaption strategies for the water supply and distribution infrastructure on condition of climatic and demographic change"). The development of the freshwater-saltwater interface is simulated for three climate and demographic scenarios.

Groundwater flow simulations are performed with the finite volume code d³f++ (distributed density driven flow) that has been developed with a view to the modelling of large, complex, strongly density-influenced aquifer systems over long time periods.

INTRODUCTION

The project NAWAK started in 2013 as an associated project of eight German institutions including research institutes as well as local water supplying companies. It was funded by the German Federal Ministry of Education and Research (033W007A). The main objective was the implementation of a planning tool that enables the local stakeholders to know the possible range of impacts caused by climate and demographic change and to identify sustainable adaption strategies.

In the BMBF-funded project "Development of sustainable adaptation strategies of water management in coastal areas under the conditions of climatic and demographic change" (NAWAK, was applied to coastal aquifer systems near the German North Sea.

Three regional 3d density-driven flow models have been set up using the code d³f++. The aim of this works was forecasting the impact of different climatic and demographic scenarios on the freshwater availability. The largest of these models will be presented here, the Sandelermöns region, covering an area of about 1000 km² and based on detailed geological and geochemical data, a subarea related recharge model (PANTA RHEI, Meon et al. 2012) as well as data from geophysics. The model contains groundwater recharge, river discharge and the well fields of three waterworks.

Based on the calibrated model three scenarios regarding the impact of sea-level rise and changing recharge- and well pumping conditions were simulated. The results were provided to the local and regional stakeholders within a special developed planning tool. Furthermore, the same modeling approach was applied to two other coastal areas near the estuary of the river Elbe.

HYDROGEOLOGICAL MODELING

The Sandelermöns model covers an area of about 1,000 km² situated at the German North sea coast as shown in Figure 1. Almost half of the model region is low-lying and formed part of the North Sea some hundred years ago so that the aquifers lead saline groundwaters. The south-western part consists of moraines with notable recharge rates and freshwater aquifers below. The three waterworks Sandelermöns, Feldhausen and Klein Horsten are situated in these areas, competing for pumping concessions and worrying about the close-by saline waters.

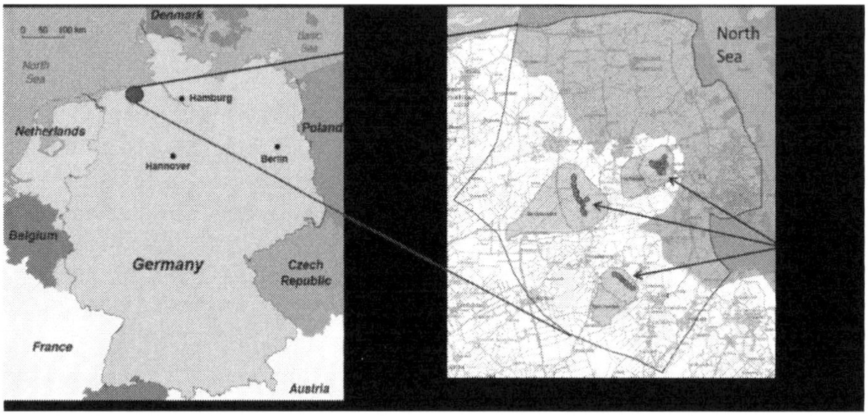

Figure 1. Situation of the Sandelermöns model area including the wells of the three waterworks and area of saline groundwater.

The hydrogeological structure model was set up by the Oldenburg-Ostfriesian Water Company (OOWV). Six formations are distinguished, starting with a thick layer of fluvial sands at the bottom, followed by small fields of Tergast clay and a continuous layer of melt-water sands. The top of the model consists of thin layers of Lauenburg clay, dune sands and silty materials. The base surface of each hydrogeological layer was read-in into the ProMesh tool (Reiter 2016) and together with ground surface data composed to a 3d model.

The south-western boundary of the model is located on a watershed and therefore assumed to be impermeable. The salt concentration is set to zero. The north-western and south-eastern boundaries are choosen to be perpendicular to the water table isohypses and therefore also regarded as impermeable for the flow. The coastal boundary is equipped with a hydrostatic pressure for seawater and a salt concentration of 34 kg/kg (35 g/l). The bottom of the model is also assumed to be impermeable because geological knowledge suggests almost impermeable clayey formations here. On the upper boundary, a Dirichlet boundary condition $p = 0$ has to be choosen for the pressure prescribed by the levelset method. The concentration is also set to zero.

The groundwater recharge was simulated by using the hydrologic model PANTA RHEI (LWI), a conceptual model with partly physic-based modules, especially for the soil water processes. 147 polygonal recharge zones were distinguished (see figure 2). Time-dependent recharge rates for the three climate scenarios WETTREG2010 B1 (moderate), WETTREG2010 A1B (long-term dry) and WETTREG2010 A2 (dry) from 2011 to 2050 were computed and coupled with the density-driven flow model d³f++. A sea-level rise of 0.25 m was assumed for this period.

The north-eastern region of the model domain is characterized by a dense net of small draining ditches and rivers conducting water to the coastal pumping stations in the dikes to keep the groundwater level below land surface. This drainage plays a crucial role in the hydraulic regime and had to be incorporated in the model. Because an explicit mapping of all ditches in a regional model is impossible, only the rivers of first and second orders were integrated, and their influence was smeared over a user defined range of surface elements.

Regrettably there exists only little information about the water levels of the receiving streams in the Sandelermöns region as well as the pumping rates of the coastal stations. Missing information was replaced by reasonable assumptions to meet the natural hydraulic regime by a careful calibration process. Furthermore, the 51 pumping wells of the waterworks were included into the model. Additionally, 36 private wells are regarded. For 2050 three different socio-economic scenarios influencing the water demand were investigated by the OOWV, a basic scenario, a green world scenario and a growth scenario. The resulting time-dependent pumping rates were used in the scenario-simulations by d³f++.

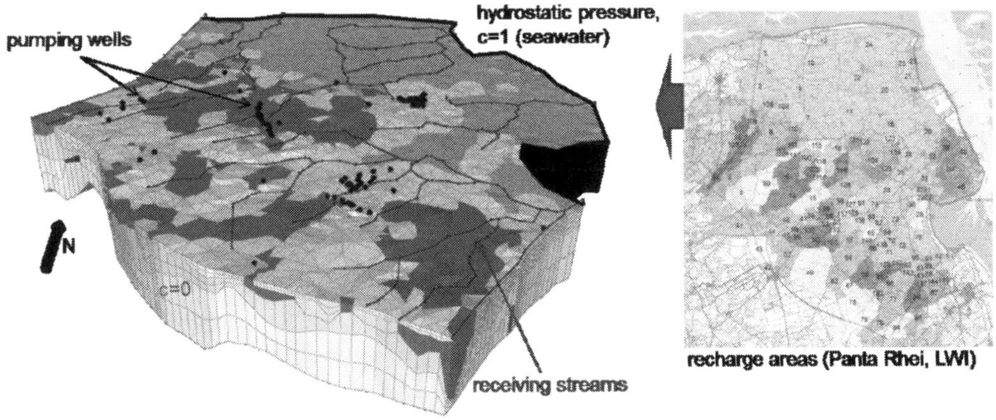

Figure 2. Sandelermöns: 3d hydrogeological model (30x vertically exaggerated) with coarse grid, rivers, pumping wells and recharge map (right).

To find an appropriate initial condition for the free groundwater table the data of 284 gauge wells were averaged over the year 2011 and interpolated. Unfortunately, only few gauges exist near the coastal line, and the model reacts very sensitive on changes in these settings. The initial condition for the salt concentration is based on geoelectromagnetic data provided by the Leibniz Institute for Applied Geology.

METHODS

For the density-driven flow model the finite volume code d³f++ is used (Fein 1999, Schneider 2016). The code may be used for groundwater flow and transport modelling in porous media as well as in fractured rock or mudstone including options for modelling salt- and heat transport as well as a free groundwater surface.

In d³f++, the equation system describing thermohaline flow is solved, consisting of the mass conservation of the fluid, the mass conservation of the brine, and energy conservation. The flow velocity follows Darcy's law. Fluid density and viscosity are depending on salt mass fraction and temperature. It should be mentioned that d³f++ solves the complete, nonlinear coupled equation system without simplifications such as the Boussinesq approximation.

d³f++ is based on the UG toolbox (Vogel et al. 2014). The use of modern numerical methods such as geometric and algebraic multigrid methods and their parallelisation enables simulations over long time periods with feasible computational effort.

In d³f++ the time dependent position of the groundwater table $\Gamma(t)$ in a fixed domain D is described as the zero set of a level set function (Frolkovič 2012). Using level set functions implicates some restrictions to these part of the model boundary: Regarding the pressure, the boundary condition $p = 0$ on $\Gamma(t)$ at any fixed time t is set directly by numerical discretization and cannot be changed. That means groundwater recharge as well as discharge may not be treated as boundary conditions, both effects have to be modelled as factors directly influencing the normal velocity S on $\Gamma(t)$.

RESULTS

The main objective of this works was predicting the situation of the freshwater-/saltwater interface up to the year 2050. Simulations started with the calibrated model in 2011 and the initial and boundary conditions as described above.

Firstly, the groundwater model had to be calibrated, based on mean values of the recharge and pumping rates in the base year of 2011. It was observed that river drainage relatively to the recharge rates is much more influencing the results than well pumping, which plays a less important role. These sensitivities form a huge problem because of the low number of gauges und the poor knowledge about the pumping rates at the overflows of the dikes. Another problem is that the results for the fluid volume may completely fail without an adequate grid refinement. Subject to the uncertainties caused by lack of data, a model calibration could be reached.

Based on the calibrated model 9 scenarios were simulated, combining both climatic and socio-economic scenarios. One difficulty was that the current distribution of saline waters con not be explained only by the hydraulic situation. The influence of the hydrostatic pressure boundary condition at the coastal line is restricted to a zone near the coastal boundary. Saline waters detected by measures outside this zone must result from former inundations hundreds of years ago, before the areas were protected by dikes. That means the simulated salt distribution is mainly determined of the initial condition. Using the airborne electromagnetic data from LIAG was therefore essential for the simulation of the scenarios.

The results of the combination of the WETTREG B1 climate scenario and the socio economic scenario basic are shown in figure 3 and figure 4. Figure 4 shows the movement of the freshwater-saltwater interface within 39 years of model time in a depth of 65 m, corresponding to the mean screen of the well field next to the saline areas.

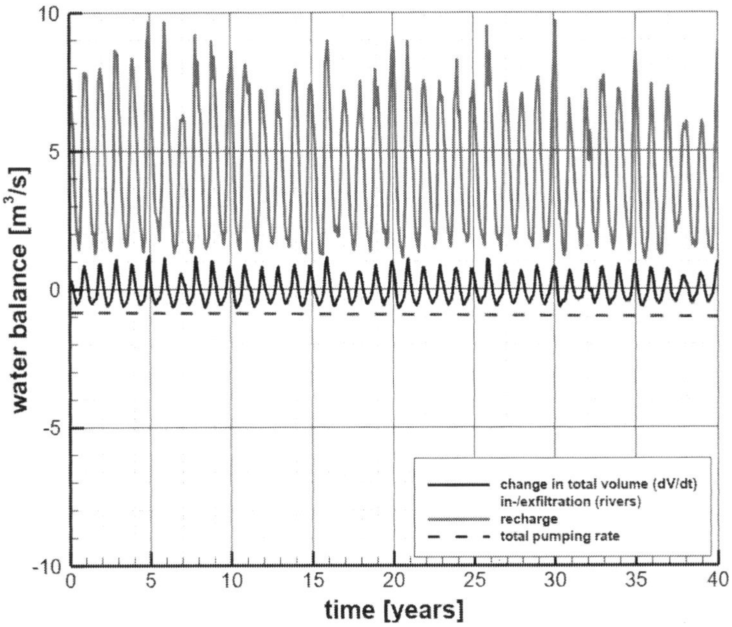

Figure 3. Water balance over the model time from 2011 to 2050, combination of the pumping rates of the basic scenario with the B1 climate scenario

Figure 4. Situation of the freshwater-saltwater interface 2015, combination of the pumping rates of the basic scenario with the B1 climate scenario

DISCUSSION AND CONCLUSIONS

One of the main results of these simulations was the predicted situation of the freshwater-saltwater interface. Besides the numerical effort a main challenge was the fact that in the modeling of this type of lowland aquifers the results are strongly controlled by the initial salt distribution and the influence of the drainage ditches. Since the initial salt distribution cannot be derived from the hydraulic regime alone, a good data base for both of these factors is essential to obtain confidential results. To be able to quantify the exfiltration by the drainage ditches in the model it is important to have more gauches and, at the other hand, that the pumping stations measure the run-off to the sea at the overflow of dikes.

Furthermore it was detected that a minimum grid refinement is crucial for getting correct results for the fluid volume or the position of the groundwater surface, respectively. Thereby it turned out again that grid convergence is an indispensable precondition for confidential simulation results, especially regarding forecasts.

In consideration of the lack of data and the still coarse grid the calibration results are rather good and plausible. Nevertheless, further grid refinements are necessary for reliable simulation results. It is desirable to achieve grid convergence, that means at least grid levels 3 or 4 (1.1 or 4.6 millions of nodes) have to be used. In the framework of the project "Implementation of strategic development goals in coastal aquifer management" (go-CAM, 02WGR1427B) the described work is continued.

REFERENCES

Fein, E., Schneider, A. (eds.): d³f—Ein Programmpaket zur Modellierung von Dichteströmungen. FKZ-02 C 0465 379, final report. Gesellschaft für Anlagen- und Reaktorsicherheit (GRS) mbH, GRS-139, Braunschweig 1999.

Frolkovič, P.: Application of level set method for groundwater flow with moving boundary, Adv. Wat. Res. 2012

Meon, G., K. Förster, M. Gelleszun and G. Riedel. 2012. Zukünftige Entwicklung von Wasserhaushalt und Hochwasser. In: NLWKN (Hg.): Globaler Klimawandel. Wasserwirtschaftliche Folgeabschätzungen für das Binnenland. Oberirdische Gewässer (33).

Reiter, S.: Mesh generation for geometric multigrid solvers -- results and roadmap, Simulation in Technology Workshop, KAUST, Thuwal, Saudi-Arabia, 10/2016.

Schneider, A. (ed.): Modelling of Data Uncertainties on Hybrid Comput-ers (H-DuR). Final report, FKZ 02 E 11062A (BMWi), Gesellschaft für Anlagen- und Reaktorsicherheit (GRS) gGmbH, GRS-392, 2016.

Vogel, A., Reiter, S., Rupp, M., Nägel, A., Wittum, G.: UG 4: A novel flexible software system for simulating PDE based models on high performance computers. Computing and Visualization in Science 16 (4), 165-179, 2014.

Contact Information: Anke Schneider, GRS gGmbH, Department Safety Assessment, Theodor-Heuss-Str., 38122 Braunschweig, Germany, Phone: +49-531-8012-248, Fax: +49-531-8012-248, Email: anke.schneider@grs.de

Comparison of manually and automatically derived fresh-saline groundwater boundaries from helicopter-borne EM data at the Jade Bay, Northern Germany

Bernhard Siemon[1], **Annika Steuer**[1], Nico Deus[2] and Jörg Elbracht[2]
[1]Federal Institute for Geosciences and Natural Resources (BGR), Hannover, Germany
[2]State Authority for Mining, Energy and Geology (LBEG), Hannover, Germany

ABSTRACT

The Federal Institute for Geosciences and Natural Resources (BGR) conducted many airborne geophysical surveys in Northern Germany during the last decades. The coastal regions of Lower Saxony were investigated by frequency-domain helicopter-borne electromagnetics (HEM) to reveal the bulk resistivity of the subsurface (sediments and pore fluids). The State Authority for Mining, Energy and Geology (LBEG) is preparing a statewide "saltwater map" for Lower Saxony with a focus on the coastal aquifers influenced by seawater intrusion. For this purpose, the HEM resistivities are used in combination with groundwater data and a geological model to derive the lower fresh-water boundary. As appropriate depth values are manually picked from vertical resistivity sections, this procedure is time consuming. Therefore, we tested an alternative, which automatically derives the fresh-saline groundwater boundary directly from the HEM resistivity models. The ambiguity between brackish/saline water and clayey sediments as source for low resistivities can somewhat be reduced by the application of gradients instead of threshold values for searching an appropriate boundary. We compare results of both methods using a dataset from a coastal region at the Jade Bay.

INTRODUCTION

A modern technique for electrical conductivity mapping is airborne electromagnetics (Paine and Minty 2005; Siemon et al. 2009). BGR operates, for example, two frequency-domain HEM systems (Resolve), which have been used to survey large parts of the coastal region of Lower Saxony in Northern Germany (Siemon et al. 2014) in cooperation with the Leibniz Institute for Applied Geophysics (LIAG). These datasets are currently used to map the seawater intrusion at the German North Sea coast (Deus et al. 2015).

As the depth of the fresh-saline groundwater boundary is an important information, e.g. for water suppliers, farmers and drainage organizations, the existing overview map (LBEG 1987) is not sufficient for many problems. In order to get a detailed status quo of the fresh-saline groundwater distribution at the coastal region of Lower Saxony, we are currently generating a model containing the lower fresh-water boundary for the entire coastline. This enables us to recognize and estimate changes related to climate change effects and to develop adaption strategies.

Survey area

The survey area Varel (319 km²) is located at the Jade Bay. The airborne survey was conducted in May 2014 (Ullmann et al. 2017). It includes the southern part of the Jade Bay and ranges from the city of Wilhelmshaven in the northwest to the Weser river in the southeast. This tidal flat area was flown during low tides in order to reduce the shielding effect of seawater.

Geology

The coastal region of the German Bight containing the East Frisian Islands, the tidal flats and the marshland are the youngest landscape elements in Lower Saxony. All sediments in this area have been accrued during the Quaternary and the accumulation is still continuing. During the last 2.6 million years, the coastal region was formed due to climate changes with at least three glacial periods, which were intercepted by two interglacials (Heunisch et al. 2007). After the Weichselian glaciation, the inland ice shields melted down and the sea level has begun to rise. The characteristic series of Holocene coastal sediments have been deposited since the Weichselian glaciation. They consist of strata of fine-grained sand, silt, clay and intercalated layers of peat (Streif 2004).

Hydrogeology

The main challenge in the study of coastal aquifers in Northern Germany is the fresh-saline groundwater boundary. Intrusion of seawater is a global challenge that could be affected by sea-level rise, changes in recharge and anthropogenic effects such as groundwater extraction or surface drainage and canalization (Werner and Simmons 2009). Low groundwater recharge rates in the marshland based on the widespread Holocene brackish and marine cohesive sediments and higher recharge rates in the moraine areas are typical for the coastal regions in Lower Saxony. The groundwater table is located very close to the surface and the upper aquifer system is between 50 and 200 m thick (LBEG 2018). The main groundwater aquifer consists of Pliocene to Saalian medium to coarse sands, in which low permeability layers are enclosed.

METHODS

Airborne EM

The penetration of the EM fields generated by HEM systems into the subsurface strongly depends on both the system parameter frequency and the ground parameter resistivity (inverse of electrical conductivity). Thus, the penetration depth is low at high frequencies or low resistivities and high at low frequencies or high resistivities. The exploration depth is thus dependent on the resistivity structure and may reach maximum depths of about 150 m for fresh-water saturated sand above saline groundwater (Siemon et al. 2011).

The inversion procedure used to model the HEM data is based on a Levenberg-Marquardt approach (Sengpiel and Siemon 2000), which applies the general matrix inversion based on singular value decomposition. The included regularisation weights the singular values. The weights are automatically determined by minimising the misfit (L1 norm). A pre-set (selectable) scaling factor (normally = 1.0) controls the model smoothness. We tested the inversion with few (6) and many (20) layers and found that smooth inversion using 20 layers and a scaling factor of 2.8 produced more stable results providing smooth but consistent resistivity models. On the other hand, extreme values generally could not be derived, i.e. resistive layers (fresh water) below or sandwiched between conductive layers (saline water or clay) appeared at somewhat lowered resistivities (compared to the few-layer inversion).

Manual picking of the bottom of fresh groundwater

The manual picking uses images of vertical resistivity sections derived from 1D HEM inversion models. These images are imported into a modelling software (Skua Gocad, Paradigm). In Lower Saxony, a chloride threshold of 250 mg/l is used to bound fresh (drinking) water. Assuming a water resistivity of 10 Ωm and a mean formation factor of 3, the lower fresh-water boundary is picked at about 30 Ωm. The application of resistivity threshold values for this picking is ambiguous, because both brackish groundwater and clayey sediments can cause similar resistivity values. Therefore, the background geological model is used to distinguish between both.

Automatic detection of the fresh-saline groundwater boundary

The ambiguity of resistivity threshold values is far more critical for the automatic detection of the fresh-saline groundwater boundary (FSB). Therefore, de Louw et al. (2011) introduced a concept using the largest ratio of consecutive layer resistivities within a pre-set resistivity range (2-5 Ωm). In extension to this ratio approach, we use a spline through the model layers followed by a calculation of steepest gradients. The spline calculated from log(r) vs. log(z) data is analysed with respect to its first and second derivatives. The spline consists of 100 values, which are equidistant in log(z). Maximum steepness of the first derivative occurs where the second derivative is zero. Generally, several sufficiently steep gradients occur. Therefore, only up to four depths assigned to the both highest (positive) and lowest (negative) gradients exceeding ±10 % are selected and stored together with the corresponding resistivities. Negative gradients occur if resistive material exists on top of conductive material, e.g. fresh water or sand above saline water or clay, and positive gradients describe the reverse case.

These steepest gradients representing the prominent layer boundaries allow following scenarios:

1) Homogeneous subsurface: Fresh (F) or saline (S) groundwater;
2) Two-layer case: Fresh above saline (FS) groundwater or the reverse case (SF); the latter requires an aquitard (clay) in-between, which generally cannot be resolved;
3) Three-layer case: A clay layer (with saline groundwater on top) is sandwiched between fresh-water aquifers (FSF) or located on top of fresh above saline groundwater (SFS);
4) Four-layer case: The latter case plus fresh groundwater on top (FSFS); the reverse case, i.e. fresh groundwater below SFS, can generally not be resolved.

In order to analyse the gradients with respect to these scenarios, the first step is to check whether an aquitard (positive gradient, scenario SF) exists or not. If no (no or one negative gradient), only the three simple cases (F, S, FS) are possible and the FSB is located below (unbounded), on top (fresh-water table) or within the resistivity model (depth of FS gradient). If yes (SF and FS gradients exist), the locations of the FS gradients with respect to the SF gradient have to be checked: Above (FSF), below (SFS) or above and below (FSFS). If two FS gradients with no intercalated SF gradient occur, i.e. the resistivity decreases stepwise, the upper one is selected, except this double boundary is located above an aquitard. Finally, the averaged resistivities of the depth ranges assigned to F have to be checked whether they are above 10 Ωm. Otherwise, the FSB has to be moved upwards.

RESULTS

The survey area Varel at the Jade Bay was selected for a comparison of manual (LBEG) and automatic (BGR) identification of the fresh-saline groundwater boundary. Figure 1 shows two resistivity maps at -3 and -30 m amsl (above mean sea level), which were derived from 1D HEM inversion models. In the southwest, a resistive feature appears on both maps indicating a thick fresh-water zone. The Jade Bay area in the north and part of the adjacent onshore areas appear conductive at shallow depth and less conductive at greater depth indicating fresh or at least less saline groundwater below the saline cover, particularly in two areas: a) The extension of the SE-NW trending fresh-water zone, and b) to the south of a former island (Arngast). The fresh water aquifer in the eastern half of the survey area seems to be thinner than in the thick fresh-water zone. Furthermore, a conductive cover occurs at the northeastern edge of the area, which is obviously caused by clayey sediments.

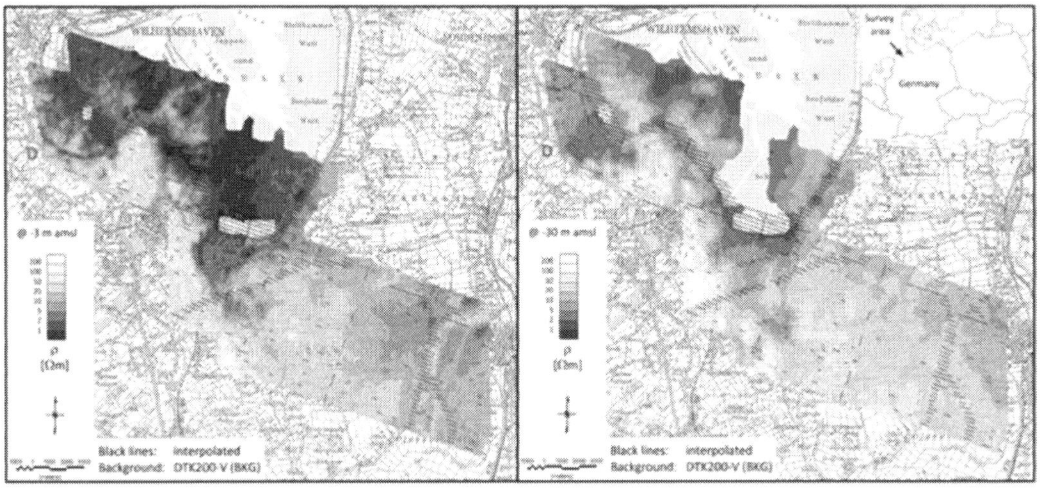

Figure 1. Resistivity at -3 m amsl (left) and -30 m amsl (right) derived from 1D HEM inversion models. The black lines mark erased and interpolated data.

Figure 2 shows the results of both LBEG and BGR approaches. They are similar, but not identical. The greatest differences occur in the southwest, where the deep fresh-water aquifer exists. There, the HEM models do not indicate a clear FSB, i.e. the fresh-water aquifer is unbounded in the models and the FSB is set onto the maximum model depth. On the other hand, the automatically derived values are generally located deeper than the manually picked values. That is not too surprising, because the manual picking searches for a boundary located at the upper level of the transition zone between fresh and saline water, whereas the automatic approach can only reveal the center of this transition zone. Thus, it depends on the thickness of the transition zone, how far apart both values are. The resulting mean difference of -16 m, however, cannot be completely explained by the thickness of the transition zone, which is assumed to be 10-15 m. The smoothness of the HEM models could be another reason for this discrepancy, because resistivity values are often lower if the corresponding resistive zone is sandwiched between or located below conductive zones. This situation affects approaches using threshold values stronger than approaches using relative values such as gradients.

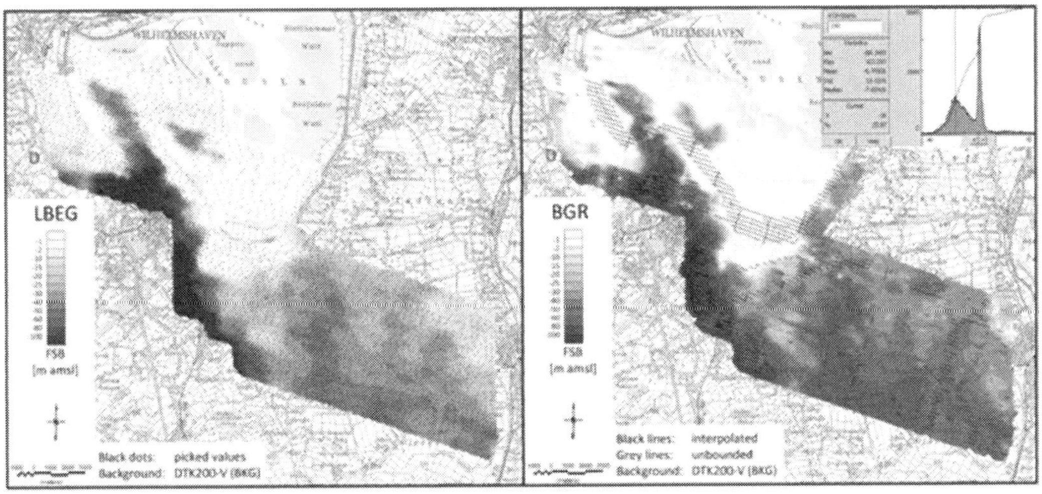

Figure 2. Elevation of fresh-saline groundwater boundaries: Manually picked (LBEG, left), automatically derived (BGR, right) from 1D HEM inversion models. A histogram of the differences (on the right map) reveals two maxima at 0 and .16 m.

DISCUSSION AND CONCLUSIONS

Airborne electromagnetics has proven to be able to reveal the fresh-saline groundwater boundary in a flat area at the Jade Bay. The HEM inversion models were used to derive that boundary by both manual picking and automatic detection. The thickness of fresh-water aquifers could be mapped onshore and offshore, even below a conductive cover such as clayey sediments saturated with saline water. Generally, the manual approach provides shallower results than the automatic approach. This is due do different targeting, because the manual approach searches for the lower drinking-water boundary and the automatic approach reveals the center of the fresh-saline transition zone. Besides this general offset, the smoothness of the inversion models has be taken into account. That particularly affects approaches applying threshold values more than approaches using relative values such as gradients.

In order to further reduce the ambiguity discussed in this paper, detailed knowledge on lithology and the influence of the diverse lithological units on the bulk resistivity is necessary. Such an approach was used to interpret the HEM data of the Province of Zeeland in the Netherlands with respect to groundwater salinity (project FRESHEM, Delsman et al. 2018). Siemon et al. (2018) discuss the differences of the FRESHEM approach and the automatic detection with respect to the fresh-saline groundwater boundary, which are rather small.

REFERENCES

de Louw, P.G.B., Eeman, S., Siemon, B., Voortman, B.R., Gunnink, J.L., van Baaren, E.S., and Oude Essink, G.H.P. 2011. Shallow rainwater lenses in deltaic areas with saline seepage. Hydrology of Earth System Sciences 15: 3659-3678, doi: 10.5194/hess-15-3659-2011.

Delsman, J.R., van Baaren, E.S, Siemon, B., Dabekaussen, W., Steuer, A., Gunnink, J.L., Karaoulis, M.C., Pauw, P.S., Vermaas, T., Bootsma, H., de Louw, P.G.B., and Oude Essink, G.H.P., 2018. Large-scale, probabilistic airborne salinity mapping for groundwater management in Zeeland, The Netherlands. In Program of 25th Salt Water Intrusion Meeting, 17.-22.6.2018, Gdańsk, Poland.

Deus, N., Elbracht, J., and Siemon, B., 2015. 3D-Modelling of the salt-/fresh water interface in coastal aquifers of Lower Saxony (Germany) based on airborne electromagnetic measurements (HEM). In Proceedings of AquaConSoil, 13th International UFZ-Deltares Conference on Sustainable Use and Management of Soil, Sediment and Water Resources, 9.-12.6.2015, Copenhagen.

Heunisch, C., G. Caspers, J. Elbracht, A. Langer, H.-G. Röhling, C. Schwarz, and H. Streif, 2007. Erdgeschichte von Niedersachsen. GeoBerichte 6: Landesamt für Bergbau, Energie und Geologie (LBEG).

LBEG, 1987. HUEK200 – Groundwater salinization, hydrogeological maps, NIBIS® Map-Server State Authority for Mining, Energy and Geology, Map accessed 29 January 2018 at http://nibis.lbeg.de/cardomap3/.

LBEG, 2018. HUEK200 – Thickness of the upper aquifer-system, hydrogeological maps, NIBIS® Map-Server State Authority for Mining, Energy and Geology, Map accessed 26 January 2018 at http://nibis.lbeg.de/cardomap3/.

Paine, J.G., and Minty, R.S. 2005. Airborne Hydrogeophysics. In Hydrogeophysics, ed Y. Rubin, and S.S. Hubbard, 333-357. Water Science and Technology Library 50, Springer.

Sengpiel, K.-P., and Siemon, B., 2000. Advanced inversion methods for airborne electromagnetic exploration. Geophysics 65: 1983-1992, doi:10.1190/1.1444882.

Siemon, B., Christiansen, A.V., and Auken, E, 2009. A review of helicopter-borne electromagnetic methods for groundwater exploration. Near Surface Geophysics 7: 629-646, doi: 10.3997/1873-0604.2009043.

Siemon, B., Steuer, A., Ullmann, A., Vasterling, M., and Voß, W., 2011. Application of frequency-domain helicopter-borne electromagnetics for groundwater exploration in urban areas. Journal of Physics and Chemistry of the Earth 36, no. 16: 1373-1385, doi:10.1016/j.pce.2011.02.006.

Siemon, B., van Baaren, E., Dabekaussen, W., Delsman, J., Karaoulis, M., Gunnink, J., Pauw, P., Meyer, U., and Vermaas, T., 2017. Automatic identification of fresh-saline groundwater interfaces from airborne EM data in Zeeland (NL). In Tagungsband zur 78. Jahrestagung der Deutschen Geophysikalischen Gesellschaft, 12.-15.2.2018, Leoben.

Siemon, B., Wiederhold, W., Steuer, A., Miensopust, M.P., Voß, W., Ibs-von Seht, M., and Meyer, U., 2014. Helicopter-borne electromagnetic surveys in Northern Germany. In Programme and Proceedings of SWIM 2014, ed Wiederhold, et al., 375-378, 23rd Salt Water Intrusion Meeting 2014, 15.-22.6.2014, Husum, Germany.

Streif, H.-J., 2004. Sedimentary record of Pleistocene and Holocene marine inundations along the North Sea coast of Lower Saxony, Germany. Quaternary International 112: 3-28.

Ullmann, A., Siemon, B., Ibs-von Seht, M., and Pielawa, J., 2017. Technischer Bericht Hubschraubergeophysik Befliegung Varel, 2014. BGR-Bericht, Archiv-Nr. 0134954, Hannover.

Werner, A.D., and Simmons, C.T., 2009. Impact of sea-level rise on sea water intrusion in coastal aquifers. Ground Water 47: 197-204.

Contact Information: Dr. Bernhard Siemon, Bundesanstalt für Geowissenschaften und Rohstoffe (BGR), Stilleweg 2, D-30655 Hannover, Germany, Phone: +49-511-6433488,
Fax: +49-511-6433662, Email: Bernhard.Siemon@bgr.de

Evaluation of groundwater potential and saline water intrusion using secondary geophysical parameters: A case study from western Maharashtra, India

N. Suneetha[1] and Gautam Gupta[1]
[1]Indian Institute of Geomagnetism, NewPanvel-410218, India

ABSTRACT

Aquifers along the coastal regions around the world are facing severe level of saline water intrusion problems. Rapid development and the associated increase in groundwater withdrawals intensify the problem. Extensive mapping of migration and extent of salt water plumes is difficult and costly. Several surficial geophysical methods have been developed for measuring salinity levels in coastal aquifers. The present study is an attempt to delineate the saline water and fresh water intrusion in parts of west coast of Maharashtra, India. A total of 86 vertical electrical soundings were carried out using the Schlumberger configuration. The contour maps for Dar-Zarrouk parameters viz. the transverse resistance (T), longitudinal conductance (S), and coefficient of anisotropy (λ) were computed at 84 sites to generate the resistivity regime of saline and fresh water bearing formations. The results exemplify that the Dar-Zarrouk parameters provide a practical elucidation in demarcating the saline and fresh water aquifers, particularly when the resistivity data interpretation encounters constraints due to intermixing of saline water aquifers, fresh water aquifers etc. Several NE-SW and NW-SE oriented major lineaments and its cris-crosses have been observed in this region.

INTRODUCTION

Saltwater intrusion can pose severe problems to coastal areas with freshwater aquifer having marine-aquifer hydraulic interaction. Saltwater intrusion happens when low density of fresh groundwater interacts with the high density of saltwater in natural conditions. Other sources of contamination include anthropogenic activity like domestic/industrial waste water and agricultural activities. In these studies, hydro-geochemistry analysis from monitoring wells and geophysical methods were used. The most widely used geophysical method to assign, particularly in salinity mapping, is geo-electrical method (Loke, 2000). Various researchers around the world have applied geo-electrical method in demarcating coastal-area hydrogeology condition (Maiti et al., 2013; Gupta et al., 2014).

The present study area is located in Sindhudurg district, western Maharashtra, India, where sand and gravel aquifers are dominant, which are favourable for constructing high-yielding wells, are scare (Fig. 1). In the overburden, the aquifers are mainly composed of clayey sand. Shear zones are expected at several places in the study area (Deshpande, 1998), where the exposed basement is fractured. The fractured zones are likely to be potential proxy indicators for groundwater prospecting. Therefore, the delineation of geologically weaker zones such as fractures is of significant societal importance.

Direct Current (DC) resistivity sounding method is one of the most popular methods that have been extensively applied for solving hydrological, geothermal, environmental and engineering problems (Zohdy et al., 1989). In the DC resistivity method, current is introduced directly into the ground through a pair of current electrodes and resulting voltage difference is measured between a pair of potential electrodes. The method provides the apparent resistivity distribution against depth which is generally found to be approximately one-third of the distance between the electrode separations. The study area is covered by the Deccan volcanic rocks, most of the soils are derived from lateritic rocks and the groundwater, is circulated through a network of voids, conduits, joints and fractures. Hence monitoring the shallow distribution of true resistivity pattern in the area is vital for mapping the faults, fractures, joints, conduits and lineaments for groundwater exploration.

In the present work, VES data from 84 stations have been analysed using secondary geophysical indicators like longitudinal conductance, transverse resistance and coefficient of electrical anisotropy, in order to understand the inhomogeneous infiltrations of fluids through pores and geologically weak zones, such as faults and fractured zones, fluid percolation pattern near the sub-surface area and the sea water intrusions effects.

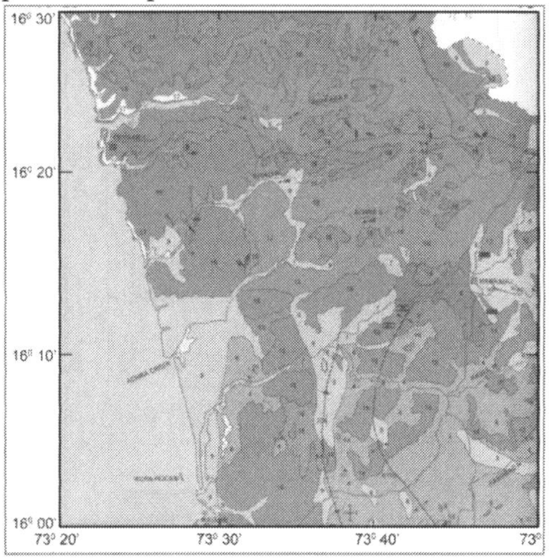

Figure 1. Geology of the area.

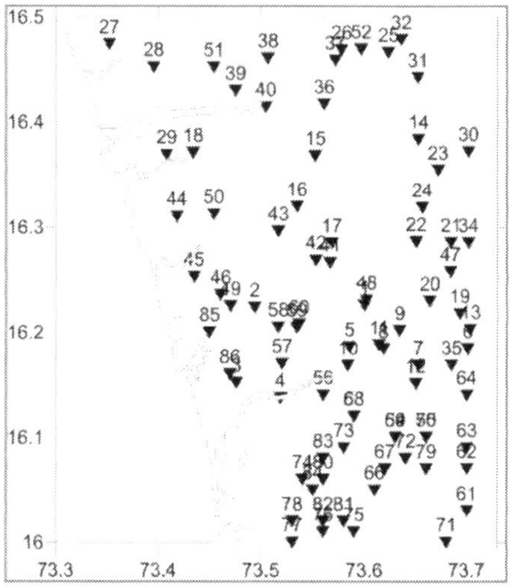

Figure 2. VES location map.

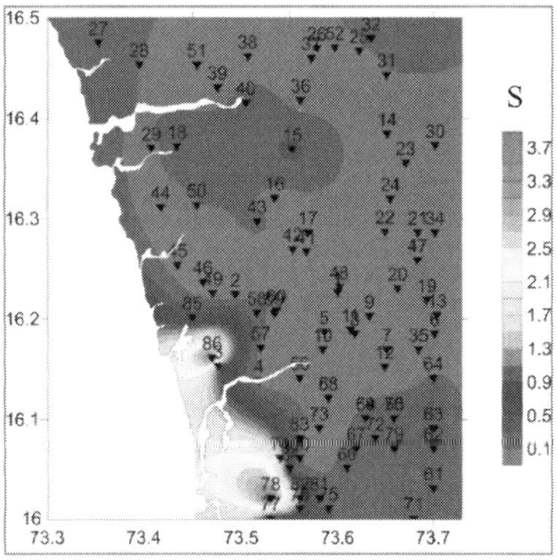

Figure 3. Longitudinal conductance map.

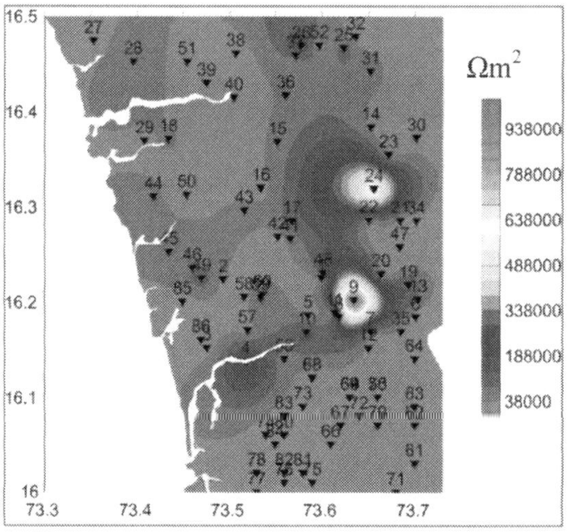

Figure 4. Transverse resistance map.

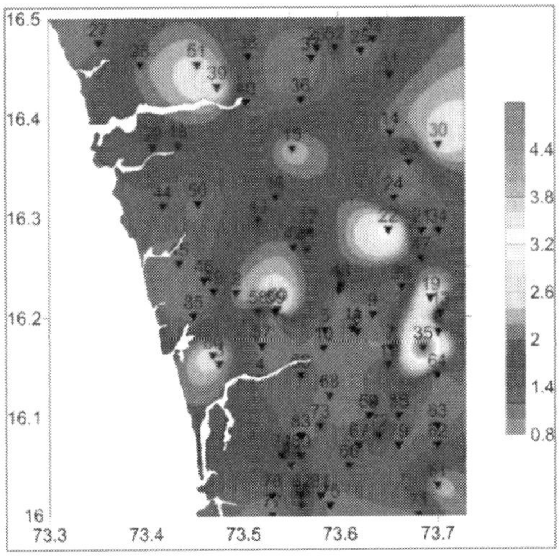

Figure 5. Electrical anisotropy map.

MATERIALS AND METHOD

The present study encompasses parts of Malvan-Vijaydurg-Kankavli, Konkan region (Fig. 2). Data were acquired with Schlumberger electrode configuration at 86 sites using IGIS made SSR-MP-AT instrument, with maximum current electrode spread (AB) of 200 m. The field data was processed and modelled using IPI2win inversion software (Bobachev 2003).

The sounding curves on log-log graph suggest 3-5 layered structure in the study area (Orellana and Mooney 1966). The resistivity and thickness values thus generated provided the primary parameters which were used to establish the secondary geoelectric indicators like transverse resistance (T), longitudinal conductance (S) and coefficient of anisotropy (λ), which helps in interpreting the subsurface lithological and structural characteristics with reduced uncertainty (Maillet 1947).

Total longitudinal conductance (S) is defined as,

$$S = \sum_{i=1}^{n} \frac{h_i}{\rho_i} \tag{1}$$

Similarly, the total transverse unit resistance (T) is defined as,

$$T = \sum_{i=1}^{n} h_i \rho_i \tag{2}$$

Where (ρi) and (hi) are the resistivity and thickness respectively and the subscript i indicates the position of the layer in the section.

Using eq. (1), the longitudinal resistivity of the current flowing parallel to the layers is given by,

$$\rho_l = \frac{H}{S} = \frac{\sum_{i=1}^{n} h_i}{\sum_{i=1}^{n} \frac{h_i}{\rho_i}} \tag{3}$$

Where H is the depth to the bottom most geoelectric layer.

Similarly, the transverse resistivity of the current flowing perpendicular to the layers is expressed using eq. (2) as,

$$\rho_t = \frac{T}{H} = \frac{\sum_{i=1}^{n} h_i \rho_i}{\sum_{i=1}^{n} h_i} \tag{4}$$

The longitudinal resistivity (ρ_l) is generally less than the transverse resistivity (ρ_t), unless the medium is uniform (Flathe 1955). Further, Keller (1982) suggested that ρ_l is dominated by the more conductive layers (in the present case, clay and weathered/fractured basalts) whereas ρt increases rapidly even if a small fraction of resistive layers are present.

Combining eq. (3) and (4), the coefficient of anisotropy (λ) is given by,

$$\lambda = \sqrt{\rho_t/\rho_l} \quad (5)$$

Here the secondary geophysical indices, namely T, S, ρ_l, ρ_t, and λ were evaluated at all the 84 VES sites so as to study the anisotropic nature and fracture geometry in the trap covered hard rock terrain for groundwater exploration. Kumar et al., (2014) suggested that high values of λ indicate different degrees of fracturing, with better water-holding ability in hard rock areas.

RESULTS AND DISCUSSION

The longitudinal conductance (S) value ranges between 0.0001 to 3.78 Ω^{-1} in the study area (Fig. 3), which help to differentiate changes in the total thickness of low resistivity materials (Galin 1979). This parameter reveals the disparity of the highly resistive basement topography, implying that high S values are indicative of deeper basement and vice versa. Also if the geologic sequence and clayey overburden is moderately thick, then high longitudinal conductance is suggestive of better protective capacity of aquifers (Oladapo and Akintorinwa 2007). Low S values of the order of 0.0001 to 0.4 Siemens is evenly spread in the central part of the study region encompassing VES stations 1, 2, 4-14, 16, 17, 19-26, 28-31, 33-57, 59-65, 67-73, 77, 79-81 and 83. In the western and southern part of the study region, high S values(0.5 to 3.78 Siemens) are observed covering the VES stations 3, 15, 18, 27, 32, 58, 66, 74-76, 78, 82, 84-86. The western part (VES stations 3, 78 and 86) is encroached by saline water presumably due to the vicinity of Arabian Sea. Geochemical analysis of groundwater in the study area (Suneetha and Gupta, 2017) reported that the electrical conductivity (EC) ranged between 174 and 9420 μS/cm (mean =686.4 μS/cm). The EC values recorded at VES points 3.78 and 86 falls beyond the acceptable level for drinking prescribed by World Health Organization (WHO, 1984), and were attributed due to the intrusion of saline water from Arabian Sea. The total dissolved salts (TDS) value at VES 3 (2,845 mg/l) also exceeds the acceptable limit prescribed by WHO (1984). Nutrient enrichment due to fertilizers and saline water intrusion could enhance TDS and, in turn, increases the EC in the study area, a fact which can be observed in other parts of the study area, reflecting high S values (VES points 3, 78 and 86). It is evident from Fig. 3 that there is a clear demarcation between saline and fresh water regions. The contour pattern and boundaries are distinct, clear and do not display any overlapping character. Due to limited depth of the investigations, these soundings end in fresh water zones. Hence the anomaly of fresh water is reflected in the S values. From the patterns of contours, it becomes easy to differentiate the region of saline water aquifers from that of the fresh water aquifers.

The transverse resistance (T) value varies from a minimum of 61 Ωm^2 to a maximum of 1055955 Ωm^2 in the entire study area (Fig. 4). Larger T values are associated with zones of high transmissivity and, hence highly permeable to fluid movement. The transverse resistance map gives a clear picture of the regions of saline and fresh aquifers. In the present scenario, saline aquifers categorize their presence by attaining T values in the range of 61-2000 Ωm^2 at VES points at south-eastern and VES 86 at the coastal side. Low T values are also observed at VES points 7, 9-10, 13, 16, 18-19, 24-25, 27-28 towards north, central and east.

Electrical anisotropy in the study area varies from 0.9-4.3 (Fig. 5). The coefficient of anisotropy is not uniform in all directions and is observed to increase from SW to NE and also from SE to NW, thus playing a major role in fracturing. More fracturing towards the NE an SW directions suggest relatively more prospective groundwater zone. Resistivity of subsurface rocks affects both the electrical anisotropy and porosity. The coefficient of anisotropy is generally 1 and rarely exceeds 2 in most of the geological settings (Zohdyet al. 1974). If the λ value varies between ~1 and up to 1.5, it is considered to be a prospective groundwater zone. Therefore low λ values are associated with lowest water table fluctuation and high λ values are related to higher water table fluctuation in the region.

CONCLUSIONS

Groundwater typically occurs in discrete aquifers in geologically intricate region and thus delineating the potential aquifer zones is often a tedious task. In the present study, the Dar Zarrouk (D-Z) parameters are particularly important and play a significant role in the construing of groundwater flow paths. Thus the D-Z parameters of the VES sounding points have been computed and the analysis shows that these parameters are useful and provide a confident solution in delineating the saline water and fresh water aquifers. This is more so when the resistivity data interpretation encounters constraints due to the intermixing of the resistivity of saline water aquifer and freshwater aquifer (Mondal et al. 2013). The behaviour of the D-Z parameters and its pattern in space over parts of Konkan region with respect to the occurrence of saline water and fresh water aquifer bodies in the coastal aquifer system has also been established. Excessive groundwater exploration may seriously affect the groundwater quality through the phenomenon of saline upconing. Therefore, it is important to maintain the balance between the fresh and saline water bodies in the coastal aquifer of Konkan region. Further these results will be useful to gain better insights of the complex geology of different intrusions in the hydrogeological system of the area.

REFERENCES

Bobachev, A.,2003. *Resistivity Sounding Interpretation*; IPI2WIN: Version 3.0.1, a 7.01.03, Moscow State University.

Flathe, H.,1955. Possibilities and limitations in applying geoelectrical methods to hydrogeological problems in the coastal area of northwest Germany.Geophysical Prospecting,3: 95–110.

Galin, D.L., 1979. Use of longitudinal conductance in vertical electricalsounding induced potential method for solving hydrogeologic problems.VestrikMoskovskogo University Geology,34:74–100.

Gupta, G.,Maiti, S., and Erram, V.C., 2014. Analysis of electrical resistivity data in resolving the saline and fresh water aquifers in west coast Maharashtra, India. Jour. Geol. Soc. India,84: 555-568.

Keller, G.V., and Frischknecht, F.C., 1966. Electrical methods in geophysical prospecting. Oxford, Pergamon Press Inc.

Maiti, S., Erram, V.C., Gupta, G., Tiwari, R.K., Kulkarni, U.D., and Sangpal, R.R.,2013a. Assessment of groundwater quality: A fusion of geochemical and geophysical information via Bayesian Neural Networks.Environ. Monit. Assessment,185:3445-3465.

 https://doi.org/10.1051/e3sconf/20185400033

Mondal, N.C., Singh, V.P. and Ahmed, S. (2013) Delineatingshallow saline groundwater zones from Southern India usinggeophysical indicators. Environ. Monit. Assess., v. 185, pp. 4869-4886.

Maillet, R., 1947. The fundamental equation of electrical prospecting. Geophysics, 12: 529–556.
Oladapo, M.I., and Akintorinwa, O.J.,2007. Hydrogeophysical study of Ogbese Southwestern, Nigeria. Global Journal of Pure and Applied Science,13(1):55-61.

Orellana, E., and Mooney, H.M., 1966. Master Tables and Curves for Vertical Electrical Sounding over Layered Structures. Interciencia, Madrid, Spain.

Todd, D.K., 1980. *Groundwater Hydrology*,In: 2nd Ed. Wiley India, 431-457.

Zohdy,A.A.R., Eaton, G.P. and Mabey, D.R. (1974) Application of surface geophysics to ground-water investigation.Techniques of water-resources investigations series of the United States Geological Survey, 2nd ed.

 https://doi.org/10.1051/e3sconf/20185400034

Hydrogeological structures as burried valley along the Eastern Pomerania Polish coast of the Baltic Sea

Anna Szelewicka[1]
[1]Marine Branch of Polish Geological Institute NRI, 5 Kościerska str., 80-328 Gdańsk

ABSTRACT

One of the characteristic feature of geological structure of the southern Baltic coast is presence of deep structures in Pleistocene sediments. Meaning of deep aquifers is unmeasurably important for water economy. The water bearing structures are situated in substantial depth and are better isolated from the surface than the water of shallow aquifers. It has important meaning for natural protection of these waters. Water resource of the structure is substantially increase in case when in water drainage zone where regional and local water circulation are focused. Simultaneously hydrogeological parameters of Pleistocene strata in the frame of deep structure are considerably advantageous rather than parameters of deeper aquifers.

INVESTIGATED AREA

Recognition and estimation of hydrogeological conditions of buried valley structure fulfilled with Pleistocene sediment was the main object of research work of Marine Branch of Polish Geological Institute, Gdańsk. The scope of research work has been attached to frame of sheets of Geological Map and Hydrogeological Map of Poland in scale 1:50 000.

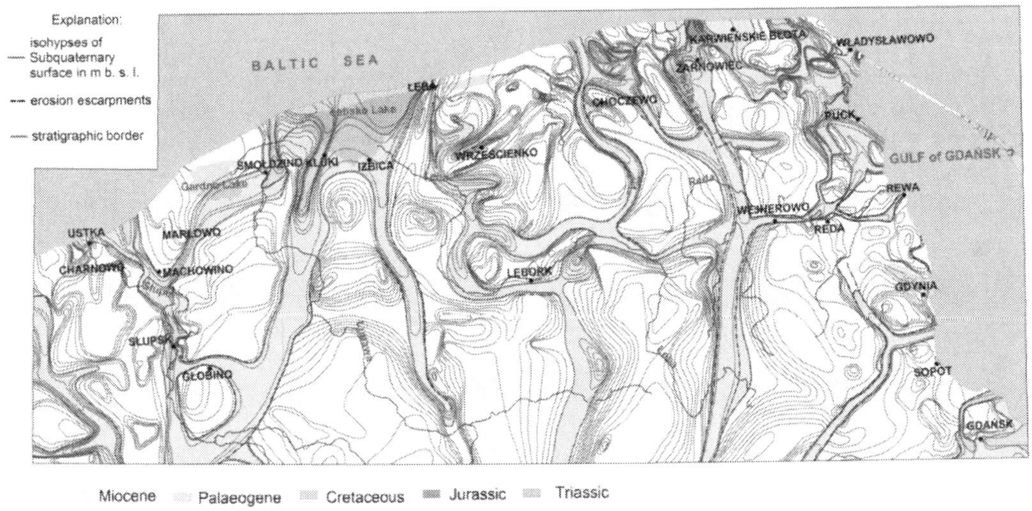

Figure 1. Subquaternary surface of the Eastern Pomerania Baltic coast (A. Szelewicka, 2011).

Elaboration of verified map of forming Subquaternary surface has allowed to singling out sequence of very often attached together long and narrow erosive structures and extensive depressions. Among this structures it is possible to differ two fundamental directions of their courses approximated for meridian and for parallel directions. Some structures like for example Łeba structure and others structures, where preferred by outflow of underground waters to Baltic sea.

GEOLOGY AND HYDROGEOLOGY

Considerable part of area is occupied by moraine plateaus and by washed out material and partly of frontal moraine hills. Most often it is sediment connected with the last Pleistocene glaciation. The Reda-Łeba glacial valley and seaside lowlands together with lake consist of loamy and peat sediments and dunes. Oligocene or Eocene formations, sometime even Cretaceous and exceptionally within Żarnowiec structure the older strata of Mesozoic are disclosed of the buried valley directly under Quaternary. Inside erosional structures has reached to Triassic formation and erosion of sediments has reached to 320 m b.s.l. Inside construction of Quaternary substratum are clearly presented three areas of Subquaternary surface which were named for the purpose of this elaboration as follows: Lowering of Łebsko Lake (I), Lowering of Puck (II) and Lowering of Vistula delta plane (III). Profile of Quaternary sediments starts at the oldest Scandinavian glaciation. Lowering of Subquaternary surface are fulfilled in general by glaciofluvial sediments and glacial tills. In the remaining area on formation of older substratum mainly Miocene sediments the Saalian and Elsterian sediments have been stated. Pleistocene sediments are ended by formation Vistulian glaciation. Lithological forms create complicated structures in plane and in geological profile where sandy deposits predominate and vertical glacial tills strata are separated and very often divided by sandy strata. Groundwater within different Quaternary aquifers occur in sands and gravel deposits of glaciations. The hydraulic headlines of deep Quaternary and Oligocene aquifer is presented on the map given on fig. 2.

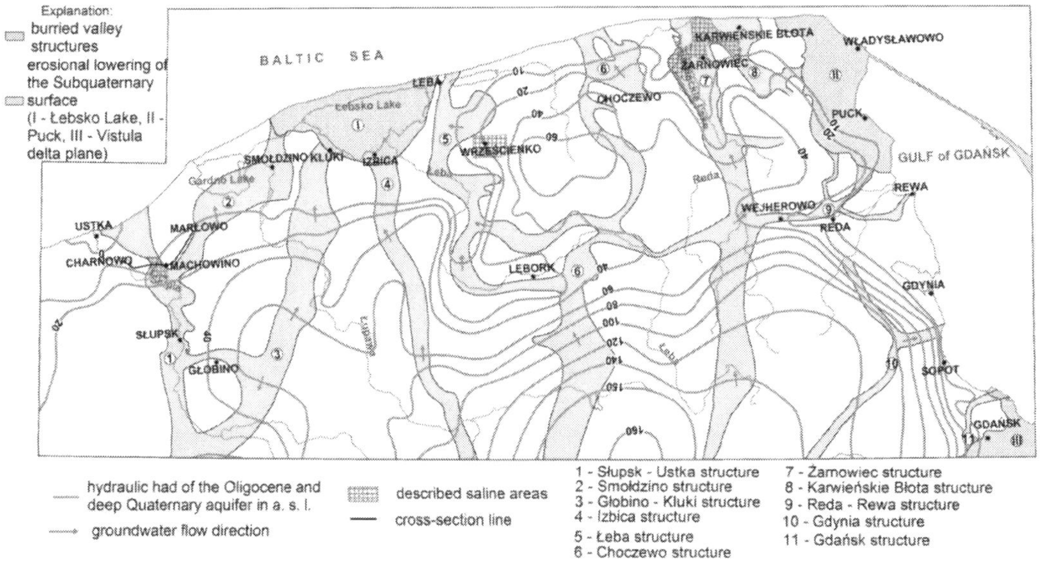

Figure 2. Scheme of the groundwater flows on the Eastern Pomerania coastal zone (A. Szelewicka, 2011).

Within structures of burried valley mainly slightly mineralized waters type HCO-Ca are observed. Salt water of Cl-Na type was stated in the areas of maximum depth of structures. The problem of salinity within Łeba structure was researched in drill hole in the area of Wrześcienko village, east direction from the structure. Concentration of chloride Ion in water was examined in this drilling hole in Wrześcienko on various depth:

- 12 mgCl/dm^3 on depth 84,0 – 96,0 (Q) – nearby drilling hole,
- 726 mgCl/dm^3 on depth 144,0 – 160,0 m (K),
- 7221 mgCl/dm^3 on depth 334,0 - 348,0 m (K),
- 79780 mgCl/dm^3 on depth 374,0 – 600,0 m (K).

Appearance of salinity in deep level of Cainozoic most probably should be linked with ascesion of salt water from Mesozoic strata. It has been stated concentration of chloride Ion in water in the area of Łeba caused by fossil water – Mesozoic brine (B. Kozerski, M. Pruszkowska 1996). Water in Słupsk – Ustka structure are low mineralized with concentration of chloride Ion from several to approx. 40 mgCl/dm^3 with mean value 18 mgCl/dm^3. Only in area where sampling from structure in the deepest aqueous layer was extracted it was observed very high concentration of chloride Ion maximum up to 855 mgCl/dm^3 in Machowino.

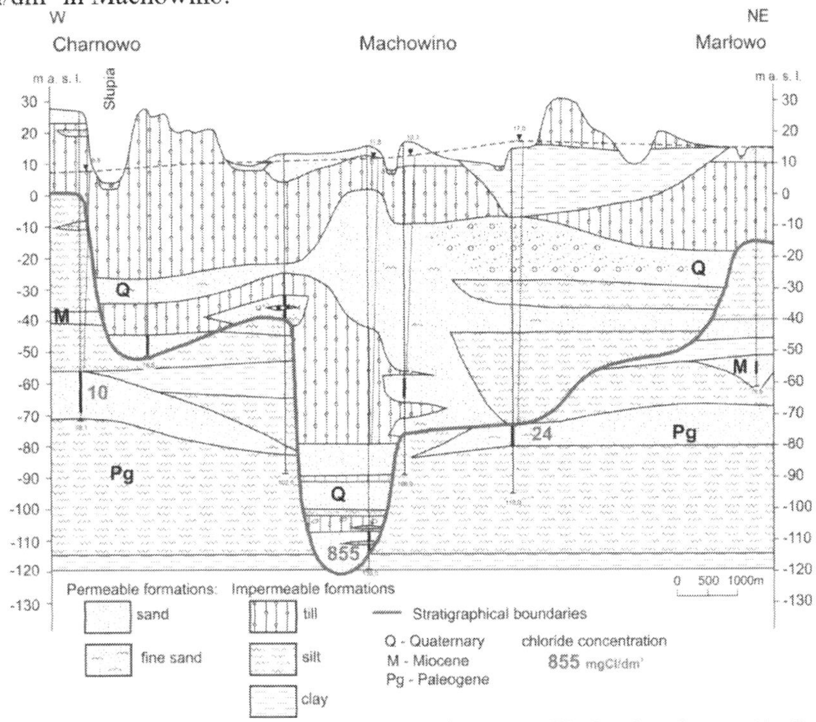

Figure 3. Hydrogeological cross-section of the Słupsk – Ustka Strukture (A. Szelewicka, 2011).

CONCLUSIONS

Groundwater intakes in the polish coastal Baltic zone cause the mixing of salt waters from ascension with the brackish waters from the intrusion of Baltic waters. Salt waters have been recognised in the deep valley structures, that enables ascension of brines from Mesozoic and Palaeozoic strata. This issue is very important from the perspective of water resources management and protection.

REFERENCES

Detailed Geological Map in scale 1:50 000 (number 1-7, 9-16, 20-27)), Polish Geological Institute – NRI. Warsaw

Kozerski B., Pruszkowska M., 1996 — Range and ingression process of brackish see water into aquifers of eastern Baltic Coast. Unpublished [in Polish], Technical University of Gdańsk.

Szelewicka A., 2011 – Recognition of water – bearing strata in burried valley structures in the northern part of Polish Lowlands. Unpublished [in Polish], Polish Geological Institute – NRI, Marine Branch of Gdańsk.

Deep submarine groundwater discharge indicated by pore water chloride anomalies in the Gulf of Gdańsk, southern Baltic Sea

Beata Szymczycha[1], Żaneta Kłostowska [1,2], Karol Kuliński [1], Aleksandra Winogradow[1], Jaromir Jakacki[1], Zygmunt Klusek[1], Miłosz Grabowski[1], Aleksandra Brodecka-Goluch [2], Bożena Graca [2], Marcin Stokowski[1], Katarzyna Koziorowska[1], Daniel Rak[1]
[1] Institute of Oceanology, Polish Academy of Sciences, Powstańców Warszawy 55, 81-712 Sopot, Poland
[2] Institute of Oceanography, University of Gdańsk, Al. Marszałka Piłsudskiego 46, 81-378 Gdynia, Poland

INTRODUCTION

Submarine groundwater discharge (SGD) is a significant pathway for material transport to the coastal zone (Burnett et al. 2006). For some elements and isotopes SGD has been thought to be the principal source (Lin et al. 2010). Therefore, the interest in coastal groundwater flow systems has increased rapidly during the last decades. Most of the studies have been focused on shallow (<20m), narrow zone (<5km) along the coastline (Lin et al. 2010). Interestingly, some deep seafloor studies (Wilson 2005; Lin et al. 2010) indicated that SGD can occur a long distance from the shoreline (~25km). In the Baltic Sea SGD has been mainly investigated in the southern part primarily at coastal zones demonstrating that groundwater seepage is comparable to river loads in case of selected chemical substances (Piekarek- Jankowska 1994; Schlüter et al. 2004; Szymczycha et al. 2012; 2014; 2016). In this study we identified deep SGD located at acoustically turbid sediments in the Gulf of Gdańsk (~70 km from the shore). Given the significance of benthic nutrients dynamics and sediment biogeochemical processes implications for the Baltic Sea environment, such as eutrophication, hypoxic and anoxic events, we aim to characterize the potential role of deep SGD in the Baltic Sea cycles of elements.

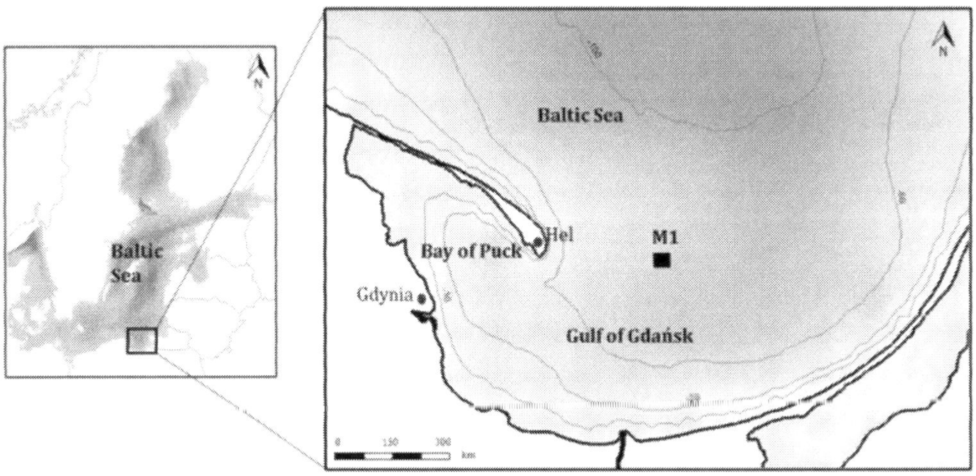

Figure 1. Location of the study area (M1).

METHODS

The study area is located in the Gulf of Gdańsk, southern Baltic Sea (Fig.1). Sea water, pore water and sediments samples were collected on board the R/V Oceania during three cruises in May 2015, January 2017 and May 2017. Seawater salinity and temperature was retrieved from CTD files while sediment cores were collected by the Gemax gravity corer. Additionally acoustic observations of the sediments were made in order to detect gas distribution. In collected sediment samples water content, calcium (Ca), total organic (TOC) and inorganic carbon (IC) were analyze while in collected pore water samples nutrients (PO_4^{3-}, NO_3^-+NO_2^-, NH_4^+), dissolved organic carbon (DOC), dissolved inorganic carbon (DIC), metals (Na, K, Mg, Ca, Al, Mn, Fe, Ni, Cr, Cu, Cd, Co, Pb) and alkalinity were analyzed. Parameters such as ORP, pH and salinity were measured *in situ*. The SGD rate was based on numerical modelling.

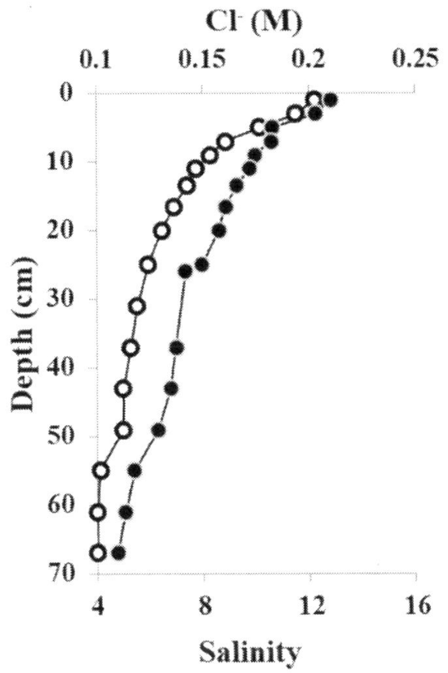

Figure 2. Examples of pore water depth profiles for chloride (hollow symbols) and salinity (solid symbols) in the study area.

RESULTS AND DISCUSSION

Seawater salinity, temperature and density were typical of the seasons when the samples were taken. The pore water depth profiles for both salinity and chloride significantly decreased with depth during every sampling campaign indicating freshwater source. The exemplary profiles are presented at Figure 2. The pore water concentrations of the main cations such as Na, Ca, Mg, K showed similar trend to chloride. Generally, pore water profiles for Cl, Na, Ca, Mg and, K, unaffected by freshwater, are constant or increase linearly with depths (Carman and Rahm 1997; Schlüter et al. 2004). The curvature profiles are characteristic for areas affected by fluid flow, in this case SGD (Schlüter et al. 2004). The general pore water trend of DIC, DOC, trace elements, PO_4^{3-}, NH_4^+, TDS and alkalinity are comparable to those observed in deep sea anaerobic sediments. Interestingly, in the deepest layers of pore water increased concentrations of Mn, Fe, Al, PO_4^{3-}, NH_4^+, DIC, DOC and alkalinity were observed most probably due to groundwater seepage.

The calculated SGD rates ranged from 0.3 to 0.7 (L m^{-2} d^{-1}). Comparable results were observed in the coastal area of the Eckernförde Bay, western Baltic Sea, where beside groundwater seepage, increased methane production and consequently methane release from sediments were detected (Schlüter et al. 2004). In our study area the sediments acoustic disturbance has been previously correlated with methane presence (Brodecka et al. 2013) while in the neighboring Bay of Puck, inner part of the Gulf of Gdańsk SGD occurrence was accompanied by methane and increased P, Si and DOC fluxes (Donis et al. 2017).

CONCLUSIONS

In this study we identified deep SGD located at acoustically turbid sediments in the Gulf of Gdańsk (~70 km from the shore). The main impacts of the anoxic deep SGD are increased efflux of chemical substances such as PO_4^{3-}, NH_4^+, DIC, DOC, trace elements (Mn, Fe, Al) and possibly methane. Therefore, SGD may significantly change their distribution both within the sediments and at the water-sediment interface.

ACKNOWLEDGMENTS

The results were obtained within the framework of the following projects: 2016/21/B/ST10/01213 sponsored by National Science Center and WaterPUCK financed by the National Centre for Research and Development (NCBR) within BIOSTRATEG program.

REFERENCES

Burnett, W.C., Aggarwal, P.K., Aureli, A., Bokuniewicz ,H.J., Cable, J.E., Charette, M.A., Kontar, E., Krupa, S., Kulkarni, K.M., Loveless, A., Moore, W.S., Oberdorfer, J.A., Oliveira, J., Ozyurt, N., Povinec, P., Privitera, A.M.G., Rajar, R., Ramessur, R.T., Scholten, J., Stieglitz, T., Taniguchi, M., Turner, J.V. 2006. Quantifying submarine groundwater discharge in the coastal zone via multiple methods. Science of the Total Environment, 367, 498-543.

Carman, R., Rahm, L. 1997. Early diagenesis sediments in and chemical characteristics of interstitial water and the deep deposition bottoms of the Baltic proper. J. Sea Res. 37, 25-47.

Donis, D., Janssen, F., Wenz, F., Dellwig, O, Escher, P., Spitzy, A., M.E. Böttcher, 2017. Biogeochemical impact of submarine ground water discharge on coastal surface sands of the southern Baltic Sea. Estuar Coast Shelf S 189, 131- 142.

Lin, T-I, Wang, C-H, You, C-F, Huang, K-F, Chen, Y-G 2010. Deep submarine groundwater discgage indicated by tracers of oxygen, strontium isotopes an barium content in the Pingtung coastal zone southern Taiwan Mar Cem 122, 51-58.

Piekarek-Jankowska H., Matciak M., Nowacki J. (1994) Salinity variations as an effect of groundwater seepage through the seabed (Puck Bay. Poland), Oceanologia, 36, 33-46.

Schlüter M., Sauter E.J., Andersen C.A., Dahlgaard H., Dando P.R., 2004. Spatial distribution and budget for submarine groundwater discharge in Eckernförde Bay (Western Baltic Sea). Limnol. Oceanogr. 49, 157–167.

Szymczycha B., Vogler S., Pempkowiak J. (2012) Nutrient fluxes via submarine groundwater discharge to the Bay of Puck, Southern Baltic, Sci. Total Environ., 438, 86-93.

 https://doi.org/10.1051/e3sconf/20185400035

Szymczycha, B., Maciejewska, A., Winogradow, A., Pempkowiak, J. 2014. Could submarine groundwater discharge be a significant carbon source to the southern Baltic Sea? In Oceanologia, 56, 327–347.

Szymczycha, B., Kroeger, K. D., Pempkowiak, J. 2016. Significance of groundwater discharge along the coast of Poland as a source of dissolved metals to the southern Baltic Sea, Marine Pollution Bulletin, 109, 151-162.

Wilson, A.M. 2005. Fresh and saline groundwater discharge to the ocean : A regional perspective. Water Resour Res 41, WO2016.

Contact Information: Beata Szymczycha, Institute of Oceanology, Polish Academy of Sciences, Powstańców Warszawy 55, 81-712 Sopot, Poland , Phone: +48 (58) 7311738: beat.sz@iopan.gda.pl

Modeling groundwater flow and salinity evolution near TSF Żelazny Most. Part I – groundwater flow

Waldemar Świdziński[1]
[1]Institute of Hydro-Engineering, Polish Academy of Sciences, IBW PAN, Gdańsk, Poland

ABSTRACT

Tailings which are by-product of the extraction of various metals (copper, gold, silver, molybdenum, etc.) are often stored in so called Tailings Storage Facilities (TSF), where they are deposited as a soil-water mixture by spigotting. In many cases the water discharged together with tailings to the TSF is rich in salts and other chemical compounds imposing negative pressure to the groundwater environment. Even in the case of total or partial lining of such facilities and well-developed drainage systems to control leaching, some portion of contaminated water often seeps either through the surrounding dams or the bed into adjacent groundwater bodies. Numerical models can be very helpful tools to assess the extent of the contamination and particularly to predict its potential development in the future. This paper and the companion one describe such a numerical model developed for Żelazny Most Tailings Storage Facility (south-west Poland), one of the world's largest tailings sites. In the first part general information about the facility is provided and a 3D hydrogeological numerical model of the structure is described. Groundwater flow pattern near the facility obtained from numerical simulations is confronted with the measurements from a comprehensively developed monitoring system. Part II will be focused on the modelling of chloride transport in groundwater.

INTRODUCTION

TSF Żelazny Most is a depository for post-flotation tailings, which are by-product of copper mined in three mines: Rudna, Lubin and Polkowice located in south-west Poland, operated by KGHM company. It belongs to one of the largest facilities of its kind in the world. Since this is only place to store tailings produced by the Polish copper mines it is a key element in this production. The tailings are transported to the TSF in the form of low concentration slurry which is discharged into the facility by spigotting along the whole perimeter of the object. Semi-liquid tailings are retained within the structure by earth dams surrounding the depository. During the spigotting the tailings undergo natural process of segregation which causes that the coarser material deposits near the dams whereas the finer one is gravitationally transported with flowing water towards the central part of the TSF, where the water is clarified forming a water pond (Figure 1). In order to accept new supplies of tailings (the annual production of which amounts 28 mln tonnes) the facility has to be continuously developed. The Żelazny Most tailings facility is being raised using the upstream construction method. Only at early stage the so-called starter dams were constructed from borrow earthen material. Afterwards the dams have been raised by 5 m high embankments utilizing coarse tailings deposited near the dams. Total area occupied by the facility amounts 14 ha with the length of perimeter 14.3 km. Due to the natural terrain relief the present height of the surrounding dams varies: 41 m for the south dam, 47 m for the north one, 57 m for west dam and over 70 m for the east dam, at the crest elevation 185 m a.s.l. Total volume of tailings stored in the TSF exceeds 600 mln m^3 and according to the future plans of copper production in KGHM it should be ready to accept next 300 mln m^3.

 https://doi.org/10.1051/e3sconf/20185400036

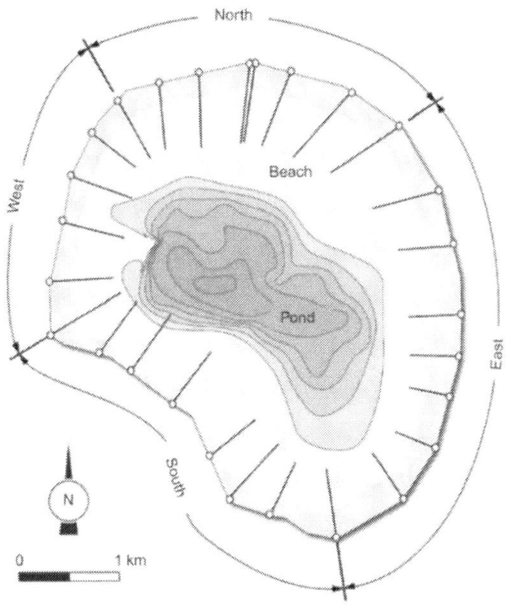

Figure 1. General scheme view of TSF Żelazny Most (Jamiolkowski 2014).

The dams as well as the bed are not sealed which means that the water discharged with tailings can almost freely infiltrate. In order to contain infiltration an extensive system of drainage has been developed. It consists of various elements, including several floors of circumferential drainage, the aim of which is to intercept the seeping waters at the highest possible elevation in order to control the position of phreatic surface within the dam body. At present there exist four floors of circumferential drainage, constructed every 10 m in height starting from the elevation 152 m a.s.l. and installed in the course of the facility development. Moreover, there is pipeline drainage of starter dam and finally the dich drains surrounding the facility at the dams' toe (Figure 2).

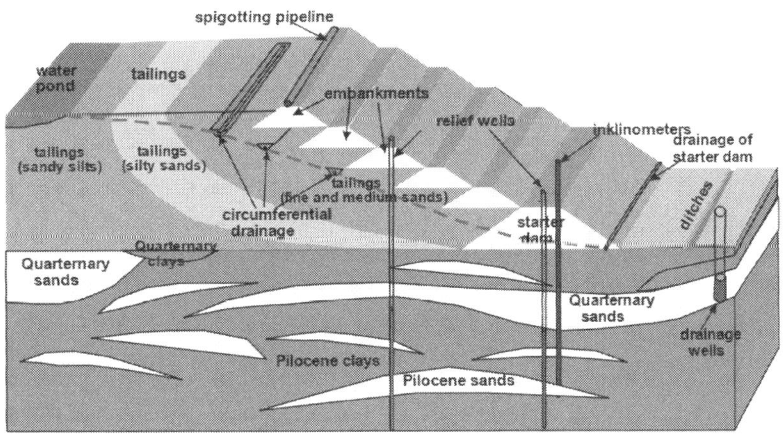

Figure 2. Schematic cross-section through the TSF Żelazny Most dam.

Despite the comprehensive system of horizontal drainage some waters still seeps into the subsoil and next downstream, infiltrating into the groundwaters. The waters discharged with tailings to TSF come from dewatering of mining fields therefore they are highly contaminated, mostly by salts. In order to reduce the negative impact of the waters seeping downstream a system of drainage wells (approximately forty wells) located around the facility outside the dams' toe has been installed. The second purpose of these wells is to reproduce the groundwater flow conditions prior to the construction of the TSF. The drainage wells continuously pump out the groundwater from the small local aquifers preventing the contaminated water from flow downstream, however small amounts of leachate still pass the hydraulic barrier. In order to recognize the existing flow paths of saline water, verify the efficiency of existing drainage system, but first of all to predict the development of salinity front with the raise of the dams and tailings, a numerical groundwater model of the TSF Żelazny Most and surrounding area has been developed.

STRUCTURE OF 3D HYDROGEOLOGICAL NUMERICAL MODEL

The 3D model, which was originally developed at the beginning of the current decade and successively updated, reproduces flow conditions in the area of TSF Żelazny Most including the following elements (Świdziński et al., 2011):

- infiltration of saline waters through the bed of water pond, the beaches and next through the mass of tailings into the groundwater,
- downstream flow of groundwater and pollution migration,
- surface watercourses and the change of their quality,
- interaction between ground and surface waters,
- drainage system of the facility,
- drinking water intake wells.

The 3D model has been developed with GMS (GroundWater Modelling System) commercial software package, initially using version 6.5 and more recently using version 8.3. The groundwater flow equations are solved by MODFLOW software integrated with GMS. The model covers the area of the current impact of the facility determined by the field measurements and the predicted area of long-term future impact. The borders of the model are defined by watercourses or local watersheds. The model domain also includes Retków-Stara Rzeka water intake located within the area of Main Groundwater Reservoir (GZWP – 314) (Figure 3). Steady-state groundwater flow is considered. The bottom boundary of the 3D model was determined based on the assumption that the most important role in the impact of the facility on groundwater is played by the shallowest multi-aquifer formations i.e. quaternary and upper tertiary aquifers. The latter is well isolated from the lower inter coal (middle and early Miocene) and under coal (Oligocene) aquifers. The elevation of the model bottom was assumed at -70,0 m a.s.l., i.e. the lowest point of the tertiary aquifer formation. The model domain has 154.8 km^2 (12.9 x 12.0 km). It was subdivided into a grid of squares and rectangles with alternate sizes. Due to essential gradients in elevation and resulting hydraulic gradients, the area of very facility was covered by square mesh with the size 25 x 25 m, zone 1 km distant from the dams toe by squares 50 x 50 m, whereas the rest of the model area with squares 100 x 100 m (Figure 3). Such discretization allowed an appropriate modelling of the work of drainage wells as well as good reproduction of the inclination of downstream slopes of the facility. The orientation of the mesh has been chosen in such a way so that the mesh line follows main direction of flow and migration of saline waters (Świdziński et al. 2011). The grid for a single calculation layer is built of 289 rows and 282 columns. Geological and hydrogeological conditions of the modelled area are extremely complex since they were significantly impacted by three glaciations which passed over Żelazny Most in Pleistocene. The various ice sheets, which are believed to have been at least 1000 m thick, have induced widespread glacio-tectonic phenomena causing many thrusts of Pleistocene deposits into the initially horizontally bedded freshwater Pliocene sediments (Jamiolkowski 2014). As a consequence it was very difficult to identify and separate single aquifers and aquitards in the **subsoil** (see Figure 2) as it is done in a standard approach in the schematisation process and generalisation of hydrogeological conditions. Moreover, despite a huge number of geological profiles (several thousands) obtained from various field investigations (standard boreholes, CPTUs, geophysical surveying) only small portion of them could have been used for identification of hydrogeological conditions since the majority of the profiles were located near the dams of the facility and a lot of other ones were too short to be included in the interpretation. Furthermore, very differentiated geological conditions which were changing very often and rapidly in plane and depth, prevented to use standard GMS tools to construct individual hydrogeological layers of similar hydraulic properties. Therefore, it was decided to construct 3D model of the subsoil in a non-standard way. First of all, for the generalization purposes four basic hydrogeological layers have been assumed:

- quaternary layers of permeable formations - $k = 10^{-4}$ m/s,
- quaternary layers of impermeable formations - $k = 10^{-9}$ m/s,
- tertiary layers of permeable formations - $k = 5x10^{-4}$ m/s,
- tertiary layers of impermeable formations - $k = 10^{-9}$ m/s.

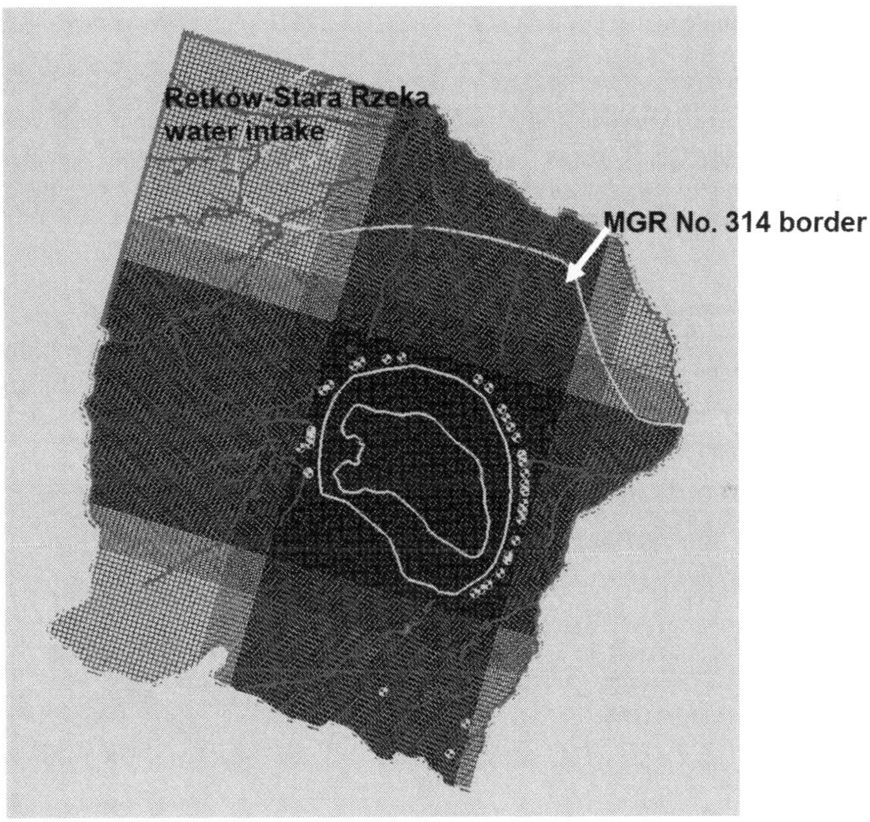

Figure 3. Finite difference mesh of 3D model for TSF Żelazny Most.

Next, the subsoil has been divided into 10 model layers of various thicknesses with the smallest thickness for uppermost layers and larger one for lower layers. The values of hydraulic conductivities, which have been assigned to generalized permeability categories, have been subsequently averaged for the given thickness of the model layers in terms of the geometric average. This procedure was applied for each hydrogeological borehole taken into account. Distributed averaged values of hydraulic conductivities were next interpolated between neighbouring boreholes for each model layer using the natural neighbour method. In such a way a fuzzy distribution of hydraulic conductivities was obtained covering the full and continuous range of permeabilities i.e.: from impermeable soils, via semi-pervious ones, and finally to fully permeable soils. The model of the foundation has been supplemented by the system of rivers and surface watercourses running within the model as well as along its borders, drainage wells around the facility, groundwater intake and farm wells in this area.

In turn, hydrogeological model for the **facility** also consists of 10 calculation layers, the lowest three of which have differentiated thickness so as to have horizontal top of the third lowest layer at the elevation of 160 m a.s.l. The subsequent 7 higher layers of the depository have equal thickness of 5 m reaching final elevation of 195 m a.s.l. Each of the calculation layers consists of external embankment and the tailings filling the depository with decreasing permeability going towards the pond. In order to reflect the change of the fines content in deposited tailings, for each calculation layer 9 zones with the same permeability have been distinguished following the decrease of permeability coefficient as a function of the distance from the embankment. Finally, based on accessible digital maps and half-tone screen of the terrain of concern the system of surface watercourses as well as all drainage elements installed in the depository together with water intakes were implemented in the model. For the majority of river sections flowing along the borders of the model the first-type (Dirichlet) boundary condition was assumed (specified head). At the north-west corner, the only artificial border of the model, the so-called specified–flux boundary was assumed which in fact is a second-type boundary condition. For all watercourses located inside the model area the third-type boundary condition (Cauchy) was specified. The work of any wells (drainage, relief and intake wells) was modeled by the second-type (Neumann) boundary condition (constant discharge). Recharge was assumed based on the long-term mean annual precipitation at the level of 597 mm reduced respectively to the effective infiltration depending on the type of the surface soils. With regard to the facility, the infiltration of saline waters throughout the mass of tailings under the pond was simulated by first-type boundary condition assuming elevation of water head whereas the infiltration of waters discharged on the facility beaches during spigotting process was simulated by third order boundary condition using MODFLOW RIVER package.

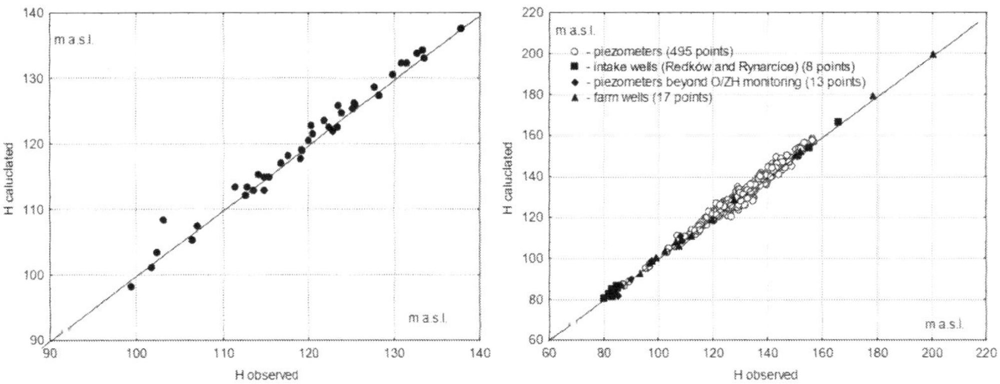

Figure 4. Observed and calculated elevations of piezometric head in a) drainage wells b) in observation points downstream the facility.

MODELLING OF GROUNDWATER FLOW REGIME

The model was calibrated so as to achieve the best fit of the simulated and real (based on the *in situ* measurements) hydrodynamic field downstream the facility. The calibration process involved adjusting local hydraulic permeability and recharge values. The comparison of calculated and measured values of the piezometric heads for drainage wells and numerous observation points downstream (piezometers, intake and farm wells - in total 533 observation points) is shown in Figure 4. It can be seen that the model reproduces accurately groundwater flow conditions within the facility and its neighborhood near and far downstream. This can be confirmed by comparing the groundwater table contours obtained from standard hydrogeological mapping with the contours obtained from the model - the match is very good (results not shown here).

DISCUSSION AND CONCLUSIONS

The groundwater flow model developed for Tailings Storage Facility Żelazny Most, Poland appeared to be a very efficient tool in reproduction of real groundwater flow regime. Taking into account the dimensions of the modeled area, complexity of the hydrogeological structure of the subsoil as well as highly differentiated flow conditions within the facility the agreement of the results of numerical simulations with the observations obtained from comprehensive monitoring system is very good. It should be stated that such a good agreement has been achieved by very careful verification, calibration and modification (if appeared necessary) of local groundwater flow conditions during calibration process which has been very much time consuming however yielding very good final results. It mostly regarded the direct vicinity of the facility, specifically near the drainage wells for which given discharge capacity has been assumed and depression simulated and compared with measured values. In some cases such modification has been carried out for intake and farm wells as well as piezometers located far downstream from the facility and for which high differences between measured and observed piezometric heads have been found. Well calibrated 3D hydrogeological model was a basis for simulation of migration of saline water into groundwater presented in Part II.

REFERENCES

Jamiolkowski M. 2014. Soil mechanics and the observational method: challenges at the Żelazny Most copper tailings disposal facility. Geotechnique, 64, No. 8, 590–619.

Świdziński W., Maciejewski St., Walter A. and Franz M. 2011. Prediction of the Żelazny Most facility impact on ground and surface waters during its operation to the elevation of 195 m a.s.l. and after its closure. Report for KGHM POLSKA MIEDŹ S.A.

Świdziński W., Maciejewski St., Walter A. 2014. Integrated assessment of the impact of the Żelazny Most facility on ground and surface waters till 2012 together with updated protection plan. Report for KGHM POLSKA MIEDŹ S.A.

Contact Information: Waldemar Świdziński, Institute of Hydro-Engineering PAS, Department of Geomechanics, Kościerska 7, 80-328 Gdańsk, Poland, Phone: +48585222945, Fax: +48585524211, Email: waldek@ibwpan.gda.pl.

Modeling groundwater flow and salinity evolution near TSF Żelazny Most. Part II – chloride transport

Waldemar Świdziński
Institute of Hydro-Engineering, Polish Academy of Sciences, IBW PAN, Gdańsk, Poland

INTRODUCTION

In Part I of this set of two papers a 3D hydrogeological numerical model (Figure 1) simulating the groundwater flow regime near the huge Tailings Storage Facility Żelazny Most, Poland was described and discussed. TSF Żelazny Most stores post-flotation tailings which are a by-product of copper mining and its extraction. The tailings are transported and discharged into the facility as a slurry containing highly saline waters which can infiltrate into groundwater strongly impacting the water environment near the TSF. Essential part of these waters is intercepted by comprehensively developed drainage system, however some small portion still passes by the drainage and enters groundwater and finally the neighboring surface watercourses. The aim of the model was to have efficient tool to simulate the groundwater flow pattern near the TSF and the transport of saline waters seeping from the facility. The results presented in Part I showed that after careful calibration and validation the model acceptably well reproduces real groundwater flow conditions within the modelled area. Next, based on the 3D model the migration of saline waters was simulated and confronted with the *in situ* measurements of the range and concentration of chlorides downstream the facility. The final step was the prediction of the change of groundwater flow regime due to increasing water level in TSF at higher elevations of the dams and stored tailings, and the associated changes of contaminated zones with the course of the facility development. In Part II the simulation of chloride transport near TSF by 3D model is described and the predicted development of salinity zones is presented.

MIGRATION OF SALINE WATERS WITHIN THE MASS OF TAILINGS

The seepage process from the pond through the mass of tailings and dam is generally determined by two factors: the water head and the permeability properties of tailings and subsoil layers. The conditions for infiltration of saline waters in the mass of tailings are dependent on its gradation which is the result of segregation and sedimentation during the spigotting process. Coarse fractions of relatively high permeability are deposited near the embankments whereas the finest fractions are deposited near the pond and, due to sedimentation process, onto its bed. Thus the infiltration of saline water throughout the tailings cannot be avoided, particularly in the close distance from the embankments. Essential portion of water infiltrates into the tailings mass via beaches of the facility during discharging of a slurry. The closer distance to the embankments the more intensive infiltration occurs. The spigotting is carried out periodically section by section which means that for a given section it effectively lasts 11% of the year, only. In the periods between spigotting the process of draining and drying takes place. However, huge reservoir of water stored in tailing mass (voids) causes that the drainage process is strongly delayed and lasts much longer than breaks between consequent discharges. It induces overlapping the effects of following discharges and equalization of the rate of water infiltrating into the subsoil. Therefore, continuous infiltration of saline waters has been assumed in the model.

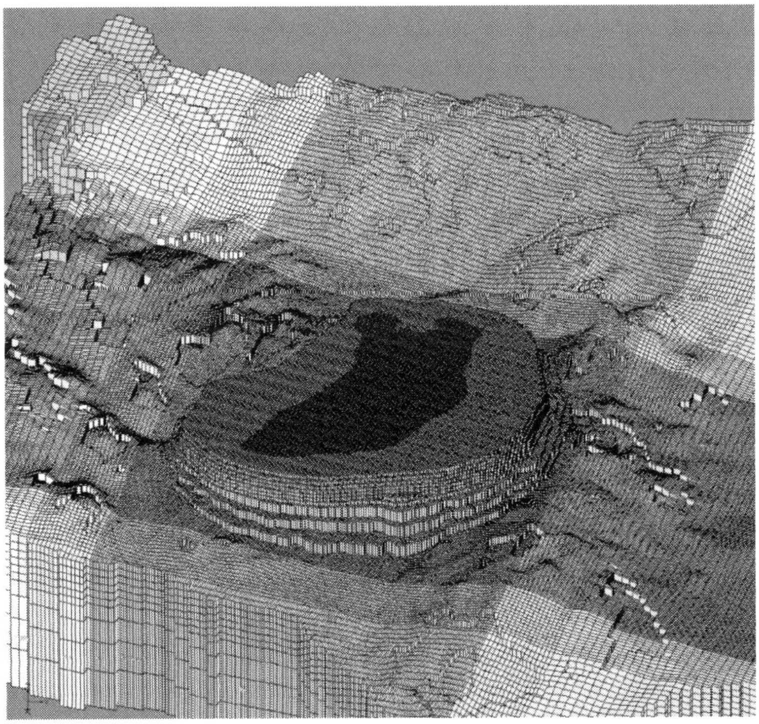

Figure 1. 3D hydrogeological model of TSF Żelazny Most area (Świdziński et al. 2012).

Segregation of tailings during discharging of a slurry as well as controlled discharge of finest material directly into the center of the facility cause large variety of hydraulic permeability of tailings both in vertical and horizontal directions and complex flow conditions within the tailings mass. *In situ* investigations of the permeability as a function of a distance from the East and West dam embankments, carried out in the past, have clearly revealed that near the pond values of coefficient of permeability are close to those corresponding to impermeable soils (10^{-9} m/s) whereas near the embankments these values represent fully permeable material (10^{-4} m/s). Moreover, the mass of tailings is a strongly anisotropic medium due to numerous laminations of semipervious layers which may partly result from the process of beach stabilization by asphalt emulsion against dusting, however correct determination of the anisotropy degree of tailings stored in the depository is very difficult. Based on infiltrometric tests the value of anisotropy can be assumed as $K_H/K_V \approx 2 \div 3$. However, the interpretation of piezometric heads observed in tailings shows that this ratio can be much higher - $K_H/K_V \geq 10 \div 15$.
The 3D model has been first calibrated for the conditions occurring at TSF at the end of 2005 and re-calibrated for 2012 implementing new information regarding the hydrogeological conditions of the subsoil. Next the model has been used to predict the impact of TSF Żelazny Most on surface and groundwaters for higher elevations related to planned lifetime of the facility. Due to some limitations, in this paper the prediction of the impact in years 2005-2012 will be presented, only.
The results of modelling of groundwater flow regime within the facility and downstream have revealed that the waters infiltrate mainly through the area of beaches during the discharging of a slurry in the proportion of 99% of all waters infiltrating from the facility whereas percolation from the pond amounts 1%, only (Świdziński, 2018). Such result can be justified taking into account drastic reduction of tailings permeability with the distance from the surrounding dams towards the pond.

MODELLING THE MIGRATION OF SALINE WATERS NEAR THE FACILITY

In order to model the migration of saline waters in groundwater MT3D (Modular Transport 3 Dimensional) software was used and to solve the advection-diffusion equation third order TVD scheme (ultimate) or MMOC (Modified Method Of Characteristics) were applied, both being an integral part of GMS package. The conservative chlorides were selected as a the modelled substance, since they do not undergo sorption process, nor react with other chemical compounds. Moreover, the chlorides, besides sulphates, are the main indicator of polluted groundwater near the facility, and thus enable a reliable assessment of changes occurring in aquatic environment caused by the facility operation.The waters discharged to the facility with tailings come from dewatering of the copper mines. They are reach in salts, the concentration of which increases with the course of exploration of new mining fields. In Figure 2 the changes of average annual values of chlorides concentration in water discharged with tailings into TSF Żelazny Most starting from 1996 till 2012 are shown.

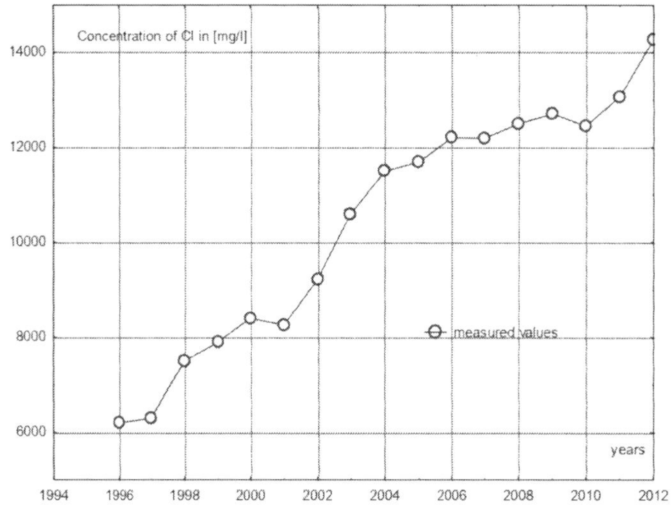

Figure 2. Mean annual concentration of chlorides in waters stored in the TSF Żelazny Most.

For the model calibrated at the end of 2005 the value of chlorides concentration in mine waters discharged into the facility was 11000 mg Cl/dm^3 and it increased at the end of 2012 up to 14280 mg Cl/dm^3. It was assumed that excess state over hydrochemical background would be modeled, only. Within the model domain woodlands and agricultural as well as meadow areas dominate, for which the average natural concentration of chlorides is 40 mg Cl/dm^3. Therefore, for the sake of simplicity as the ultimate value for groundwater pollution by chlorides 210 mgCl/dm^3 was assumed so as the total chlorides concentration would not exceed the ultimate value assigned to the III class clean water (250 mg Cl/dm^3), the last class corresponding to good quality water according to the recent Polish regulations. Consequently, the initial condition in the form of given concentration of salts was applied only for the areas polluted by saline waters from the facility identified by the *in situ* measurements. Such measurements are being carried out annually in the frame of operational monitoring. The distribution of concentration of salts in the groundwater downstream the facility, introduced to the model as initial condition, corresponded to that observed in 2005, the year the model was originally calibrated and validated. For the rest of areas the concentration of chlorides was assumed to be 0. In turn, in order to model the intrusion of saline waters during the spigotting the slurry over the beaches and from the pond first-type boundary condition was assumed. The condition relies on providing the value of chlorides concentration for specified grid blocks. As it is shown in Figure 2 the concentration of chlorides increases with time, thus such change had also to be incorporated into the prognostic calculations for 2012. The dispersion process was modelled using two hydrodynamic dispersion constants i.e. longitudinal (α_L) and transverse (α_T). For the analysed case 20 m and 6 m for longitudinal and transverse dispersion constants were assumed, respectively. Moreover, the effective porosity of the foundation soils varied depending on their permeability, while for the tailings a constant value was assumed. Taking into account that the rate of facility development is relatively slow (increase of water level in the pond 1.5 m/year) as well as that the *in situ* observations show small fluctuations of groundwater level near the TSF, steady-state flow conditions have been incorporated. Little change of groundwater regime near the TSF is mostly caused by the work of drainage system overtaking the essential portions of water infiltrating into the subsoil in the near downstream whereas far downstream groundwater level is mostly impacted by precipitation. Despite the stationarity of flow process assumed, in long-term predictions quasi non-steady-state approach was applied by execution of calculations for higher elevations of the dams in several time steps. The calculation stages of groundwater flow regime were coupled with the simulation of chloride transport in groundwater. The latter was modelled as fully non-steady-state process based on the stationary hydrodynamic field being the solution for a given time step (assumed to be one year starting from 2005). The elevations of the water in the pond for consequent years were introduced into the model based on the real measurements at the end of the year. The initial condition for calibration year (2005) was the distribution of the concentration of chlorides in the groundwater downstream the facility based on *in situ* measurements. In turn, the initial condition for the concentration of chlorides in subsequent years was the distribution of this concentration downstream the facility being the result of numerical calculations for previous year as well as average concentration of chlorides in the pond based on the *in situ* observations.

Table 1. Predicted budget of groundwater near TSF Żelazny Most .

	Inflow m^3/s	Inflow m^3/d	Outflow m^3/s	Outflow m^3/d
Infiltration from the pond	0.00248	214	-	-
Infiltration through the beaches	0.37910	32754	-	-
Precipitation through the dams	0.01650	1426	-	-
Discharge of drainage wells	-	-	0.14474	12506
Discharge of horizontal drainage system	-	-	0.23393	20212
Total	0.39808	34394	0.37867	32718
Difference between inflows and outflows			**0.01940**	**1676**

THE RESULTS OF NUMERICAL SIMULATIONS

The groundwater budget for the area of the facility and its closest vicinity (including the location of drainage wells) calculated by the numerical model for the end of 2012 is presented in Table 1. According to the results given in Table 1, at the end of 2012 the amount of saline waters not captured by the drainage system flowing out of the facility downstream was 1676 m^3/day and it was lower comparing to 2005 (2078 m^3/day, (Świdziński et al. 2011). It can be also seen that the amount of waters seeping downstream is equivalent to approximately 5% of total amount of waters infiltrating through the mass of tailings into the subsoil whereas 95% of it is captured by drainage system. Modelled range of groundwater with concentration of chlorides higher than limit level (> 250 mg Cl/dm^3 – solid line) together with the pollution zones based on *in situ* measurements (filled areas) for 2012 is shown in Figure 3.

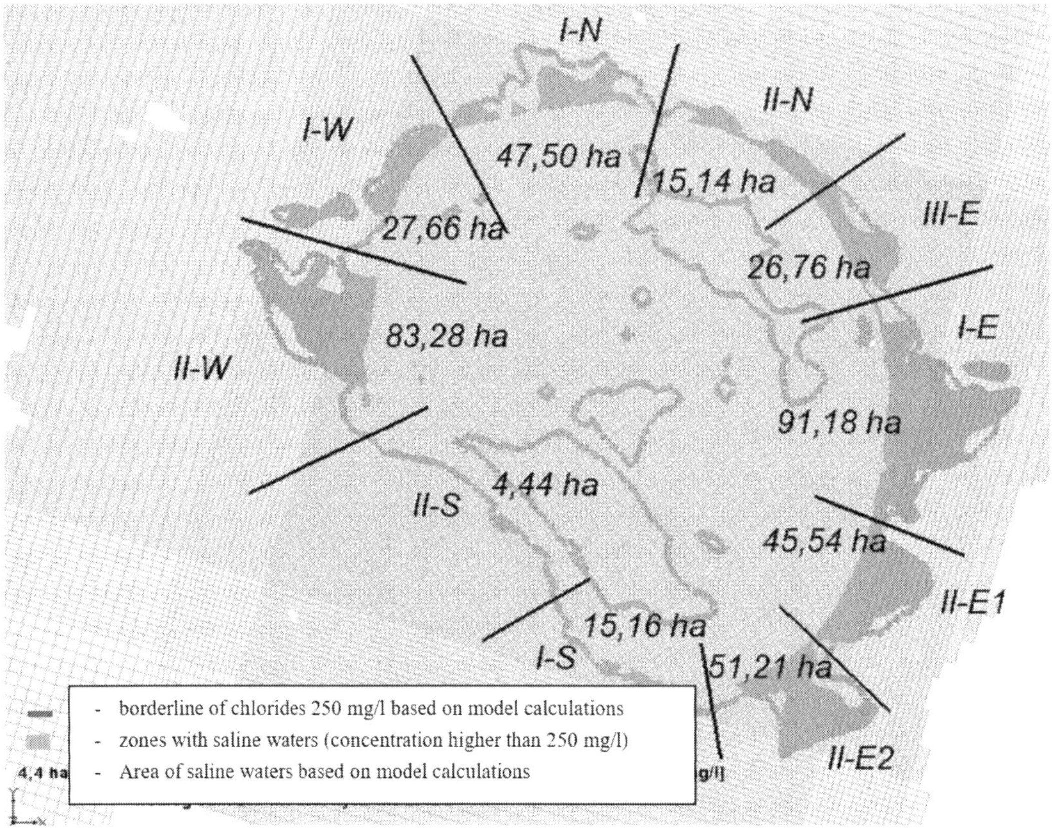

Figure 3. Measured and calculated ranges of polluted groundwater downstream TSF Żelazny Most at the end of 2012.

Simulated total polluted area is approximately 92 ha larger than resulting from *in situ* measurements (~316 ha), however both the maximum ranges as well as the shape of polluted zones designated by the 250 Clmg/dm^3 contour, are similar. There are some regions where modelling results show the presence of chlorides higher than the limit value which is not confirmed by the results of operational monitoring e.g. larger range of saline waters in zone I-N downstream north dam as well as opposite situations e.g. zone I-E in the central part of east dam. There are also some separated regions where field measurements do not show higher concentration of chlorides whereas they are simulated by the numerical model (e.g. the polluted „island" in zone I-E). Best quantitative agreement with regard to the polluted area has been achieved for eastern part for which the difference between measured and simulated results was around 9% what proves good calibration of the model for this particular region. Somewhat worse agreement was obtained for the rest of the dams. In general, the numerical results overestimate the *in situ* observations whereas quite good reproduction of the shapes of polluted zones is observed. The largest difference with regard to the polluted area is for northern part (100%) and 50% for western area downstream the facility, however it should be noted that the total contribution of these areas is lower than the eastern part.

DISCUSSION AND CONCLUSIONS

In the set of two papers a 3D numerical model which simulates groundwater flow and its chemical changes caused by the largest Tailings Storage Facility Żelazny Most, Poland has been presented. The model serves mainly for prediction of the potential future changes of groundwater flow regime and the extent of contamination by salinity downstream the facility in the course of its further development. The results of predictive simulations for the period 2005-2012, confronted with *in situ* observations, have proved that the model quite well reproduces the shape, range and area of polluted zones. Long-term *in situ* measurements of the salinity of groundwater, carried out every year in hundreds of the observation points near the TSF, show that the polluted areas do not change too much. It was confirmed by predictive simulations for the whole planned lifetime of the facility (results not presented here). The results of numerical calculations have shown that the situation should not essentially change in future and a decrease of the amount of saline waters flowing out of the facility can be expected. It means that the hydraulic barrier created by drainage wells installed around the facility efficiently prevents the outflow of saline waters downstream the facility.

REFERENCES

Świdziński W. 2018 Modelling of saline waters transport in groundwater near Tailing Storage Facility (TSF). Part I – hydrodynamics of groundwater. SWIM proceedings (to be supplemented later on).

Świdziński W., Maciejewski St., Walter A. and Franz M. 2011. Prediction of the Żelazny Most facility impact on ground and surface waters during its operation to the elevation of 195 m a.s.l. and after its closure. Report for KGHM POLSKA MIEDŹ S.A.

Świdziński W., Maciejewski St., Walter A. 2014. Integrated assessment of the impact of the „Żelazny Most" facility on ground and surface waters till 2012 together with updated protection plan. Report for KGHM POLSKA MIEDŹ S.A.

Contact Information: Waldemar Świdziński, Institute of Hydro-Engineering PAS, Department of Geomechanics, Kościerska 7, 80-328 Gdańsk, Poland, Phone: +48585222945, Fax: +48585524211, Email: waldek@ibwpan.gda.pl.

Laboratory Scale Investigation of Dispersion Effects on Saltwater Movement due to Cutoff Wall Installation

Masahiro Takahashi[1], Kazuro Momii[2] and Roger Luyun Jr.[3]
[1]Nippon Koei Co. Ltd., Tokyo, Japan
[2]Kagoshima University, Kagoshima, Japan
[3]University of the Philippines Los Baños, Laguna, Philippines

ABSTRACT

In the numerical investigation of saltwater transport in coastal aquifers, we need to correctly evaluate the hydrodynamic dispersion in the flow field. In this study, we focused on the role of dispersivity in the removal process of residual saltwater in a laboratory scale cutoff wall experiment. From a pulse-type fluorescent tracer injection experiment in a saturated porous media of glass beads with a mean diameter of 0.088 cm, the estimated longitudinal and transverse dispersivities were found to be 0.07 cm and 0.0025 cm, respectively. Numerical analysis of the saltwater intrusion and subsequent removal after cutoff wall installation using SEAWAT and the generated dispersivity ratio (α_L/α_T) of 28 reproduces well the measured salt concentration changes with time. Whereas, if a dispersivity ratio of 10 is used in the numerical simulation, transverse dispersion in the saltwater and freshwater mixing zone becomes large and the residual saltwater is removed faster than the laboratory experiment. Inversely, if 100 was used, the residual saltwater removal time took longer. The transverse dispersion is a key parameter in the mechanical dispersion of saltwater in the mixing zone after cutoff wall installation.

INTRODUCTION

Among several countermeasures to prevent saltwater intrusion in coastal aquifers, artificial subsurface barriers store and control groundwater and ensure a consistent extraction of freshwater without causing seawater intrusion. Luyun et al. (2009) showed experimentally and numerically that residual saltwater trapped upon installation of these cutoff walls gradually retreated before being completely removed from the reservoir behind the cutoff wall. However, they observed slight differences in the time for complete removal between the experimental and numerical results. To numerically investigate the seawater movement, we need to correctly evaluate the dispersion parameters in the flow field. In this study, we focused on the role of dispersivity in the saltwater transport in the laboratory scale cutoff wall experiment. The objectives of this study are to determine the longitudinal and transverse dispersivities of a porous medium using fluorescein tracer test and to demonstrate their influence on saltwater movement after cutoff wall installation.

MATERIALS AND METHODS

Experimental setup

We conducted our experiments in a transparent flow tank with internal dimensions of 60 cm length, 30 cm height, and 1 cm thickness, sufficiently thin to approximate quasi 2-dimensional system (Figure 1). A 5-mm diameter injector tube used for pulse tracer injection is located 10 cm from the left saltwater reservoir and 5 cm from the bottom of the flow tank. Beside it, 5 cm from the left reservoir and 5 cm from the bottom, a 2-mm diameter electric conductivity (EC) sensor is used to measure the salt concentration change with time during cutoff wall experiments. We packed the main flow tank with homogeneous glass beads with a nominal diameter of 0.88 mm to model an unconfined aquifer. The freshwater and saltwater reservoirs were separated by fine mesh screens to each side of the glass bead section. The heads in the reservoirs were controlled by adjustable drainage pipes.

Tracer calibration and pulse experiments

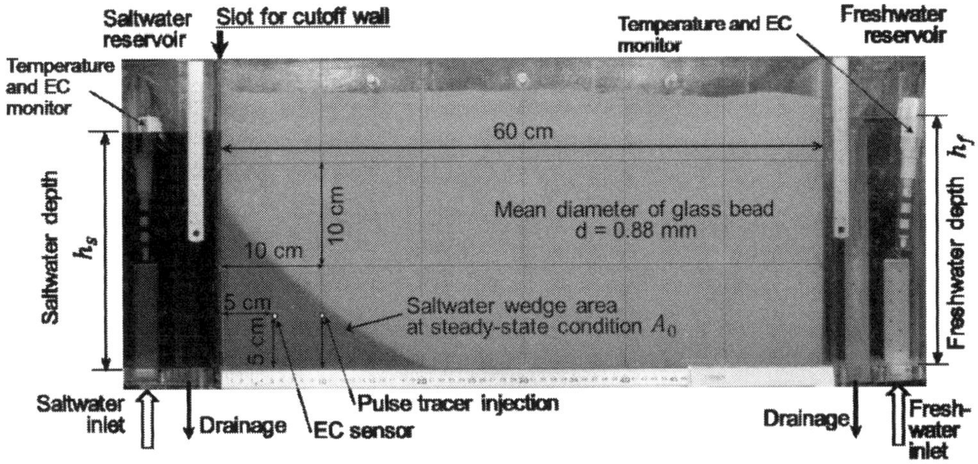

Figure 1. Experimental set-up.

We conducted tracer experiments to determine the longitudinal dispersivity, α_L, and transverse dispersivity, α_T, of the porous medium. We chose the fluorescein sodium salt ($C_{20}H_{10}Na_2O_5$) as a tracer. When a fluorescent dye tracer is exposed to ultra-violet light, the dye emitted visible light in proportion to the tracer concentration (Huang et al. 2002). We used two 40W Black Light Blue ultraviolet (UV) tubes as a light source at a distance of 80 cm inside a dark room to excite the fluorescent tracer. Images were recorded using a Nikon D5100 digital camera.

A tracer calibration test of the setup was first conducted to determine the relationship between the tracer concentration and luminosity. We saturated the porous medium with known concentrations of fluorescein salt (C = 0, 0.2, 2, 5, 10, 15 mg L^{-1}) and took photographs under UV light in the darkroom. We used the image processing software, Image J (U. S. National Institutes of Health, USA) to split the RGB (Red, Green, and Blue color bands) components and determine the luminosity values of the different concentrations using pixel value statistics. One pixel is equivalent to 0.024 cm. The Green color band shows very good correlation with tracer concentration, so we used this relationship to convert luminosity to actual concentration in the pulse injection tests.

We then performed a pulse tracer test with the water level at the left and right reservoirs set to 23.5 cm and 23.0 cm, respectively. The tracer solution with 15 mg L^{-1} concentration was introduced over the full thickness of the porous medium by using the injector. We injected a slug of 1.5 mg fluorescein salt for a period of 60 s using an automatic syringe pump to ensure a constant injection rate. As the tracer plume moves in the direction of flow, images were then captured by the digital camera. All these were done inside the dark room. We processed the captured images using Image J and with the derived relationship between G luminosity values and concentration, we determined the concentration profiles of the tracer plume at times equal to 0 min, 10 min, and 29 min after the 1-min tracer injection.

Cutoff Wall Experiments

We prepared saltwater by dissolving commercial salt in water and dyed it with a red edible pigment (Food dye red No. 102 New Coccine). Small peristaltic pumps were used to constantly supply saltwater and freshwater from supply buckets below the setup to the left and right reservoirs, respectively. Diluted saltwater overflowing from the water level adjustment cylinder on the drains into an independent channel while freshwater falls back into the freshwater bucket and recirculated into the system by the pumps. The cutoff wall experiment procedure was similar to Luyun et al. (2009). Initially, the porous medium and reservoirs were filled with freshwater. The water levels were set to a constant head of h_f =24.0 cm on the freshwater side and h_s =23.0 cm on the saltwater side (Figure 1). When the flow stabilized, a shutoff wall was inserted to isolate the saltwater reservoir and the freshwater in it was then replaced with the red saltwater solution. The saltwater intrusion process was initiated with the removal of the shutoff wall. Density driven flow progressed until steady state was achieved. After steady state was established, a 20-cm cutoff wall was inserted into the slot and the movement of the residual saltwater wedge was recorded by a digital camera. We measured the areas of the attenuating saltwater wedge at different time intervals. The experiment ended when the saltwater wedge was completely flushed out.

Numerical Simulations

In the pulse tracer experiments, we used MODFLOW and MT3DMS (Zheng and Wang 2008) to simulate the concentration profiles developed from the image analysis. A uniform grid spacing of 0.48 cm was used. Using all parameters from the experiments, we assumed several paired values of longitudinal and transverse dispersivities. The paired values that give the minimum root mean squared difference (RMSD) between concentration profiles determined from image analysis and those from the simulation, were considered to be the valid values of the longitudinal and transverse disperivities in our experiment.

We used the SEAWAT Ver4 (Langevin et al. 2008) to simulate the 20-cm cutoff wall experimental condition and using the estimated longitudinal and transverse dispersivities. Additionally, we performed the simulations using dispersivity ratio α_L/α_T of 10 and 100. We then compared the changes in concentration of the saltwater wedge as measured at the EC probe location during the experiment and the computed values at the same location from the simulation results. The purpose of the simulation is not only to reproduce the experimental results but also to demonstrate the effects of dispersivity on the saltwater removal process due to the cutoff wall installation.

RESULTS AND DISCUSSION

Pulse Tracer Experiments

Figure 2 shows a comparison of the tracer concentration profiles along the horizontal axis from the injection point. The longitudinal dispersivity, α_L, and transverse dispersivity, α_T, values that gave the minimum RMSD to the plotted concentration profiles from image analysis, were determined to be 0.07 cm and 0.0025 cm, respectively. The concentration profiles (black lines) calculated using MODFLOW and MT3DMS from these derived values are shown to be in good agreement with the experimental values. The dipersivity ratio then is α_L/α_T =28, which is within the range of 10 and 100 given by Todd and Mays (2005).

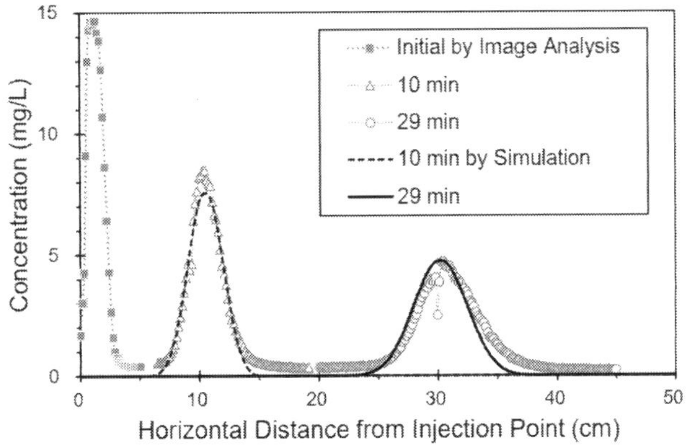

Figure 2. Comparison of the tracer concentration profiles along the horizontal axis at t = 0 min (initial), 10 min and 29 min after the pulse tracer injection.

Cutoff Wall Experiments

Figure 3 presents the comparison of the experimental and the numerical results at (a) initial steady state, and (b) 12 h after installation of the 20-cm cutoff wall. The numerical model predicted well the overall outline of the saltwater wedge. Dispersive flux of salt is carried with the freshwater as it flows along the mixing zone and over the cutoff wall. Saltwater is transported by diffusion and dispersion and then carried to the outlet by the freshwater flow. Since the cutoff wall prevented additional saltwater supply, there is a net flux of salt flowing out over the wall. The saltwater wedge area slowly but continuously decreased from A_0=138 cm^2 at initial steady state to A_{12}= 38 cm^2 and A_{24}= 7 cm^2 after 12 and 24 h, respectively. In the experiment, all the saltwater was flushed out after t_{max} =33 h (Figure 4).

The freshwater velocity profiles (Figure 3) near the freshwater reservoir at the initial steady state and after the cutoff wall installation did not change. This supported the experimental measurement that the freshwater discharge towards the saltwater reservoir remains constant during the saltwater removal process after the cutoff wall installation. In general, the freshwater velocity near the discharge zone is smaller than that above the saltwater wedge toe location because the cross-sectional area of freshwater flow decreases gradually towards the seaside discharge zone from the toe location.

Comparing the attenuating saltwater wedge versus time, Figure 4 shows that the experimental and numerical results are in good agreement. The reduction rate of the saltwater wedge area gradually becomes slower with time. As the saltwater wedge area reduces due to the saltwater removal, the cross-sectional area of freshwater flow above it increases and the freshwater velocity decreases. Smaller velocity results in smaller dispersion and the dilution of the saltwater across the mixing zone between the saltwater and freshwater is weakened. Thus, it takes more time to remove the residual saltwater with time after the cutoff wall installation. Dispersion is a key parameter to the dilution of the saltwater in the mixing zone during saltwater removal.

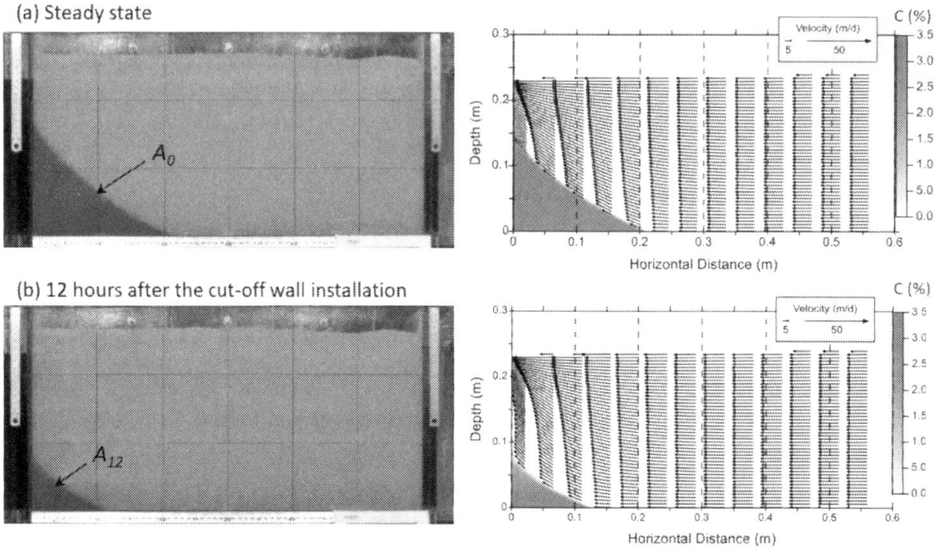

Figure 3. Comparison between the experiments and numerical simulations.

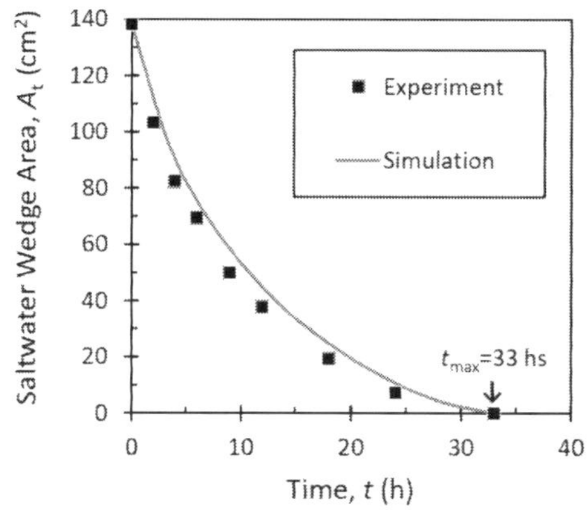

Figure 4. Experimental and numerical results of reduction in saltwater wedge area.

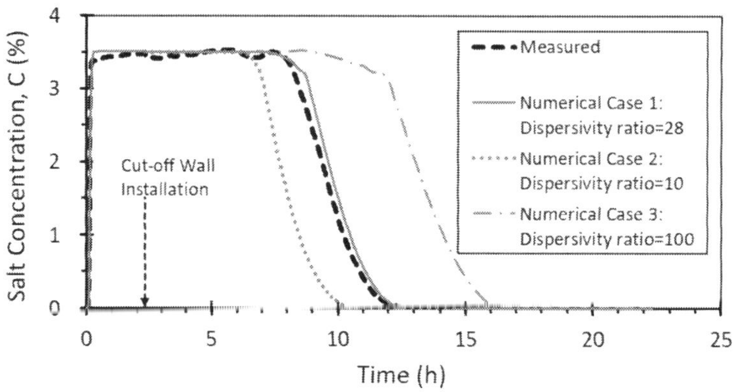

Figure 5. Measured and numerical results of salt concentration changes during saltwater intrusion process and subsequent removal after cutoff wall installation.

Figure 5 compares the measured concentration of the attenuating saltwater wedge with three cases of model computations using different dispersivity ratios: (a) the experimental Case 1, $\alpha_L/\alpha_T = 28$ ($\alpha_L = 0.07$ cm, $\alpha_T = 0.0025$ cm); (b) the hypothetical Case 2, $\alpha_L/\alpha_T = 10$ ($\alpha_L = 0.07$ cm, $\alpha_T = 0.007$ cm), and; (c) the hypothetical Case 3: $\alpha_L/\alpha_T = 100$ ($\alpha_L = 0.07$ cm, $\alpha_T = 0.0007$ cm). Note that the value of longitudinal dispersivity, α_L, is fixed. During the intrusion process, there is no apparent difference in the salt concentration. Upon installation of the cutoff wall, however, it can be observed that only the concentration change in Case 1 agreed well with that of the experiment. For Case 2, the saltwater wedge attenuated and was completely flushed out faster than in the experiment. For a constant longitudinal dispersivity, the lower the ratio the higher is the transverse dispersivity and the broader is the mixing zone of the saltwater wedge. This means that it will be easier for the dispersed flux of salt to be carried away by the freshwater discharge as it flows along the mixing zone and over the cutoff wall. It is the opposite in Case 3 where the saltwater wedge was completely flushed out much later than in the experiment. The higher the dispersivity ratio, the lower is the transverse dispersivity and the thinner is the mixing zone leading to a longer time of complete removal of the saltwater wedge.

CONCLUSIONS

In our laboratory-scale study in a saturated porous media made of glass beads with a mean diameter of 0.088 cm, the longitudinal and transverse dispersivities, and their ratio were estimated using a pulse-type fluorescent tracer injection experiment to be 0.07 cm, 0.0025 cm, and 28 cm, respectively. By using these values, the numerical calculation of saltwater intrusion and removal processes after cutoff wall installation, reproduce well the experimental salt concentration changes with time. Our numerical simulation illustrates the fact that accurate estimation of dispersivity is important and the dispersivity ratio $\alpha_L/\alpha_T = 10$ usually employed is not always applicable. Although mechanical dispersion in the transverse direction is a much weaker process than in the longitudinal direction, our study demonstrates that the transverse dispersion plays an important role in the reduction of residual saltwater after the cutoff wall installation.

REFERENCES

C. D. Langevin, D. T Thorne, Jr, A. M. Dausman, M. C. Sukop, and W. Guo 2008. SEAWAT Version 4: A Computer Program for Simulation of Multi-Species Solute and Heat Transport. US Geological Survey Techniques of Water Resources Investigations 6-A22.

Huang, W., C. Smith, D. Lerner, S. Thornton, and A. Oram 2002. Physical modeling of solute transport in porous media: evaluation of an imaging technique using UV excited fluorescent dye. Water Research 36: 1843-1853.

C. Zheng and P. P. Wang 2008. MT3DMS Documentation and User's Guide. U.S. Army Corps of Engineers.

Luyun Jr., R., K. Momii, and K. Nakagawa 2009. Laboratory-scale saltwater behavior due to subsurface cutoff wall. Journal of Hydrology 377: 227-236.

Todd, D.K., and L. W. Mays 2005. Groundwater Hydrology Third Edition. John Wiley & Sons, Inc.

Contact Information: Roger A. Luyun Jr., Land and Water Resources Division, College of Engineering and Agro-industrial Technology, University of the Philippines Los Baños, Phone/Fax: +6349-536-2387, Email: raluyun1@up.edu.ph

MANTRA-O18: An Extended Version of SUTRA Modified to Simulate Salt and $\delta^{18}O$ Transport amid Water Uptake by Plants

Su Yean Teh[1], Donald L. De Angelis[2], Clifford I. Voss[3], Leonel Sternberg[4] and Hock Lye Koh[5]
[1]School of Mathematical Sciences, Universiti Sains Malaysia, Penang 11800, Malaysia
[2]U.S. Geological Survey, Wetland and Aquatic Research Center, Gainesville, FL, USA
[3]U.S. Geological Survey, Menlo Park, CA, USA
[4]Department of Biology, University of Miami, Coral Gables, FL, USA
[5]Jeffrey Sachs Center on Sustainable Development, Sunway University, Jalan Universiti, Bandar Sunway, 47500 Selangor, Malaysia

ABSTRACT

Sea level rise and the increasing landward intrusion of storm surges pose the threat of replacement of salinity-intolerant vegetation of important coastal habitats by salinity-tolerant vegetation. Therefore, a means is needed to better understand the processes that influence this vegetation shift and to aid in the management of coastal resources. For this purpose, a hydrology–salinity–vegetation model known as MANTRA was developed by coupling a spatially explicit model (MANHAM) for simulation of vegetation community dynamics along coastal salinity gradients with SUTRA, a USGS groundwater flow and transport model. MANTRA has been used to project possible future changes in Coot Bay Hammock in southern Florida under conditions of gradually rising sea level and storm surges. The simulation study concluded that feasibility exists of a regime shift from hardwood hammocks to mangroves subject to a few conditions, namely severe damage to the existing hammock after a storm surge and a sufficiently persistent high salinity condition and high input of mangrove seedlings. Early detection of salinity stress in vegetation may facilitate sustainable conservation measures being applied. It has been shown that the $\delta^{18}O$ value of water in the xylem of trees can be used as a surrogate for salinity in the rooting zone of plants, which is difficult to measure directly. Hence, the model MANTRA is revised into MANTRA-O18 by including the $\delta^{18}O$ of the tree xylem dynamics. A simulation study by MANTRA-O18 shows that effects of increasing salinization can be detected many years before the salinity-intolerant trees are threatened with replacement.

INTRODUCTION

Climate change and resulting sea level rise (SLR) will inflict changes that may be irreversible in coastal ecosystems, particularly those of low lying landscapes and atoll islands. Sustained by a fragile balance of freshwater and seawater interactions, the Everglades ecosystem is especially susceptible to sea level rise as documented in Ross et al. (2000, 2009). Along coastal southern Florida, the freshwater marsh has been observed to be replaced by mangroves (Gleason et al. 1974; Willard et al. 1999; Williams et al. 1999). Many such shifts from salinity-intolerant vegetation to salinity-tolerant vegetation have been attributed to sea level rise (Alexander and Crook 1974; Lara et al. 2002; Kirwan and Megonigal, 2013). The pace of such shifts may be affected positively or negatively by the self-reinforcing positive feedback between the vegetation and salinity (Passioura et al. 1992; Sternberg et al. 2007), as well as the frequent and intensified salinity pulses associated with the increasing impact of storm surges as a consequence of sea level rise (Scheffer et al. 2001; Teh et al. 2008). There is a need to understand, predict, and prepare for the consequences of climate change-related impacts, in particular the effects of SLR and storm surges on both the short-term dynamics of salinity in the soil and groundwater and the long-term effects on vegetation. For this purpose, a hydrology–salinity–vegetation model known as MANTRA (Teh et al. 2013) was developed by coupling a spatially explicit model (MANHAM) for simulation of vegetation community dynamics along coastal salinity gradients with SUTRA, a USGS's groundwater flow and transport model. MANTRA has been applied to a Coot Bay Hammock along the southwestern coast of Everglades National Park to project the possible future changes in such coastal hammocks under sea level rise and storm surges (Teh et al. 2015). This simulation study underscores that three conditions are necessary for a hardwood hammock to undergo a regime shift leading to a mangrove community; sufficiently severe damage to the existing hammock to open a gap to allow growth of invading propagules, a large input of salinity persisting for a long enough period of time to favor growth of mangrove propagules in competition with remaining freshwater vegetation, and an input of enough mangrove propagules to allow mangroves to be present in sufficient number to influence the future soil salinity. It is desirable to have an early indicator of impending shifts in vegetation due to salinity stress. Water salinity of the vadose zone, salinity of xylem water and predawn water potential are some of the potential indicators of critical transition from salinity-intolerant vegetation to salinity-tolerant vegetation but there are uncertainties and limitations in the measurements of these indicators (Zhai et al. 2016). It has been shown that the oxygen isotope composition ($\delta^{18}O$ value) of plant stem water may be an indicator of salinity stress (Vendramini and Sternberg 2007). Hence, the model MANTRA is revised into MANTRA-O18 by including the $\delta^{18}O$ of the tree xylem dynamics. A brief overview of MANTRA-O18 is given in the following section.

MANTRA-O18

MANTRA-O18 is an extended version of SUTRA (Voss and Provost 2002) that is capable of simulating (i) vegetation community dynamics and (ii) variable-density flow and transport of two solutes; i.e., salt and ^{18}O, through variably to fully saturated porous media. So this extended version named MANTRA-O18 is:

(a) An improvement of SUTRA in that the U.S. Geological Survey's spatially explicit model of vegetation community dynamics along coastal salinity gradients (MANHAM) is integrated, and,
(b) A simplified version of SUTRA-MS (Hughes and Sanford, 2015) in that the number of solutes is limited to two with one solute (salt) effecting fluid density and the other solute (^{18}O isotope) not affecting fluid density.

In MANTRA-O18, the water uptake rates of hammock and mangrove are determined based on the solute (salt) concentration *S* calculated by the SUTRA module. The total water uptake by plants then affects the fluid density and fluid pressure, which consequently change the salinity.

Further details regarding MANHAM and MANTRA can be found in Teh et al. (2008) and Teh et al. (2013). Changes were made to the fluid and solute mass balance equations so that the pure water fluid uptake by the salinity-excluding plants is properly accounted for, including the incorporation of a second solute mass balance equation for ^{18}O isotope; hence the name MANTRA-O18. Internally in MANTRA-O18, the proportion of ^{18}O isotope ($^{18}O/^{16}O$) is tracked numerically and then converted to $\delta^{18}O$ for a resulting output using

$\delta^{18}O = \left[{}^{18}O / {}^{18}O_{standard} - 1 \right] \times 1000$ with ${}^{18}O_{standard}$ as the standard mean ocean water (SMOW).

TEST CASE: PURE FLUID OUTFLOW

To illustrate the effect of pure fluid outflow on salinity and ${}^{18}O$ isotope, a domain as illustrated in Figure 1 used. It should be noted that domain setup and the parameter values are used for testing purposes so their scale and magnitude may not be realistic. The saline seawater is marked with $\delta^{18}O$ value of +4‰ while the fresh pure water is marked with $\delta^{18}O$ value of −3‰.

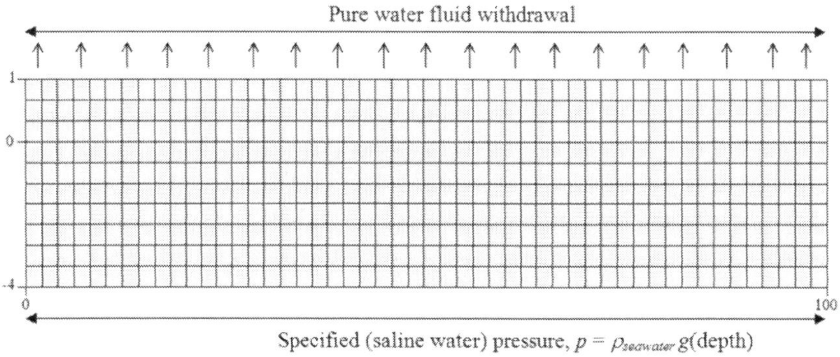

Figure 1. Domain setup of simple fluid outflow case.

The aquifer is assumed to sit on top of saline groundwater so a constant specified pressure is defined at the bottom boundary. At the top boundary, a constant outflow of pure water fluid is imposed. Note that there is no recharge of freshwater so the continuous uptake of water will unreasonably cause a severe build-up of salt. This is certainly not a realistic setup but the objective of this set up is to demonstrate the build-up of salt and transport of ${}^{18}O$ isotope in MANTRA-O18 so the results will be illustrated at a fixed time at t = 100 day. Figure 2(a) illustrates the build-up of salt in the presence of pure water fluid outflow from the surface cells. The corresponding simulated $\delta^{18}O$ concentration is shown in Figure 2(b), showing no build-up of $\delta^{18}O$, as the $\delta^{18}O$ isotope is withdrawn together with the pure water fluid. As the fresh water fluid ($\delta^{18}O$ = −3‰) available in the domain is being continually withdrawn by the plants, the saline groundwater ($\delta^{18}O$ = +4‰) at the bottom boundary gradually infiltrates into the upper layers and mixes with the fresh water fluid. The infiltration of saline groundwater from the bottom boundary and the withdrawal of $\delta^{18}O$ isotope by the plants causes the $\delta^{18}O$ isotope concentration to decrease from the bottom to the top of the domain.

SLR SIMULATION

The simulation results of MANTRA-O18 for a coastal transect subjected to SLR rate of 3 mm/year are illustrated here. Figure 3 shows the steady-state hammock and mangrove biomass, salinity and $\delta^{18}O$ isotope profiles simulated by means of MANTRA-O18 before the SLR event. The changes in hammock and mangrove distribution, as well as salinity and $\delta^{18}O$ isotope profiles for 100 years after SLR, are shown in Figure 4.

 https://doi.org/10.1051/e3sconf/20185400039

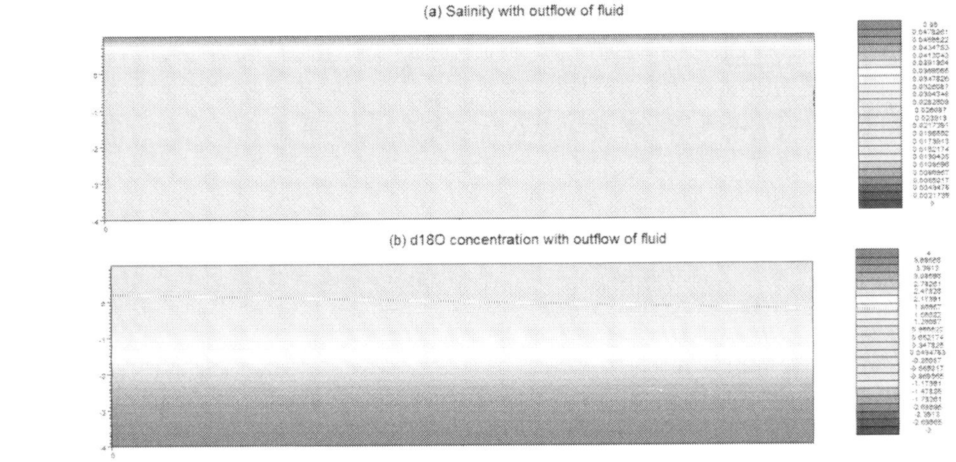

Figure 2. Simulated (a) salinity and (b) δ^{18}O concentration with outflow of fluid.

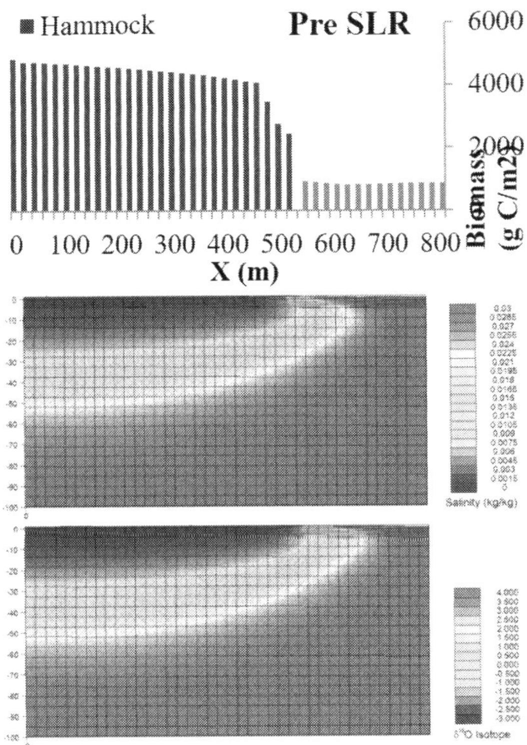

Figure 3. Simulated hammock and mangrove biomass (top), salinity (middle) and δ18O isotope (bottom) profiles without sea level rise (SLR).

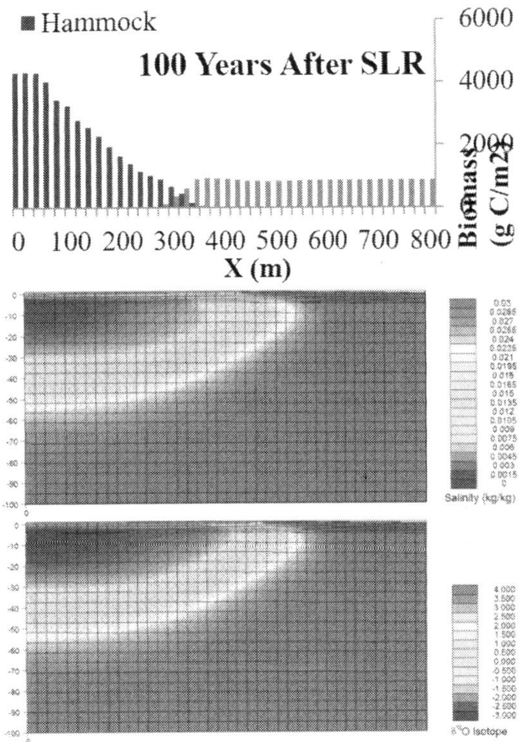

Figure 4. Simulated hammock and mangrove biomass (top), salinity (middle) and δ18O isotope (bottom) profiles 100 years after SLR begins.

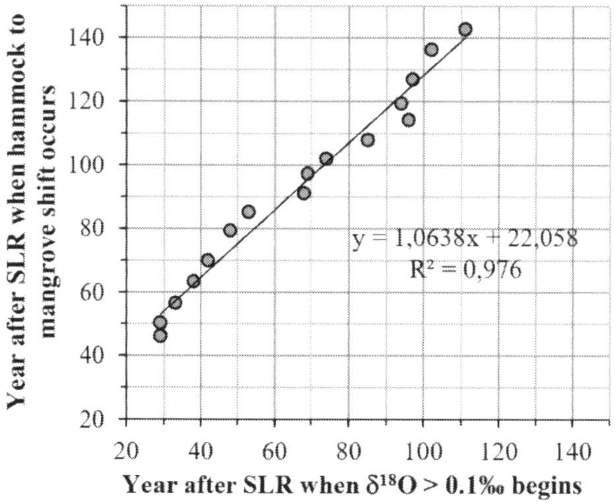

Figure 5. Number of years after SLR initiation when hammock to mangrove shift occurs plotted against the number of years after SLR initiation, when $\delta^{18}O$ value of these cells exhibit a difference of greater than 0.1‰.

DISCUSSION AND CONCLUSION

This paper introduces the model MANTRA-O18 in which $\delta^{18}O$ of the tree xylem is coupled with the modeling of hydrology, salinity, and the responses of both salinity-intolerant and their competing salinity-tolerant trees. The results of a test case are presented to illustrate the features of the simulated salinity and $\delta^{18}O$. MANTRA-O18 has also been used to assess the potential of tracking the yearly $\delta^{18}O$ of plant stem water to predict the shifting of vegetation as a result of SLR. As shown in Figure 5, simulation results by MANTRA-O18 on a test case indicate a linear relation between the year when the shift from hammock to mangrove occurs and the year when the $\delta^{18}O$ difference is greater than 0.1‰ (the current precision of measurement of the isotopic composition of stem water). This suggests that for this test case of SLR, the annually-averaged $\delta^{18}O$ could start to exhibit values greater than 0.1‰ at least 20 years (mean ≈ 26 years, standard deviation ≈ 5 years, min ≈ 17 years, max ≈ 35 years) before the hammock to mangrove shift occurs. The correlation obtained here is for this particular problem setup, so the value could be different for other setups. However, the notion that the $\delta^{18}O$ difference substantially precedes the shift in vegetation is likely transferrable to other situations.

ACKNOWLEDGEMENT

Financial support provided by USGS grant for this study is gratefully acknowledged. TSY gratefully acknowledge the Loreal-UNESCO Women in Science Fellowship 2017 and FRGS grant 203/PMATHS/6711569.

REFERENCES

Alexander, T.J. and A.G. Crook. 1974. Recent vegetational changes in southern Florida. In: Gleason, P.J. (Ed.), Environments of Southern Florida: Present and Past. Memoir 2: Miami Geological Society, Miami, Florida, pp. 61–72.

Gleason, P.J., A.D. Cohen, H.K. Brooks, P. Stone, W.G. Smith and W. Spackman Jr. 1974. The environmental significance of Holocene sediments from the Everglades and saline tidal plain. In: Gleason, P.J. (Ed.), Environments of Southern Florida: Present and Past. Memoir 2: Miami Geological Society, Miami, Florida, pp. 61–72.

Hughes, J.D. and W.E. Sanford. 2005. SUTRA-MS a Version of SUTRA Modified to Simulate Heat and Multiple-Solute Transport: U.S. Geological Survey Open-File Report 2004-1207, 141 p.

Kirwan, M.L. and J.P. Megonigal. 2013. Tidal wetland stability in the face of human impacts and sea-level rise. Nature 504: 53–60.

Lara, R., C. Szlafsztein, M. Cohen, U. Berger and M. Glaser. 2002. Implications of mangrove dynamics for private land use in Braganc,a, North Brazil: a case study. Journal of Coastal Conservation 8: 97–102.

Passioura, J.B., M.C. Ball and J.H. Knight. 1992. Mangrove may salinize the soil and in so doing limit their transpiration rate. Functional Ecology 6: 476–481.

Ross, M.S., J.F. Meeder, J.P. Sah, P.L. Ruiz and G.J. Telesnicki. 2000. The southeast saline Everglades revisited: 50 years of coastal vegetation change. Journal of Vegetation Science 11:101–112.

Ross, M.S., J.J. O'Brien, R.G. Ford, K. Zhang and A. Morkill. 2009. Disturbance and the rising tide: the challenge of biodiversity management for low island ecosystems. Frontiers in Ecology and Environment 9: 471–478

Scheffer, M., S. Carpenter, J.A. Foley, C. Folke, and B. Walker. 2001. Catastrophic shifts in ecosystems. Nature 413: 591–596.

Sternberg, L.D.L., S.Y. Teh, S.M.L.Ewe, F. Miralles-Wilhelm and D.L. DeAngelis. 2007. Competition between hardwood hammocks and mangroves. Ecosystems 10: 648–660.

Teh, S.Y., D.L. DeAngelis, L.D.L. Sternberg, F.R. Miralles-Wilhelm, T.J.I. Smith and H.L. Koh. 2008. A simulation model for projecting changes in salinity concentrations and species dominance in the coastal margin habitats of the Everglades. Ecological Modelling 213: 245–256.

Teh, S.Y., H.L. Koh, D.L. DeAngelis and M. Turtora. 2013. Interaction between salinity intrusion and vegetation succession: A modeling approach. Theoretical & Applied Mechanics Letters 3: 1-032001.

Teh, S.Y., M. Turtora, D.L. DeAngelis, J. Jiang, L. Pearlstine, T.J. Smith and H.L. Koh. 2015. Application of a Coupled Vegetation Competition and Groundwater Simulation Model to Study Effects of Sea Level Rise and Storm Surges on Coastal Vegetation. Journal of Marine Science and Engineering 3: 1149–1177.

Vendramini P.F. and L.S.L. Sternberg. 2007. A faster plant stem-water extraction method. Rapid Communications in Mass Spectrometry 21: 164-168.

Voss, C. I., and A.M. Provost. 2002. (Version of September 22, 2010), SUTRA, A model for saturated-unsaturated variable-density ground-water flow with solute or energy transport. U.S. Geological Survey Water-Resources Investigations Report 02-4231, 291 p.

Willard, D.A., C.W. Holmes, W.H. Orem, L.M. Weimer. 1999. Plant communities of the Everglades: a history of the last two millenia. In: Gerould, S., Higer, A. (compilers). U. S. Geological Survey Program on the South Florida Ecosystem—Proceedings of South Florida Restoration Science Forum, May 17–19, 1999, Boca Raton, Florida. U. S. Geological Survey Open-File Report 99-181, Tallahassee, Florida, pp. 118–119.

Williams, K., Z.S. Pinzon, R.P. Stumpf and E.A. Raabe. 1999. Sea level rise and coastal forests on the Gulf of Mexico. Open-File Report 99-441. U.S. Geological Survey, Center for Coastal Geology, St. Petersburg, Florida.

Zhai, L., J. Jiang, D.L. DeAngelis and L.S.L. Sternberg. 2016. Prediction of plant vulnerability to salinity increase in a coastal ecosystem by stable isotopic composition (δ18O) of plant stem water: a model study. Ecosystems 19: 32–49,

Contact Information: Su Yean Teh, Universiti Sains Malaysia, School of Mathematical Sciences, 11800 USM, Pulau Pinang, Malaysia, Phone: 604-6534770,
Fax: 604-6570910, Email: syteh@usm.my

Groundwater salinity mapping of the Belgian coastal zone to improve local freshwater storage availability

Dieter Vandevelde[1], Esther van Baaren[2], Joost Delsman[2], Marios Karaoulis[2], Gualbert Oude Essink[2,9], Perry de Louw[2,8], Tommer Vermaas[2], Pieter Pauw[2], Marco de Kleine[2], Sara Thofte[3], Rasmus Teilmann[3], Kristine Walraevens[4], Marc Van Camp[4]; Huits Dominique[5], Willem Dabekaussen[6], Jan Gunnink[6], Alexander Vandenbohede[7]
[1]Flanders Environment Agency (VMM), Brussels, Belgium
[2]Deltares, Utrecht, The Netherlands
[3]SkyTEM, Aarhus, Denmark
[4]Ghent University, Department of Geology, Belgium
[5]Inagro Knowledge Centre Water, Roeselare, Belgium
[6]Geological Survey of The Netherlands (TNO), The Netherlands
[7]De Watergroep, Brussel, Belgium
[8]Wageningen University & Research, Wageningen, The Netherlands
[9]Utrecht University, Utrecht, The Netherlands

ABSTRACT

In the European TOPSOIL project, countries around the North Sea are searching for solutions for climate related threats. They explore the possibilities of using the topsoil layer to solve current and future water challenges. The main objective is to improve the climate resilience of the water management of the topsoil and shallow aquifers in the North Sea region. TOPSOIL is supported by the Interreg VB North Sea Region program in line with priority 3 of the program: 'Sustainable North Sea Region, protecting against climate change and preserving the environment'.

The Belgian part of this project, called FRESHEM for GO-FRESH Vlaanderen ('FREsh Salt groundwater distribution by Helicopter ElectroMagnetic survey for Geohydrological Opportunities FRESH water supply'), focuses on mapping the salinity distribution of groundwater using airborne electromagnetics and aims to look into a number of measures that could increase the availability of freshwater for agriculture in the polder area. Two pilot projects will evaluate the possibilities for freshwater storage and aims to specify what measures can be taken to achieve this. Together with the other water users and water managers, The Flanders Environment Agency wants to prepare a plan for the realization of one or more pilot projects that can improve the availability of freshwater.

INTRODUCTION

Shallow coastal aquifers are vulnerable to climate change and sea-level rise (Cliwat, 2011; Ferguson & Gleeson, 2012; Oude Essink et al., 2010; Vandenbohede et al., 2008). Some coastal regions in Europe are already dealing with an increased threat of drought and flooding (Faneca Sànchez et al., 2012; Giambastiani et al., 2007; Pauw et al., 2012; Rasmussen et al., 2013; Yang et al., 2015). Climate change and sea-level rise disturb the fragile balance between fresh and salt water and can lead to a more saline environment. The summer of 2017 for instance showed in the Belgian coastal plain how urgent water scarcity can be and how fast salinization of the fresh water surface water system can occur in polder areas.

Farmers in West-Flanders (Belgium) make significant use of fresh groundwater from the confined Paleocene aquifer for their water supply. For years, the groundwater level of the Paleocene aquifer has been decreasing, compromising the security of water supply. In addition, shallow water resources are scarce because of the occurrence of brackish and saline water in the top aquifer near the coast, and the shallow occurrence of aquitards further inland. The TOPSOIL project will explore the possibilities of using the shallow aquifer to increase the availability of fresh groundwater.

The salinity distribution in the coastal plain was mapped by De Breuck et al. (1974 and 1989). To get an overview of the current situation of shallow water resources, the salinization of the coastal aquifer is reinvestigated by airborne electromagnetics. These results are used in a regional water system analysis where physical conditions, measures and potentials to improve the freshwater availability are investigated. This water system analysis will lead to regional maps showing potentials to apply specific measures to improve the freshwater availability.

AIRBORNE SURVEY

The fresh-salt water distribution of the coastal aquifers is mapped using the SkyTEM system (Figure 2) which is an airborne transient electromagnetic sensor that measures ground conductivity using electromagnetic waves (Auken et al., 2008; Schaars & Rolf, 2014; Steuer et al., 2009; Viezzoli et al., 2009). The SkyTEM system delivers high resolution data of the subsurface resistivity thanks to a fast turn-off time and early time data, and a high signal to noise ratio. About 2400 line km are flown which covers the main part of the Belgian polder area (Figure 1). The inversion process produces bulk resistivity profiles for each flight line (Figure 3). With the aid of a lithological model, bulk resistivity will be converted into a 3D image of pore water resistivity and chloride concentrations. Finally, a map showing the distribution of porewater salinity of the area will be available.

Figure 1. Flight lines (color coded on daily production).

The processed and inverted airborne electromagnetic data are verified by results of other hydrogeological research like recent groundwater analysis, borehole logs and SlimFlex measurements (flexible borehole electromagnetic soundings). Comparison will take place with data of vertical electrical soundings (VES).

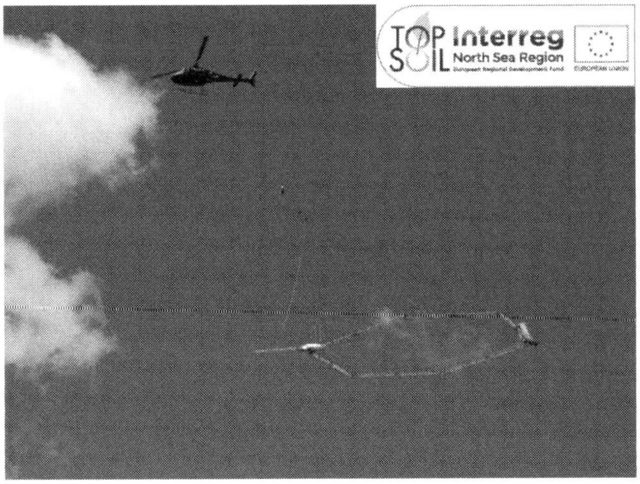

Figure 2. SkyTEM system

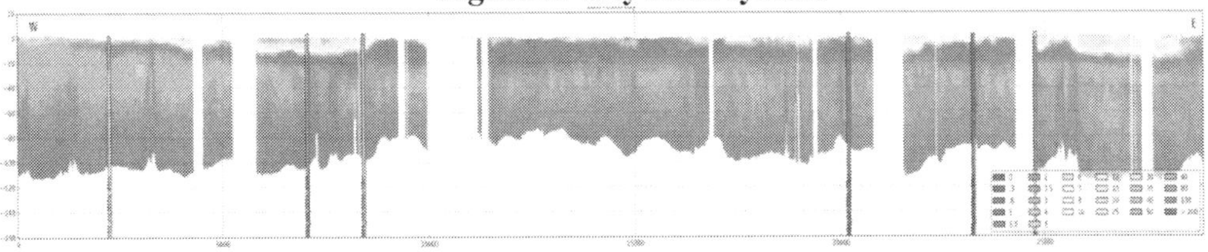

Figure 3. Bulk resistivity profile of flight line 103001 (Westcoast).

IMPROVING THE FRESH-WATER AVAILABILITY

The results of the airborne survey will be used to investigate the potential to improve the freshwater availability in the western polder area. The consortium (Deltares, Ghent University, TNO, Inagro, De Watergroep) has already a lot of experience with these kind of measures (van Baaren et al., 2015, 2018; Bironne et al., 2017; Delsman et al., 2015, 2017, 2018; Oude Essink et al., 2014; Pauw et al., 2015). This experience will be used to select the most appropriate measures for the study area. The following steps are distinguished:

Water system analysis

The current fresh-salt water distribution and the possibilities to increase the existing freshwater lenses, are determined by physical conditions (hydrogeology, topography) as well as human conditions (drainage systems, groundwater extractions). The water system analysis will collect all necessary data and will determine relationships between those conditions. Possible measures will be listed. For each measure boundary, conditions will be defined together with the benefits and disadvantages. The water system analysis will finally lead to a map which shows the potential to apply a certain kind of measure.

Stakeholder participation

By consulting local stakeholders about their current and future needs we will get an idea of the water availability, water use and water demand today and in the near future. It will also give insight into how realistic the implementation of a certain measure will be. The key to a successful implementation of measures lies in fostering support through participation.

Pilot areas

For a selected area, the most appropriate measures will be withheld. For these measures, a financial analysis will be made (costs and benefits). In consultation with stakeholders, two pilot areas will be selected and an implementation plan will be made. The realization of these pilot projects will serve as model projects for future investigations.

CONCLUSIONS

The Belgian part of the TOPSOIL project explores the possibilities of using the shallow aquifer to increase the availability of fresh groundwater for agriculture. An airborne electromagnetic survey (SkyTEM) updates information about the current situation of fresh groundwater reserves in the coastal aquifer. The output of this airborne survey will serve as basic information for a water system analysis. This analysis leads to a map which shows the potential to apply a certain kind of measure to improve the freshwater availability. Together with stakeholders, pilot areas will be selected and will serve as model projects to future investigations.

REFERENCES

Auken, E., Christiansen, A. V., Jacobsen, L. H., & Sørensen, K. I. 2008. A resolution study of buried valleys using laterally constrained inversion of TEM data. Journal of Applied Geophysics, 65(1), 10–20. https://doi.org/10.1016/j.jappgeo.2008.03.003

Van Baaren, E. S., Delsman, J. R., Karaoulis, M., Pauw, P. S., Vermaas, T., Bootsma, H., ... Meyer, U. 2018. FRESHEM Zeeland - FREsh Salt groundwater distribution by Helicopter ElectroMagnetic survey in the Province of Zeeland. Deltares report 1209220. Utrecht, Netherlands.

van Baaren, E. S., De Lange, G., Vermeulen, P. T. M., & Vonhögen - Peeters, L. 2015. Exploring groundwater resources on Jurong Island, Singapore; Monitoring, modelling and assessing risks (PUB report).

Bironne, A., Saha, A., Lee, W. K., Babovic, V., Vonhogen, L., van Baaren, E. S., ... Hoogendoorn, B. 2017. Utilization of reclaimed land as groundwater reservoir. In 14th Annual Meeting Asia Oceania Gesciences Society, 6-11 Augustus.

De Breuck, W., De Moor, G., Maréchal, R., & Tavernier, R. 1974. Diepte van het grensvlak tussen zoet en zout water in de freatische laag van het Belgische kustgebied (1963-1973). Verziltingskaart. Brussel, Militair Geografisch Instituut.

De Breuck, W., De Moor, G., Maréchal, R. & Tavernier R. 1989. Diepte van het grensvlak tussen zoet en zout water in de freatische watervoerende laag van noordelijk Vlaanderen (1974-1975). Cliwat. (2011). Groundwater in a Future Climate.

Delsman, J. R., De Paz, J. M., De Klerk, M., Stuyt, L., De Louw, P. G. B., Visconti, F., ... Pecora, S. 2015. Application of the FWOO method in the Vega Baja Segura catchment Application of the FWOO method in the Vega Baja Segura catchment. Deltares Rapport 1220325, 98.
Delsman, J. R., Van Baaren, E. S., Karaoulis, M., Pauw, P. S., Vermaas, T., Bootsma, H., ... Revil, A. 2018. Large-scale, probabilistic airborne EM mapping provides detailed view of groundwater salinity in the province of Zeeland, Netherlands. To Be Submitted to Environmental Research Letters.

Delsman, J. R., Boekel, E. Van, Reinhard, S., Van Loon, A., Bartholomeus, R., Mulder, M., ...

Schasfoort, F. 2017. Regioscan Zoetwatermaatregelen, Verkennen van het perspectief van kleinschalige zoetwatermaatregelen voor de regionale zoetwateropgave.

Faneca Sànchez, M., Gunnink, J. L., van Baaren, E. S., Oude Essink, G. H. P., Auken, E., Elderhorst, W., & De Louw, P. G. B. 2012. Modelling climate change effects on a Dutch coastal groundwater system using airborne electromagnetic measurements. Hydrology and Earth System Sciences, 16(12), 4499–4516. https://doi.org/10.5194/hess-16-4499-2012

Ferguson, G., & Gleeson, T. 2012. Vulnerability of coastal aquifers to groundwater use and climate change. Nature Climate Change, 2(5), 342–345. https://doi.org/10.1038/nclimate1413

Giambastiani, B. M. S., Antonellini, M., Oude Essink, G. H. P., & Stuurman, R. J. 2007. Saltwater intrusion in the unconfined coastal aquifer of Ravenna (Italy): A numerical model. Journal of Hydrology, 340(1–2), 91–104. https://doi.org/10.1016/j.jhydrol.2007.04.001

Oude Essink, G. H. P., van Baaren, E. S., & De Louw, P. G. B. 2010. Effects of climate change on coastal groundwater systems: A modeling study in the Netherlands. Water Resources Research, 46(May), 1–16. https://doi.org/10.1029/2009WR008719

Oude Essink, G. H. P., Van Baaren, E. S., Zuurbier, K. G., Velstra, J., Veraart, J., Brouwer, W., … Schroevers, M. 2014. GO-FRESH: Valorisatie kansrijke oplossingen voor robuuste zoetwatervoorziening, KvK rapport (in Dutch).

Pauw, P. S., De Louw, P. G. B., & Oude Essink, G. H. P. 2012. Groundwater salinisation in the Wadden Sea area of the Netherlands: quantifying the effects of climate change , sea-level rise and anthropogenic interferences. Netherlands Journal of Geosciences - Geologie En Mijnbouw, 91–3(3), 373–383. https://doi.org/10.1017/S0016774600000500

Pauw, P. S., Van Baaren, E. S., Visser, M., De Louw, P. G. B., & Oude Essink, G. H. P. 2015. Increasing a freshwater lens below a creek ridge using a controlled artificial recharge and drainage system: a case study in the Netherlands. Hydrogeology Journal. https://doi.org/10.1007/s10040-015-1264-z

Rasmussen, P., Sonnenborg, T. O., Goncear, G., & Hinsby, K. 2013. Assessing impacts of climate change, sea level rise, and drainage canals on saltwater intrusion to coastal aquifer. Hydrology and Earth System Sciences, 17(1), 421–443. https://doi.org/10.5194/hess-17-421-2013

Schaars, F., & Rolf, H. 2014. SkyTEM PWN: Eindrapportage van het airborne geofysisch onderzoek Noordhollands Duinreservaat PWN.

Steuer, A., Siemon, B., & Auken, E. 2009. A comparison of helicopter-borne electromagnetics in frequency- and time-domain at the Cuxhaven valley in Northern Germany. Journal of Applied Geophysics, 67(3), 194–205. https://doi.org/10.1016/j.jappgeo.2007.07.001

Vandenbohede, A., Luyten, K., & Lebbe, L. C. 2008. Effects of Global Change on Heterogeneous Coastal Aquifers: A Case Study in Belgium. Journal of Coastal Research, 24, 160–170. https://doi.org/10.2112/05-0447.1

Viezzoli, A., Auken, E., & Munday, T. 2009. Spatially constrained inversion for quasi 3D modelling of airborne electromagnetic data an application for environmental assessment in the Lower Murray Region of South Australia. Exploration Geophysics, 40(2), 173–183. https://doi.org/10.1071/EG08027

Yang, J., Graf, T., & Ptak, T. 2015. Impact of climate change on freshwater resources in a heterogeneous coastal aquifer of Bremerhaven, Germany: A three-dimensional modeling study. Journal of Contaminant Hydrology, 177–178, 107–121. https://doi.org/10.1016/j.jconhyd.2015.03.014

Contact Information: Dieter Vandevelde, Flanders Environment Agency, Koning Albert II-laan 20, 1000 Brussels, Belgium, Phone: +32 2 214 21 27,
Email: di.vandevelde@vmm.be

Saltwater intrusion under climate change in North-Western Germany - mapping, modelling and management approaches in the projects TOPSOIL and go-CAM

Helga Wiederhold[1], Wolfgang Scheer[2], Reinhard Kirsch[3], M. Azizur Rahman[1] and Mathias Ronczka[1]
[1]Leibniz Institute for Applied Geophysics (LIAG), Hannover, Germany
[2]State Agency for Agriculture, Environment and Rural Areas Schleswig-Holstein (LLUR), Flintbek, Germany
[3]Institut of Geosciences, University of Kiel, Germany

ABSTRACT

Climate change will result in rising sea level and, at least for the North Sea region, in rising groundwater table. This leads to a new balance at the fresh–saline groundwater boundary and a new distribution of saltwater intrusions with strong regional differentiations. These effects are investigated in several research projects funded by the European Union and the German Federal Ministry of Education and Research (BMBF). Objectives and some results from the projects TOPSOIL and go-CAM are presented in this poster.

INTRODUCTION

Predicted results of climate change in the North Sea region (DK, GE, NL, BE, UK) are warmer and dryer summers and increased precipitation in the colder seasons leading to enhanced groundwater recharge and rising groundwater table (CLIWAT Working Group 2011). Additionally, a rise of sea level up to 1 m is predicted. At the coastline, this will lead to a new balance between seawater and freshwater with consequences for the saltwater intrusions. Rise in sea level will also extend the reach of saltwater in the River Elbe towards the inland. Hence, the area of influence of saltwater intrusion from the river will also extend. Size and sign of this effect can have strong regional variation and must be quantified to enable action plans to ensure sufficient freshwater for the human population, agriculture and industry. Investigations of future development of saltwater intrusions in the North Sea regions started 2008 with the EU INTERREG IVB North Sea project CLIWAT and 2013 with the German government funded project NAWAK. Focus areas were the coastal region of Lower Saxony and the North Sea islands of Föhr and Borkum. Results were, e.g., enhanced demand for drainage and a new saltwater intrusion pattern (Burschil et al. 2012, Sulzbacher et al. 2012). Based on these results, two new projects TOPSOIL and go-CAM started in 2015 and 2017 respectively.

THE TOPSOIL PROJECT

Focus of the INTERREG VB North Sea project TOPSOIL (www.topsoil.eu) is the interaction of soil and groundwater under climate change. The TOPSOIL project considers 16 pilot areas that are located in Belgium (2 pilot areas), Denmark (4 pilot areas), Germany (5 pilot areas), the Netherlands (3 pilot areas) and the UK (2 pilot areas). In Northern Germany, aspects of seawater intrusion are investigated in 2 corresponding pilot areas in Schleswig-Holstein (GE-1: Störmarsch) and Lower Saxony (GE-2: Elbe-Weser region)

https://doi.org/10.1051/e3sconf/20185400041

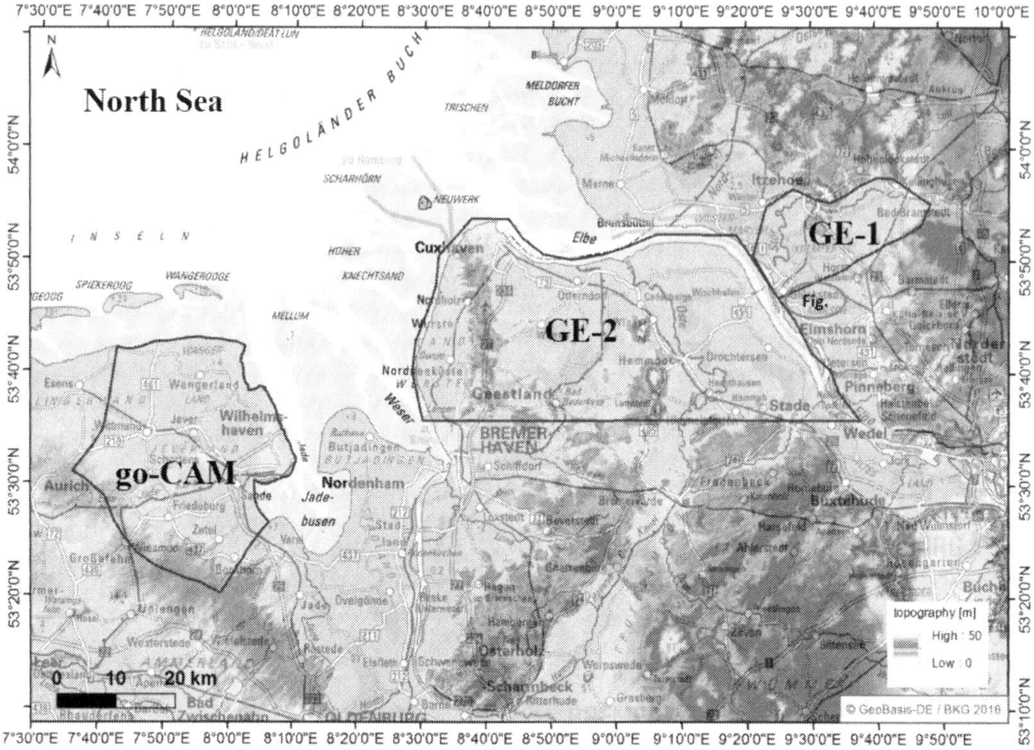

Figure 1. Location map of pilot areas GE-1 and GE-2 concerning the TOPSOIL project and the go-CAM project area. For nearly all areas HEM data are available from BGR (Siemon et al. 2014).

situated on both sides of the River Elbe (Figure 1). At these regions, saline water intrusion from the sea and from the River Elbe has been considered as the main impact on the groundwater catchment due to climate change. The drainage and irrigation channels might also play a vital role on the fresh–saline groundwater boundary development and possible upconing of saltwater into the drainage channels due to change in rainfall, groundwater recharge and groundwater gradients. These phenomena will be investigated at two different scales: local and regional. These impacts will be investigated by developing density driven groundwater models for both pilot areas. A local scale (ca. 300 km^2) groundwater model has been considered for GE-1 where the process of saltwater upconing to the drainage channels will be investigated. Additionally, the intrusion of brackish water from the River Elbe due to sea level rise and river bed deepening will be analyzed. Earlier observations indicate that the deepening of the River Elbe was followed by an increased salt load of the river water. As a consequence, the groundwater in the region near the river is already brackish and might increase further towards the inland. To support the modeling activity and to monitor river-aquifer interaction, installation of a monitoring station using the vertical electrode chain SAMOS is planned (Grinat 2018). The monitoring station will be installed near the River Elbe in Glückstadt to observe the fluctuation in resistivity in the subsurface due to change in salt content of the water and water level at the river. The groundwater model will be developed by Leibniz institute for Applied Geophysics (LIAG) with support from the Geological Survey of Schleswig-Holstein (LLUR).

A regional scale (ca. 1700 km^2) groundwater model has been planned for GE-2 to investigate the following: (i) saltwater intrusion from the sea due to sea level rise, (ii) saltwater intrusion from the River Elbe due to sea level rise and river bed deepening, (iii) possibility of managed aquifer recharge to store and transport water at the Geest area (near the western coast), (iv) role of drainage and irrigation channel on the development of or change in fresh–saline groundwater boundary. The geological and hydrogeological condition of this is complex and salinity distribution in vertical and horizontal direction is non-uniform. Groundwater quality is also heterogeneous (Rahman et al. 2018). Therefore, a development of a density driven groundwater flow and transport model is a big challenge here. The groundwater model will be developed by LIAG in collaboration with the Geological Survey of Lower Saxony (LBEG).

In these regions, helicopter-borne electromagnetic (HEM) surveys were performed by the Federal Institute for Geosciences and Natural Resources (BGR) (Siemon et al. 2014). HEM data provide useful information about the subsurface properties (such as presence of clay) and salinity. Due to high data coverage and high resolution of horizontal and vertical information, HEM data is considered as one of the main source of salinity information for the groundwater model. But it is a big challenge to transform the resistivity (Ωm) information to salinity (in mg/l), due to presence of clay and saltwater who both have the property of low resistivity. Therefore, a systematic analysis technique will be developed to transform the HEM data to salinity. Challenging is also the lateral heterogeneity of the near surface covering layers (Figure 2).

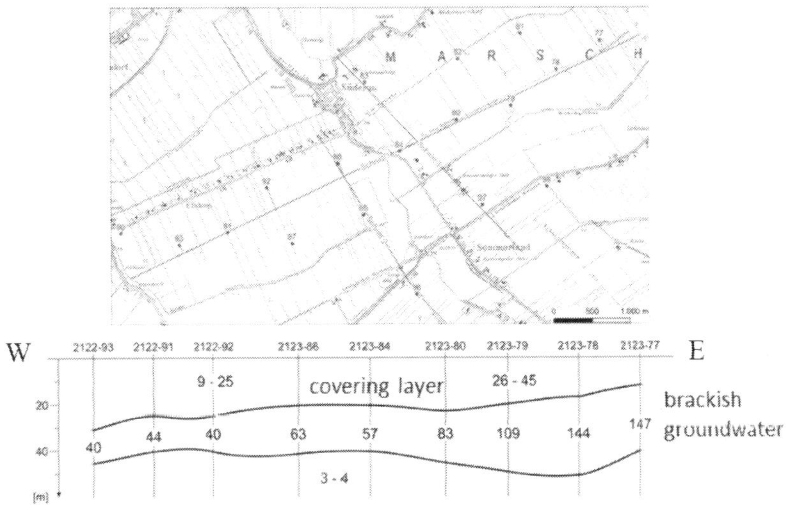

Figure 2. Vertical resistivity section (resistivity in Ωm) from vertical electrical soundings show the resistivity increase of the brackish groundwater (= decrease of salt content) with increasing distance to the river Elbe. Different electrical resistivities of the covering layer (Klei, clayey organic sediment) are caused by different clay contents with consequences for the hydraulic conductivities of this layer.

To develop the groundwater models, an integrated approach has been presented by Wiederhold et al. (2017) and a modified version is shown in Figure 3. An interdisciplinary approach, considering geological, geophysical and geochemical information has been formulated for groundwater catchment characterization (see details in Rahman et al. 2018). These information together with hydrological and meteorological information from several organizations (e.g., Germany's National Meteorological Service (DMD), Lower Saxony department of water, coastal and nature conservation (NLWKN) etc.) facilitate to analyze and assess the coastal groundwater catchment. These information will be fed to the groundwater model that will be used afterwards to achieve the planned objectives mentioned before.

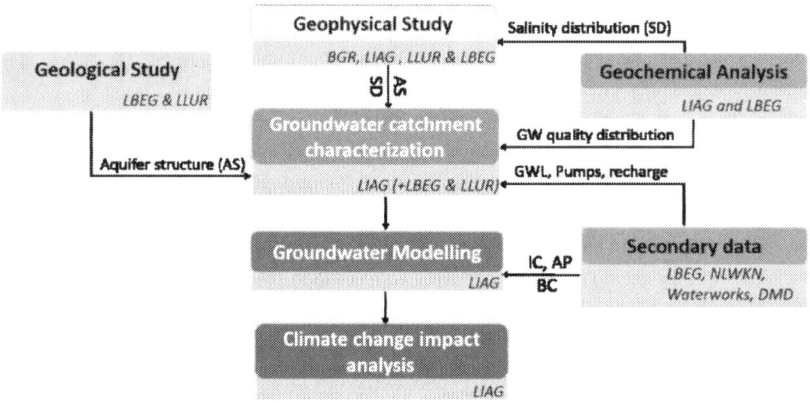

Figure 3. Integrated concept of coastal zone studies in the two pilot areas of the TOPSOIL project.

THE GO-CAM PROJECT

Globally, integrated coastal zone water resources planning and management targets to develop and preserve the coastal zone as an ecologically intact and economically flourishing habitat for human taking environmental, ecological, economic, social and geo-political conditions into consideration. Active involvement of transdisciplinary actors to the planning and management of coastal zone makes the entire formal planning and decision making procedure complex. Since a few decades, the conflicts among the different actors became worse in the coastal region where climate change, sea level rise, and salinization play very vital role on the limited availability of fresh groundwater resources. Hence, proper decision making requires a contribution from transdisciplinary sciences and engineering areas (e.g., natural sciences, social science, economic analysis etc.). A multicriteria decision analysis (MCDA) tool has the ability to combine all these information and contribute to the intelligent and optimal planning for sustainable use of water resources at the coastal area.

Whereas the TOPSOIL project focuses on the development of approaches jointly to analyze the climate change impact due to sea level rise and increase in groundwater table by field investigation and model development, go-CAM project aims at the development and implementation of a multi-criteria steering instrument (coastal aquifer management, CAM) for the sustainable use of water resources in coastal areas (https://www.tu-braunschweig.de/lwi/hywa/forschung projekte/gocam). This action can be considered as one step further to implementation of coastal zone groundwater management strategies.

The specific objectives of the project is: (i) development of groundwater model for some pilot area in Lower Saxony, (ii) development of MCDA tool for coastal zone water management, (iii) transfer of the dialog platform CAM developed for the project region

Northern Germany to international partner regions (such as Brazil, Turkey and South Africa).

LIAGs part are field investigations including mapping and monitoring of saltwater intrusions in the coastal region of Niedersachsen (Figure 1) in cooperation with TU Braunschweig and water supply companies (OOWV). Mapping of saltwater occurrence will be carried out with geophysical methods. As in the TOPSOIL region GE-1, monitoring of the fresh–saline groundwater boundary by the system SAMOS is planned at two locations. Density driven groundwater flow and transport modeling will be done by Gesellschaft für Anlagen- und Reaktorsicherheit (GRS) (Schneider et al. 2018).

ACKNOWLEDGMENT

The TOPSOIL project is co-funded by the Interreg North Sea Region Programme, project J-No. 38-2-27-15. The go-CAM project is funded by the German Federal Ministry of Education and Research under grant 02WGR1427E.

REFERENCES

Burschil T, W. Scheer, R. Kirsch, and H. Wiederhold. 2012. Compiling geophysical and geological information into a 3D model of the glacially affected island of Föhr. Hydrol. Earth Syst. Sci. 9(4), 5085-5119, DOI: 10.5194/hessd-9-5085-2012.

CLIWAT WORKING GROUP (2011): Groundwater in a Future Climate - The CLIWAT Handbook. ISBN: 87-7788-265-2, Central Denmark Region, Aarhus. http://cliwat.eu/xpdf/groundwater_in_a_future_climate.pdf

Grinat M (2018): Long-time resistivity monitoring of a freshwater/saltwater transition zone using vertical electrode systems. 25th Saltwater Intrusion Meeting, Gdansk, this volume.

Rahman, A.M., E. González, H. Wiederhold, N. Deus, J. Elbracht, and B. Siemon. 2018. Characterization of a regional coastal zone aquifer using interdisciplinary approach – an example from Weser-Elbe region, Lower Saxony, Germany. 25th Saltwater Intrusion Meeting, Gdansk, Poland, this volume.

Schneider, A. et al. *titel not known yet.* 25th Saltwater Intrusion Meeting, Gdansk, this volume.

Siemon, B., H. Wiederhold, A. Steuer, M.P. Miensopust, W. Voß, M. Ibs-von Seht, and U. Meyer, 2014. Helicopter-borne electromagnetic surveys in Northern Germany. In Programme and Proceedings of SWIM 2014, ed Wiederhold, et al., 375-378, 23rd Salt Water Intrusion Meeting 2014, Husum, Germany.

Sulzbacher H., H. Wiederhold, B. Siemon, M. Grinat, J. Igel J, T. Burschil, T. Günther, and K. Hinsby. 2012. Numerical modelling of climate change impacts on freshwater lenses on the North Sea Island of Borkum using hydrological and geophysical methods. Hydrol. Earth Syst. Sci., 16, 3621-3643, DOI10.5194/hess-16-3621-2012.

Contact Information: Helga Wiederhold, Leibniz-Institut für Angewandte Geophysik (LIAG), Stilleweg 2, Hannover, Germany, Phone: +49-511-6433520,
Fax: +49-511-6433665, Email: helga.wiederhold@liag-hannover.de

AUTHOR INDEX

Adil Sbaï M. 161
America Ilja 1
Amraoui N. 161
Ataie-Ashtiani B. 70, 104
Azizur Rahman M. 223
Baïsset M. 7, 110
Belfort B. 70
Bier George 1
Bikše J. 13, 144
Bootsma Huite 36
Böttcher Michael E. 99
Brodecka-Goluch Aleksandra 188
Chang Yawen 19
Claus Jasper 88
Custodio Emilio 24
Dabekaussen W. 217
Damrat Mateusz 30
De Angelis Donald L. 210
De Kleine M. 217
De Louw P. 1, 36, 217
Dellwig Olaf 99
Delsman J. 217
Depret Devlin 88
Desens A. 42
Deus Nico 137, 173
Devriese Gert Jan 88
Dominique H. 217
Dzierzbicka-Glowacka L. 119
Elbracht Jörg 137, 173
Eley M. 167
Epping Dieter 47
Erkens Gilles 36
Escher Peter 99
Essink G. O. 217
Fahs M. 70, 104
Gelleszun M. 167
Gomula-Kawecka Anna 125
González Eva 137
Grabowski Milosz 188
Graca Bozena 188
Graf Thomas 42
Grinat Michael 47
Gumula-Kawecka Anna 119
Gunnink J. 217
Gupta Gautam 179
Hoc Ryszard 51
Houben G. J. 42, 56, 115
Howahr M. 167
Hu Bill X. 19
Jakacki Jaromir 188
Jaworska-Szulc Beata 119, 125, 131
Jenner Anna-K. 99
Kaczor-Kurzawa Dorota 59
Karaoulis M. 217
Kirsch Reinhard 223
Klostowska Zaneta 64, 188
Klusek Zygmunt 188
Koh Hock Lye 210
Koohbor B. 70
Kooi Henk 36
Kordalski Zbigniew 76
Kozerski Bohdan 131
Koziorowska Katarzyna 188
Kramer Mark 36
Krawiec Arkadiusz 81
Kuhn T. 42
Kulinski Karol 64, 188
Kwaterkiewicz Andrzej 131
Lebbe Luc 88
Leczynski Leszek 64
Lengier Monika 64
Li Xue 19
Lidzbarski Miroslaw 94
Lipka Marko 99
Logashenko D. 167
Lovighi H. 110
Luyun Roger 204
Meyer Robert 47
Momii Kazuro 204
Mozafari B. 104
Neyens D. 7, 110
Pauw P. 217
Polemio M. 150
Post V. E. A. 42, 115
Potrykus Dawid 119, 125
Pruszkowska-Caceres Malgorzata 119, 125, 131
Przewlócka Maria 125, 131
Rahman Mohammad Azizur 137
Rak Daniel 188
Reiter S. 167
Retike I. 13, 144
Rizzo Enzo 150
Robakiewicz Malgorzata 155
Ronczka Mathias 223
Sadurski Andrzej 51, 76, 81
Santaloia F. 150
Scheer Wolfgang 223
Schmiedinger Iris 99
Schneider A. 167
Siemon Bernhard 137, 173
Simmons C. T. 70, 104
Sternberg Leonel 210
Steuer Annika 173
Stokowski Marcin 188
Struck Ulrich 99
Sültenfuß Jürgen 99
Suneetha N. 179
Swidzinski Waldemar 192, 198
Szelewicka Anna 185

AUTHOR INDEX

Szymczycha Beata 64, 188
Szymkiewicz Adam 119, 125, 131
Takahashi Masahiro 204
Tarnawska Ewa 94
Teh Su Yean 210
Teilmann R. 217
Thofte S. 217
Van Baaren E. 217
Van Camp M. 217
Van Der Zee Sjoerd 1
Vandenbohede A. 217
Vandevelde D. 217
Vermaas T. 217
Voss Clifford I. 210
Walraevens K. 217
Walther M. 42
Warumzer Rafal 30
Westphal Julia 99
Wiederhold H. 137, 167, 223
Winde Vera 99
Winogradow Aleksandra 188
Wisniowski Zenon 51
Wolf J. 167
Wu Zijun 99
Younes Rafic 104
Zhao H. 167
Zuffianò L. E. 150